Stochastic Mechanics
Random Media
Signal Processing and Image Synthesis
Mathematical Economics
Stochastic Optimization and Finance
Stochastic Control

Applications of Mathematics

Stochastic Modelling and Applied Probability

44

Springer
New York
Berlin
Heidelberg
Barcelona
Hong Kong
London
Milan
Paris
Singapore
Tokyo

Applications of Mathematics

1 Fleming/Rishel, **Deterministic and Stochastic Optimal Control** (1975)
2 Marchuk, **Methods of Numerical Mathematics,** Second Ed. (1982)
3 Balakrishnan, **Applied Functional Analysis,** Second Ed. (1981)
4 Borovkov, **Stochastic Processes in Queueing Theory** (1976)
5 Liptser/Shiryayev, **Statistics of Random Processes I: General Theory** (1977)
6 Liptser/Shiryayev, **Statistics of Random Processes II: Applications** (1978)
7 Vorob'ev, **Game Theory: Lectures for Economists and Systems Scientists** (1977)
8 Shiryayev, **Optimal Stopping Rules** (1978)
9 Ibragimov/Rozanov, **Gaussian Random Processes** (1978)
10 Wonham, **Linear Multivariable Control: A Geometric Approach,** Third Ed. (1985)
11 Hida, **Brownian Motion** (1980)
12 Hestenes, **Conjugate Direction Methods in Optimization** (1980)
13 Kallianpur, **Stochastic Filtering Theory** (1980)
14 Krylov, **Controlled Diffusion Processes** (1980)
15 Prabhu, **Stochastic Storage Processes: Queues, Insurance Risk, Dams, and Data Communication,** Second Ed. (1998)
16 Ibragimov/Has'minskii, **Statistical Estimation: Asymptotic Theory** (1981)
17 Cesari, **Optimization: Theory and Applications** (1982)
18 Elliott, **Stochastic Calculus and Applications** (1982)
19 Marchuk/Shaidourov, **Difference Methods and Their Extrapolations** (1983)
20 Hijab, **Stabilization of Control Systems** (1986)
21 Protter, **Stochastic Integration and Differential Equations** (1990)
22 Benveniste/Métivier/Priouret, **Adaptive Algorithms and Stochastic Approximations** (1990)
23 Kloeden/Platen, **Numerical Solution of Stochastic Differential Equations** (1992)
24 Kushner/Dupuis, **Numerical Methods for Stochastic Control Problems in Continuous Time** (1992)
25 Fleming/Soner, **Controlled Markov Processes and Viscosity Solutions** (1993)
26 Baccelli/Brémaud, **Elements of Queueing Theory** (1994)
27 Winkler, **Image Analysis, Random Fields, and Dynamic Monte Carlo Methods: An Introduction to Mathematical Aspects** (1994)
28 Kalpazidou, **Cycle Representations of Markov Processes** (1995)
29 Elliott/Aggoun/Moore, **Hidden Markov Models: Estimation and Control** (1995)
30 Hernández-Lerma/Lasserre, **Discrete-Time Markov Control Processes: Basic Optimality Criteria** (1996)
31 Devroye/Györfi/Lugosi, **A Probabilistic Theory of Pattern Recognition** (1996)
32 Maitra/Sudderth, **Discrete Gambling and Stochastic Games** (1996)

(continued after index)

Richard Serfozo

Introduction to Stochastic Networks

Springer

Richard Serfozo
School of Industrial and Systems Engineering
Georgia Institute of Technology
Atlanta, GA 30233
USA
rserfozo@isye.gatech.edu

With 9 figures.

Mathematics Subject Classification (1991): 60K25, 60G55, 90B22

Library of Congress Cataloging-in-Publication Data
Serfozo, Richard.
Introduction to stochastic networks / Richard Serfozo.
p. cm. — (Applications of mathematics ; 44)
Includes bibliographical references and index.
ISBN 0-387-98773-8 (hardcover : alk. paper)
1. Queuing theory. 2. Stochastic analysis. I. Title.
II. Series
QA274.8.S47 1999
519.8′2—dc21 98-17363

Printed on acid-free paper.

Production managed by Robert Bruni; manufacturing supervised by Jeffrey Taub.
Camera-ready copy prepared from the author's TeX files.
Printed and bound by Maple-Vail Book Manufacturing Group, York, PA.
Printed in the United States of America.

9 8 7 6 5 4 3 2 1

ISBN 0-387-98773-8 Springer-Verlag New York Berlin Heidelberg SPIN 10712659

To Joan and Kip

Preface

The term stochastic network has several meanings. Here it means a system in which customers move among stations where they receive services; there may be queueing for services, and customer routing and service times may be random. Such a system is often called a queueing network.

Typical examples of stochastic networks are as follows:

Computer and Telecommunications Networks: Data packets, read/write transactions, files, or telephone calls move among computers, buffers, operators, or switching stations.

The Internet: Queries, e-mail, advertisements, purchase orders, news, and zillions of other e-messages move among host computers, PCs, people, and mail-order stores.

Manufacturing Networks: Parts, orders, or material move among work stations, inspection points, automatically-guided vehicles, or storage areas.

Equipment Maintenance Networks: Parts or subsystems move among usage sites and repair facilities.

Logistics and Supply-Chain Networks: Parts, material, personnel, trucks, or equipment move among sources, storage depots, and production facilities.

Parallel Simulation and Distributed Processing Systems: Messages, data packets and signals move among buffers and processors.

Stochastic networks also arise in many other areas such as in biology, physics, and economics.

Issues concerning the operation of a stochastic network include the following: Where are its bottlenecks or major delays? How does one network design compare with another? What are good rules for operating the network (e.g., customer priorities or routings)? What is a least-cost network (e.g., numbers of machines, tools, or workers). Examples of performance objectives of a network are as follows. The probability of a busy signal in a telecommunications network should be less than one percent. The expected waiting times in a computer system should be less than certain values. The probability of meeting manufacturing deadlines should be above ninety percent.

To address such issues requires an understanding of the behavior of the network in terms of the equilibrium (or stationary) probability distribution of the numbers of units at the nodes. These distributions are used to evaluate a variety of performance measures such as throughputs on arcs and at nodes, expected costs, and percentage of time a node is overloaded. The equilibrium distribution is a basic ingredient for constructing objective functions or constraints used in mathematical programming algorithms to select optimal network designs and protocols. The quality of a network is also determined by the duration of travel and sojourn times in it, such as the time for a unit to travel from one sector to another or the amount of time it takes for a unit to visit a certain set of nodes. Equilibrium distributions are used in describing the means or distributions of such travel times.

This book describes a number of stochastic network models that have been developed over the last thirty years. The focus is on Markov process models, whose equilibrium distributions and performance parameters are analytically tractable via closed form expressions or computational algorithms. The network models can be categorized as follows:

- Classical Jackson and multiclass BCMP and Kelly networks. The development of these networks in Chapters 1 and 3 is under a unified framework of a Whittle network.
- Reversible networks. A self-contained description of these networks and the related theory of reversible Markov processes is in Chapter 2.
- Networks with string transitions. These are extensions of Whittle networks to batch movements and more intricate transitions involving strings of events; see Chapter 7.
- Networks with product form stationary distributions. Chapter 8 characterizes these networks, which include quasi-reversible networks.
- Spatial queueing systems in which customers move in a general space where they obtain services (e.g., mobile phones moving in a region). The space–time Poisson models in Chapter 9 are generalizations of the classical $M/G/\infty$ infinite-server system; they are characterized via random transformations of Poisson processes. The models in Chapter 10 are spatial analogues of Whittle networks.

In addition to describing network models, a major aim of this book is to provide introductions to Palm probabilities for stationary systems and to Little laws for queues and utility processes. These topics are the subjects of Chapters 4–6, which also address network issues concerning customer travel times, flows between nodes and network sojourn times. To emphasize its simplicity and usefulness, the subtheory of Palm probabilities for stationary Markov processes comes before the theory for general stationary processes. Palm probabilities for Markov processes are simply ratios of rates of certain events, where the rates are obtained by an extended Lévy formula. The classical Lévy formula is for expectations of functionals of Markov process, and the extended formula in Chapter 4 applies to functionals that may include information about the entire sample path of the Markov process. Chapter 5 gives a rather complete development of Little laws that describe a variety of sample path averages of waiting times and other performance parameters of queues and general stochastic systems.

Many properties of networks are represented by point processes. Chapters 4 and 6 use point processes as a framework for counting events over time, and Chapters 9 and 10 use point processes to represent customer locations in a region. These applications of point processes are self-contained and understandable without a knowledge of the theory of point processes. Aside from the last chapter, most of the point process material concerns Poisson processes. Chapter 4 presents necessary and sufficient conditions for a point process functional of a Markov process to be a Poisson process. These conditions establish, for instance, that the departure times from a stationary Jackson network form a Poisson process. Another topic in Chapter 9 concerns random transformations of Poisson processes (e.g., translations and partitions) that result in new Poisson processes. These transformations are useful for representing particle movements in space and time.

The book is intended for engineers, scientists, and system analysts who are interested in stochastic network models and related fundamentals of queueing theory. My aim was to write a monograph that would be useful as a reference and for teaching as well. All or parts of Chapters 1–6 and sections 1–7 in Chapter 9 could be used in graduate courses related to network modeling or applied probability. The more advanced models discussed in Chapters 7–10 would be suitable for seminars. A prerequisite for the first eight chapters is an introduction to stochastic processes (not using measure theory) covering Markov chains, Poisson processes, and continuous-time Markov processes. Knowledge of measure theory is needed for the spatial models discussed in the last two chapters.

Finally, I would like to express my appreciation to those who helped create and perfect this book. First, I thank the taxpayers of this country who have supported the NSF, which funded part of my research. I am grateful to Karl Hinderer for inviting me to present my initial crude network notes in a short course in Karlsruhe. Many thanks go to Bingyi Yang and Xiaotao Huang for doing their Ph.D. research with me that resulted in Chapters 7 and 10, respectively. Chapter 8 is based on joint work with Xiuli Chao, Masakiyo Miyazawa, and H. Takada. I thank them for the insightful wrestling matches with notation we had via numerous e-mail exchanges. My last thanks go to Bill Cooper, Christian Rau, and German Riano for their superb proofreading, which eliminated many errors.

Contents

1

Jackson and Whittle Networks

This chapter describes the equilibrium behavior of Jackson and Whittle networks. In such a network, the numbers of discrete units or customers at the nodes are modeled by multidimensional Markov processes. The main results characterize the equilibrium distributions of the processes. These distributions yield several performance parameters of the networks including throughput rates, expected customer waiting times, and expected busy periods for servers.

1.1 Preliminaries on Networks and Markov Processes

In this section, we present the framework we will use for modeling a stochastic network as a Markov process. Included is a review of some basics of Markov processes.

We will consider a network that operates as follows. The network consists of m nodes, labeled $1, 2, \dots, m$, where m is finite. Discrete units or customers move among the nodes where they are processed or served. We will often use the word "unit" instead of customer because it is shorter and has a broader connotation. For example, in a computer or telecommunications network, a node might be a computer, data file, or switching station; and a unit might be a data packet, message (batch of packets), telephone call, or transaction. In a manufacturing network, a node might be a work station, storage area, inspection point, source of demands, or station for automatically-guided vehicles; and a unit might be a part, group of parts, request for a product, or a message.

In our m-node network, randomness may be present in the servicing or routing of the units—it may emanate from the units' characteristics or the nodes' structures or a combination of both. The evolution of the network is represented by a continuous-time stochastic process $\{X_t : t \geq 0\}$ whose states are vectors $x = (x_1, \ldots, x_m)$ in a finite or infinite state space $\mathbb{E}$, where x_j denotes the number of units at node j. Chapters 3 and 8 discuss processes with more general states that include information other than the quantities of units at the nodes.

The network is *closed* with ν units in it if the total number of units $|x| = x_1 + \ldots + x_m$ is always equal to ν. Then $\mathbb{E} = \{x : |x| = \nu\}$. Otherwise, the network is *open*—it is *open with finite capacity* ν if $\mathbb{E} = \{x : 0 \leq |x| \leq \nu\}$, and it is *open with unlimited capacity* if $\mathbb{E} = \{x : 0 \leq |x| < \infty\}$.

Assume that X is a continuous-time Markov jump process (or continuous-time Markov chain). Then its probability distribution is determined by its *transition rates*

$$q(x, y) \equiv \lim_{t \downarrow 0} t^{-1} P\{X_t = y | X_0 = x\}, \quad y \neq x,$$

and $q(x, x) \equiv 0$. We adopt the standard convention that the process X is regular in the sense that it cannot take an infinite number of jumps in a finite time interval. Also, to avoid degeneracies, we assume the process does not have any absorbing states. To model an actual network by this process, one must translate the operational features of the nodes and the rules of routing units into a specific transition rate function q. We will study several networks in this framework. We call X a *Markov network process* that represents the numbers of units at the nodes of an m-node network.

Since X is a Markov jump process, its sojourn time in any state is exponentially distributed. Specifically, whenever X enters a state x, it remains there for a time that is exponentially distributed with rate $q(x) \equiv \sum_y q(x, y)$. Then it jumps to a state y with probability $p(x, y) \equiv q(x, y)/q(x)$. These exponential sojourns and transitions continue indefinitely. The resulting sequence of states X visits forms a Markov chain with transition probabilities $p(x, y)$.

A standard way of defining the transition rates $q(x, y)$ is to specify the exponential sojourn rates $q(x)$ and probabilities $p(x, y)$ and then determine q by setting $q(x, y) = q(x)p(x, y)$. The following example illustrates how this is done for a Markov process that may have transitions from a state back to itself. Such situations arise in networks where a unit exiting a node may be instantaneously fed back to the same node for another service.

Example 1.1. *Construction of a Markov Process.* Suppose X is a stochastic process on a countable state space $\mathbb{E}$ such that the sequence of states it visits is a Markov chain with transition probabilities $\bar{p}(x, y)$, where the probability $\bar{p}(x, x)$ of a transition from the state x back to itself may be positive. In addition, whenever the process is in state x, the time to the next transition is exponentially distributed with rate $\lambda(x)$. Now, the sequence of "distinct" states visited by X is clearly a Markov chain with transition probabilities $p(x, y) = \bar{p}(x, y)/(1 - \bar{p}(x, x))$. Also, if $\bar{p}(x, x) > 0$, the "entire" sojourn time in a state x is the sum of successive exponential times with rate $\lambda(x)$ until a transition takes it to a new state with probability $1 - \bar{p}(x, x)$. Consequently, the sojourn time is exponentially distributed with rate $q(x) = \lambda(x)(1 - \bar{p}(x, x))$ (see Exercise 1). Then by the discussion above, X is a Markov process with transition rates $q(x, y) = q(x)p(x, y) = \lambda(x)\bar{p}(x, y)$. □

Much of our focus will be on the network process X with *single-unit movements* described as follows. Envision the units as moving within the set of nodes $M = \{1, \dots, m\}$ if the network is closed or $M = \{0, 1, \dots, m\}$ if the network is open. Here node 0 denotes the outside of the network. This node 0 is only a source or sink; the network state x does not record any population size for it. A typical transition of X will be triggered by the movement of one unit from some node j to some node k in M. Specifically, when X is in state x and a unit moves from j to k, then the next state of the network is $T_{jk}x$, which is the vector x with one less unit at node j and one more unit at node k. For example $T_{30}x$ is the vector x with x_3 replaced by $x_3 - 1$. We will sometimes write $T_{jk}x = x - e_j + e_k$, where $e_0 = 0$ and e_j is the unit vector with 1 in component j and 0 elsewhere, for $j = 1, \dots, m$.

The exponential sojourn time in state x is usually formulated as follows. For each pair j, k in M, one assumes that the time to the next "potential" movement of a unit from j to k, or potential transition from x to $T_{jk}x$ is exponentially distributed with rate $q(x, T_{jk}x)$, and these times are independent. The form of $q(x, T_{jk}x)$ depends on the network being modeled. Then the sojourn time in state x, being the minimum of these independent exponential times, is also exponential with rate $q(x) = \sum_j \sum_k q(x, T_{jk}x)$. Such sums are for all j and k in the node set M unless specified otherwise. Moreover, $q(x, T_{jk}x)/q(x)$ is the probability that the jump is triggered by the j-to-k movement. This interpretation of the transition rates in terms of exponential times to potential movements is often used as a guide for formulating the rate function q for particular networks.

Later chapters cover networks with more general concurrent or multi-unit movements in which a typical transition is from x to $x + a - d$, where $a = (a_1, \dots, a_m)$ and $d = (d_1, \dots, d_m)$ denote the numbers of arrivals to and departures from the respective nodes. In these instances, the natural assumption is that, whenever the network is in state x, the time to the next potential transition to state $x - d + a$ is exponentially distributed with rate $q(x, x - d + a)$ and these times are independent for the possible vectors d and a.

To describe the equilibrium behavior of Markov processes, we will use the following notation. Assume that $\{X_t : t \geq 0\}$ is a Markov jump process as described above on a countable state space $\mathbb{E}$ with transition rates $q(x, y)$. A positive measure π on $\mathbb{E}$ is an *invariant measure for X* (or for q) if it satisfies the *balance equations*

$$\pi(x) \sum_y q(x, y) = \sum_y \pi(y) q(y, x), \quad x \in \mathbb{E}. \tag{1.1}$$

The measure may be infinite and the process may be reducible or null recurrent. If X is irreducible and positive recurrent, then there is a unique positive probability measure π that satisfies the balance equations. In this case, X is called an *ergodic process*, and π is called the *stationary* or *equilibrium distribution* of X. For simplicity, we will often present an invariant measure for an ergodic process and not take the extra step to normalize the measure to be a stationary distribution.

When the process X is ergodic, its stationary distribution π is also the *limiting distribution* in the sense that

$$\lim_{t\to\infty} P\{X_t = x\} = \pi(x).$$

A stochastic process is *stationary* if its finite-dimensional distributions are invariant under any shift in time. Because X is a Markov process, a necessary and sufficient condition for it to be stationary (or in equilibrium) is that $P\{X_t = x\} = \pi(x)$ for each x and t.

A variety of costs and performance parameters of Markov processes are expressible in terms of the following functionals. Suppose that a value (e.g., a cost or utility) is incurred continuously at the rate of $f(x)$ per unit time whenever the process X is in state x. Then the total value incurred in the time interval $(0, t]$ is

$$\int_0^t f(X_s)\,ds.$$

One may also be interested in values associated with the transitions of X. Suppose $h(x, y)$ is a value associated with each transition of X from x to y. Then the total value for the transitions in $(0, t]$ is

$$\sum_n h(X_{\tau_{n-1}}, X_{\tau_n})1(\tau_n \in (0, t]),$$

where $0 \equiv \tau_0 < \tau_1 < \tau_2 \ldots$ are the transition times of X. Note that X_{τ_n} is the value of X at the nth transition. Here, 1(statement) denotes the indicator function that is 1 or 0 depending on whether the "statement" is true or false.

The ergodic theory for Markov processes justifies that the limiting averages of the preceding functionals exist. Furthermore, these limits are expected values of the functionals when the process X is stationary. These properties are summarized in the following result. We say that a sum $\sum_n a_n$ *exists* (or is absolutely convergent) if $\sum_n |a_n| < \infty$.

Theorem 1.2. (Law of Large Numbers) *If the Markov process X is ergodic with stationary distribution π, then with probability one (w.p.1)*

$$\lim_{t\to\infty} t^{-1} \int_0^t f(X_s)\,ds = \sum_x \pi(x) f(x),$$

$$\lim_{t\to\infty} t^{-1} \sum_n h(X_{\tau_{n-1}}, X_{\tau_n})1(\tau_n \in (0, t]) = \sum_x \pi(x) \sum_y q(x, y)h(x, y),$$

provided the sums exist. These limit statements also hold when the random functions are replaced with their expectations. Furthermore, if X is stationary, then the preceding limits are the respective expected values

$$E\int_0^1 f(X_s)\,ds, \qquad E\sum_n h(X_{\tau_{n-1}}, X_{\tau_n})1(\tau_n \in (0, 1]).$$

This theorem yields the following properties. First, the average number of transitions of X from a set A to a set B per unit time is

$$\begin{aligned}\pi q(A, B) &\equiv \sum_{x \in A} \pi(x) \sum_{y \in B} q(x, y) \\ &= \lim_{t \to \infty} t^{-1} \sum_n 1(X_{\tau_{n-1}} \in A, X_{\tau_n} \in B, \tau_n \in (0, t]).\end{aligned}$$

This is also the expected number of such transitions in a unit time interval when X is stationary. The quantity $\pi q(A, B)$ is sometimes called the *probability flux between A and B*. In particular, $\pi q(x, y)$ is the average or equilibrium rate of transitions from x to y (the $q(x, y)$ is the "infinitesimal" transition rate). In light of this, the total balance equations (1.1) are $\pi q(x, \mathbb{E}) = \pi q(\mathbb{E}, x)$, $x \in \mathbb{E}$. That is, in equilibrium, the average number of transitions per unit time from x to all the other states equals the average number of transitions from the other states into x. Or, loosely speaking, the rate of flow out of x equals the rate of flow into x.

More generally, summing the total balance equations on $x \in A$ yields

$$\pi q(A, \mathbb{E}) = \pi q(\mathbb{E}, A).$$

Also, subtracting $\pi q(A, A)$ from this equation yields

$$\pi q(A, A^c) = \pi q(A^c, A), \tag{1.2}$$

where A^c denotes the complement of A. This says the rate of flow out of A equals the rate of flow into A, which is what one would anticipate for a stable system.

The rate of flow into A is related to the number of entrances of X into A in the time interval $(0, t]$, which is

$$N_A(t) = \sum_n 1(X_{\tau_{n-1}} \in A^c, X_{\tau_n} \in A, \tau_n \in (0, t]).$$

The *rate at which the process X enters A* is defined by

$$\lambda(A) \equiv \lim_{t \to \infty} t^{-1} N_A(t) \quad \text{w.p.1.}$$

Then it follows that

$$\lambda(A) = \pi q(A^c, A) = \sum_{x \in A^c} \pi(x) \sum_{y \in A} q(x, y).$$

This rate is also related to the time T_n of the nth entrance of X into A. Namely, by the law of large numbers for point processes (see Theorem 5.8),

$$\lim_{n \to \infty} n^{-1} T_n = \lambda(A)^{-1} \quad \text{w.p.1.} \tag{1.3}$$

Another quantity of interest for the Markov process X is the average sojourn or waiting time in A defined by

$$W(A) = \lim_{n \to \infty} n^{-1} \sum_{i=1}^{n} W_i(A) \quad \text{w.p.1,}$$

where $W_i(A)$ is the time X spends in A on its ith visit.

Theorem 1.3. *If the Markov process X is ergodic with stationary distribution π, then the limit $W(A)$ exists and*

$$W(A) = \lambda(A)^{-1} \sum_{x \in A} \pi(x).$$

If in addition X is stationary, then $W(A)$ is also the expected waiting time in A with respect to the Palm probability that X enters A at time 0.

PROOF. The first assertion follows since applications of (1.3) and Theorem 1.2 yield

$$\begin{aligned} W(A) &= \lim_{n\to\infty} n^{-1} T_{n+1} \lim_{n\to\infty} T_{n+1}^{-1} \int_0^{T_{n+1}} 1(X_t \in A)\, dt \\ &= \lambda(A)^{-1} \sum_{x \in A} \pi(x) \quad \text{w.p.1.} \end{aligned}$$

The second assertion follows from Theorem 4.31 in Chapter 4, where Palm probabilities are first introduced. The second assertion is also a special case of the inversion formula for Palm probabilities in Corollary 6.16. □

The following notion of reversibility plays an important role in network modeling.

Definition 1.4. The Markov process X is *reversible* if there is a positive measure π on $\mathbb{E}$ that satisfies the *detailed balance equations*

$$\pi(x)q(x, y) = \pi(y)q(y, x), \quad x, y \in \mathbb{E}. \tag{1.4}$$

The π is an invariant measure since it also satisfies (1.1), which are the detailed balance equations summed over y. We also say that *q is reversible with respect to π*.

By the law of large numbers for Markov processes, the detailed balance equation (1.4) says that, for an ergodic process, the average number of transitions of the process from state x to state y is equal to the average number of transitions in the reverse direction from y to x. And if the process is stationary, then the expected number of transitions x to y is equal to the expected number of y to x transitions.

A distinguishing feature of reversible transition rates is that they have the following simple, canonical form.

Theorem 1.5. *The transition rate q is reversible if and only if it is of the form*

$$q(x, y) = \gamma(x, y)/\pi(x), \quad x \neq y \in \mathbb{E}, \tag{1.5}$$

for some positive function π on $\mathbb{E}$ and some nonnegative function γ on $\mathbb{E} \times \mathbb{E}$ such that $\gamma(x, y) = \gamma(y, x)$, $x, y \in \mathbb{E}$. In this case, π is an invariant measure for q.

PROOF. If q is reversible with respect to π, then (1.5) is satisfied with $\gamma(x, y) = \pi(x)q(x, y)$. Conversely, any transition function of the form (1.5) satisfies the detailed balance equations, and hence q is reversible. □

The canonical representation (1.5) is useful as a quick check for determining whether a process is reversible: just find a symmetric function γ such that $q(x, y)/\gamma(x, y)$ is independent of y. The preceding representation is the only property of reversibility needed in this chapter. We will resume the discussion of reversible Markov processes and reversible networks in the next chapter.

1.2 Tandem Network

This section gives a glimpse of what lies ahead. It describes the equilibrium behavior of a tandem network, which is an example of an open Jackson network.

Consider a network consisting of m nodes in series as shown in Figure 1.1 below. Units enter node 1 according to a Poisson process with intensity λ. Each unit is served at nodes $1, \ldots, m$ in that order, and then it exits the system. Each node is a single server that serves the units one at a time on a first-come, first-served basis, and a unit's service time at node j is exponentially distributed with rate μ_j, independent of the arrival process and other services. When a unit arrives to a node and the server is busy, the unit joins the queue at that node to wait for its service.

The state of the network is represented by a vector $x = (x_1, \ldots, x_m)$ in the set $\mathbb{E} \equiv \{x : |x| < \infty\}$, where x_j denotes the number of units at node j. Let X_t denote the state of the network at time t. The process $X = \{X_t : t \geq 0\}$ is an open, unlimited-capacity network process that evolves as follows. Upon entering a state x, it remains there until a new unit arrives to node 1, or there is a service completion at one of the nodes. In other words, typical transitions of the process are from x to $T_{01}x = x + e_1$ (an arrival into node 1), or from x to $T_{j,j+1}x = x - e_j + e_{j+1}$ (a service completion at node j), provided $x_j \geq 1$. Here $m + 1 \equiv 0$. The time until such a transition is exponentially distributed, and so X is a Markov process. Its transition rates are

$$q(x, T_{01}x) = \lambda, \qquad q(x, T_{j,j+1}x) = \mu_j 1(x_j \geq 1),$$

and otherwise the rates are 0.

The balance equations that an invariant measure π must satisfy are

$$\pi(x) \sum_{j=0}^{m} q(x, T_{j,j+1}x) = \sum_{j=1}^{m+1} \pi(T_{j,j-1}x) q(T_{j,j-1}x, x) 1(x_j \geq 1), \qquad x \in \mathbb{E}.$$

Since each node j resembles an $M/M/1$ queueing process with input rate λ and service rate μ_j, one might conjecture that the stationary distribution of the process

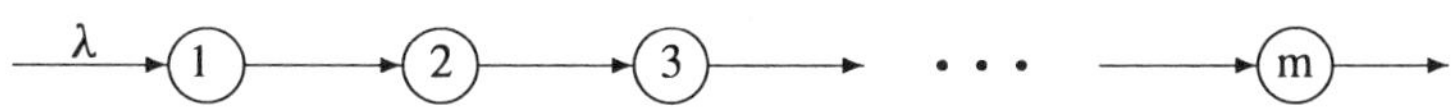

FIGURE 1.1. Tandem Network

is a product of stationary distributions of $M/M/1$ systems of the form

$$\pi(x) = \prod_{j=1}^{m}(1-\rho_j)\rho_j^{x_j},$$

where $\rho_j \equiv \lambda/\mu_j$. We assume $\rho_j < 1$ for each j.

To prove this conjecture, note that from the definition of q and $m+1=0$, it follows that π defined above satisfies

$$\begin{aligned}
\pi(x)q(x, T_{01}x) &= \pi(x)\lambda = \pi(T_{0m}x)q(T_{0m}x, x),\\
\pi(x)q(x, T_{j,j+1}x) &= \pi(x)\mu_j 1(x_j \geq 1)\\
&= \pi(T_{j,j-1}x)q(T_{j,j-1}x, x)1(x_j \geq 1), \quad 1 \leq j \leq m.
\end{aligned}$$

Summing these equations, we see that π satisfies the balance equations above. The preceding are *partial balance equations* that say the average number of movements of units per unit time from node j to node $j+1$ that takes the network out of state x is equal to the average number of movements from $j-1$ to j that takes the network into state x.

For the rest of this section, suppose the tandem network process X is stationary. Because its stationary distribution π has a product form, the numbers of units at the nodes at any fixed time t are independent, and the number of units at each node j has a geometric distribution $(1-\rho_j)\rho_j^{x_j}$, just as if it were an $M/M/1$ queueing system operating in isolation. A typical item of interest is the probability distribution of the total number of units in the network. This distribution is $P\{|X_t| = n\} = \sum_{|x|=n}\pi(x)$. From a result we will prove later (Proposition 1.31), it follows that

$$P\{|X_t| = n\} = \prod_{j=1}^{m}(1-\rho_j)\sum_{i=1}^{m}\rho_i^{n+m-1}\prod_{\ell\neq i}(\rho_i - \rho_\ell)^{-1},$$

when the ρ_j's are distinct. For nondistinct ρ_j's, Proposition 1.32 applies.

This distribution is useful for optimization problems such as the following. Suppose there is a cost c_j per unit time of having a service rate μ_j at node j. Then the problem is

$$\begin{aligned}
&\min_{\mu_1,\ldots,\mu_m} \sum_{j=1}^{m} c_j\mu_j\\
&\text{s.t.} \quad P\{|X_t| > b\} \leq p.
\end{aligned}$$

Here b is a desired upper bound on the total system quantity, and p is a probability representing the quality of service.

A related quantity for the system is the average time $W(A)$ the system spends in the set $A = \{x : |x| > b\}$. The average rate at which X enters this set is

$$\lambda(A) = \sum_{|x|=b}\pi(x)q(x, T_{01}x) = \lambda P\{|X_t| = b\}.$$

Then by Theorem 1.3,

$$W(A) = \pi(A)/\lambda(A) = \frac{P\{|X_t| > b\}}{\lambda P\{|X_t| = b\}}.$$

Next, consider the point process $N_j(t)$ that denotes the number of units that move between node j and node $j + 1$ in time t. The rate of flow between j and $j + 1$ is

$$EN_j(1) = \sum_x \pi(x)q(x, T_{j,j+1}x) = \lambda.$$

By a result we prove later (Theorem 4.22), it follows that N_j is a Poisson process with intensity λ. A key ingredient for this is that each unit can make at most one visit to node j.

For a unit that enters the system in equilibrium at time 0, let W_j denote the time the unit spends in node j. We will show in Theorem 4.43 that $W_1, \ldots, W_m$ are independent exponential random variables and W_j has a rate $\mu_j - \lambda$. These exponential waiting times are with respect to the Palm probability that a unit enters the system at time 0. We discuss Palm probabilities in Chapters 4 and 5. This result is useful for addressing issues concerning the total time $W_1 + \cdots + W_m$ a unit spends in the network.

1.3 Definitions of Jackson and Whittle Processes

In this section, we define Jackson and Whittle processes. They are Markov processes that represent networks in which units move among the nodes according to independent Markovian routing and their service rates depend on the congestion. In a Jackson network, the service rate at each node depends only on the number of units at that node, whereas in a Whittle network, the service rate at each node is a function of the numbers of units at all the nodes.

Throughout this section, we assume that $\{X_t : t \geq 0\}$ is a stochastic process that represents the numbers of units at the nodes in an m-node network with single-unit movements. It is convenient to consider closed and open networks at the same time. Accordingly, assume the network may be any one of the following types:

- Closed network with ν units and state space $\mathbb{E} = \{x : |x| = \nu\}$.
- Open network with unlimited capacity and state space $\mathbb{E} = \{x : |x| < \infty\}$.
- Open network with finite capacity ν and state space $\mathbb{E} = \{x : |x| \leq \nu\}$.

Think of the units moving in the *node set*

$$M \equiv \begin{cases} \{1, \ldots, m\} & \text{if the network is closed} \\ \{0, 1, \ldots, m\} & \text{if the network is open.} \end{cases}$$

In Chapter 3, we discuss how our results apply to networks with multiple types of units.

The major assumption we make is that whenever the network is in state x, the time to the next movement of a single unit from node j to node k (i.e., a transition

from x to $T_{jk}x = x - e_j + e_k$) is exponentially distributed with rate $\lambda_{jk}\phi_j(x)$. The λ_{jk} are nonnegative with $\lambda_{jj} = 0$, and $\phi_j(x)$ is positive except that $\phi_j(x) = 0$ if $x_j = 0$ and $j \neq 0$. This assumption of exponential times to movements is satisfied under the following conditions:
(i) Whenever the network is in state x, the time to the next departure from node j is exponentially distributed with rate $\phi_j(x)$.
(ii) Each departure from j is routed to node k with probability λ_{jk}, independently of everything else.

For our development, we follow the standard convention that λ_{jk} may either be a routing probability or a nonnegative intensity of selecting the nodes j and k (like intensities in birth–death processes) and call it the j-to-k *routing intensity* or *routing rate*. Think of λ_{jk} as the transition rates of a continuous-time Markov jump process that depicts the movement of one unit in the node set M—this is an artificial *routing process* separate from the network process. With no loss in generality, we assume the routing process does not have transient states and it need not be irreducible (further comments on this are in the next section).

The $\phi_j(x)$ is the *service rate* or *departure intensity* at node j when the network state is x. If the network is open, units enter it at node k according to a system-dependent Poisson process with intensity $\lambda_{0k}\phi_0(x)$. The $\phi_0(x)$ is the "arrival intensity" from the outside. When $\phi_0(\cdot) \equiv 1$, the arrivals from outside into the respective nodes are independent Poisson processes with intensities $\lambda_{01}, \ldots, \lambda_{0m}$ (a zero intensity for a node means it does not have arrivals from outside). With a slight abuse of notation, we refer to λ_{jk} and $\phi_j(x)$ individually as rates or intensities, even though they are only parts of the compound rate $\lambda_{jk}\phi_j(x)$. Also, we call them "routing" and "service" rates, but they may have other interpretations.

Under the preceding assumptions, X is a Markov network process with single-unit movements and its transition rates are

$$q(x, y) = \begin{cases} \lambda_{jk}\phi_j(x) & \text{if } y = T_{jk}x \in \mathbb{E} \text{ for some } j \neq k \text{ in } M \\ 0 & \text{otherwise.} \end{cases} \tag{1.6}$$

We sometimes express these transition rates compactly as $q(x, T_{jk}x) = \lambda_{jk}\phi_j(x)$, where it is understood that $T_{jk}x$ is in $\mathbb{E}$.

In addition to the assumption on exponential times to movements, we assume the service intensities are balanced as follows.

Definition 1.6. The service intensities ϕ_j are Φ-*balanced* if Φ is a positive function on $\mathbb{E}$ such that, for each $x \in \mathbb{E}$ and $j, k \in M$ with $T_{jk}x \in \mathbb{E}$,

$$\Phi(x)\phi_j(x) = \Phi(T_{jk}x)\phi_k(T_{jk}x).$$

This is a natural condition on the service intensities under which the process has a tractable stationary distribution. More insights on Φ-balance are in Section 1.13. Here is an important example.

Example 1.7. *Independently Operating Nodes.* Consider the case in which each $\phi_j(x)$ is a function $\phi_j(x_j)$ of only x_j—there are no additional restrictions on the form of these functions. We sometimes refer to these functions as being *node-dependent* service rates. An easy check shows that these rates are balanced by

$$\Phi(x) = \prod_{j \in M} \prod_{n=1}^{x_j} \phi_j(n)^{-1}.$$

Here and below, we use the convention that $\prod_{n=1}^{x} a_n = 1$ if $x = 0$. □

This completes the description of the network processes we will study. We name them as follows.

Definition 1.8. The Markov network process X with transition rates (1.6) and Φ-balanced service intensities is a *Whittle process*. It is a *Jackson process* if the service intensity $\phi_j(x)$ is a function $\phi_j(x_j)$ only of x_j, for each $j = 1, \dots, m$, and $\phi_0(\cdot) \equiv 1$ when the network is open.

Jackson and Whittle network processes are prominent because their stationary distributions have closed-form expressions. It is convenient to study these processes together since they have many features in common. Note that in a Jackson process, the service intensities $\phi_j(x_j)$ are "node-dependent" (a function of x_j), indicative of independently operating nodes. In a Whittle process, however, the service intensities are "system-dependent" (a function of x), indicative of dependently operating nodes. Jackson processes were named after Jackson who introduced them in 1957. Special cases of Whittle processes have been studied, but not under this name. We introduce the name Whittle processes to recognize his major contributions to the understanding of networks with system-dependent transitions.

Throughout the rest of this chapter, we will assume that X is either a Jackson process or a Whittle process as defined above. We will make it clear when results apply specifically to a Jackson process.

The following are some observations about the sample paths, services and routings in the Whittle process X. Because it is a Markov process, each of its sojourn times in state x is exponentially distributed with rate

$$\sum_j \sum_k q(x, T_{jk}x) = \sum_j \phi_j(x) \sum_k \lambda_{jk}.$$

Also, when the network is in state x, the time until a "potential" departure from node j (the minimum of the departures times to nodes $k \neq j$) is exponentially distributed with rate $\phi_j(x) \sum_k \lambda_{jk}$. This follows because the minimum of independent exponential variables is also exponential with rate being the sum of the rates of the variables. Upon ending a sojourn in state x, the process jumps to state $T_{jk}x \in \mathbb{E}$ with probability

$$p_{jk} = q(x, T_{jk}x) / \sum_\ell q(x, T_{j\ell}x) = \lambda_{jk} / \sum_\ell \lambda_{j\ell}, \quad j, k \in M. \tag{1.7}$$

Note that this probability is independent of $\phi_j(x)$ and the state x. The $\{p_{jk}\}$ is a Markov chain matrix with $p_{jj} = 0$. We refer to p_{jk} as the *routing probabilities* of

X. The p_{jk} is the conditional probability that a unit moves from j to k given that it does move out of j. Since expression (1.7) is independent of x, one can view the units departing from node j as being routed independently and identically according to the probabilities p_{jk}, $k \in M$.

The convention $\lambda_{jj} = 0$ does not rule out the possibility that a unit exiting node j may be fed back to j for another service. Such feedbacks are modeled as follows.

Example 1.9. *Networks with Feedbacks at Nodes.* Consider the Whittle process X under the assumption that, whenever it is in state x, the time to the next departure from node j is exponentially distributed with rate $\phi_j(x)$. But now, assume that a unit departing node j enters node k with probability $\bar{p}_{jk}$, independently of everything else, where the probability $\bar{p}_{jj}$ of a feedback may be positive. Then it follows, from the construction of Markov processes with feedbacks described in the last section, that the process X is a Whittle process with transition rates $q(x, T_{jk}x) = \bar{p}_{jk}\phi_j(x)$. In this case, a transition from x to $T_{jk}x$ occurs with probability $\bar{p}_{jk}$, and the exponential sojourn time in state x has the rate $\sum_j \phi_j(x)(1 - \bar{p}_{jj})$. □

In a transition from x to $T_{jk}x$, we refer to a "single unit" moving from j to k. However, more than one unit may actually move in the transition, as long as the node populations before and after the transition are x and $T_{jk}x$, respectively. For instance, in a manufacturing network, a part exiting a certain node j may be considered as a completed part that actually exits the network and triggers another unit outside the network to take its place and enter node k.

1.4 Properties of Service and Routing Rates

This section gives more insight into service intensities. It also shows how the routing rates determine the irreducibility of Jackson and Whittle processes.

The service intensities of a Jackson or Whittle network have various interpretations. The following are standard examples of node-based intensities; more intricate system-dependent intensities are discussed later. Viewing the node-dependent $\phi_j(x_j)$'s as relative service intensities, we say that node j *consists of s exponential servers* with rate μ_j if $\phi_j(x_j) = \mu_j \min\{x_j, s\}$, $x_j \geq 1$. This typically represents the case in which there are s independent servers, $1 \leq s \leq \infty$, and their service times (or the service times required by the units) are independent and exponentially distributed with rate μ_j. This node therefore operates independently of the other nodes in that its rate does not depend on x_k, for $k \neq j$. A number of service disciplines are allowable since the units are indistinguishable (e.g., first-come, first-served, service in random or arbitrary order, or last-come-last-served). The case $\phi_j(x_j) = \mu_j x_j$ (when $s = \infty$) implies that node j is simply a delay point for each unit, and each delay is independent of everything else. Such a delay might be a time for a self-processing operation (e.g., a think time, maturation time, storage period, or self-maintenance time).

Another interpretation of ϕ_j is that it represents a *processor-sharing* scheme in which units are processed as follows. At any instant when x_j units are present, the time to the next "potential" departure of the ith unit there is exponentially distributed with rate $\mu_i(x_j) > 0$ such that $\sum_{i=1}^{x_j} \mu_i(x_j) = \phi_j(x_j)$. That is, node j works on the ith unit with intensity or rate $\mu_i(x_j)$, and all the units receive service simultaneously. This is *egalitarian processor-sharing* when $\mu_i(x_j) = \phi_j(x_j)/x_j$: each unit gets the same share of the $\phi_j(x_j)$. Regardless of the particular processing rule, the total departure intensity is simply $\phi_j(x_j)$. Keep in mind that a processor-sharing intensity function ϕ_j can have any form. For example, $\phi_j(x_j) = \mu_j$ can be a processor-sharing intensity even though it can also represent a single server with a first-come, first-served discipline.

The following result is a criterion for the network process X to be irreducible. Recall that the *routing process* of X is a Markov process with transition rates λ_{jk}. The sequence of states this routing process visits forms a Markov chain, whose transition probabilities are the routing probabilities $p_{jk} = \lambda_{jk}/\sum_{k'} \lambda_{jk'}$.

Proposition 1.10. *The Jackson or Whittle process X is irreducible if and only if its routing process is irreducible.*

PROOF. First, assume X is irreducible. To prove the routing process is irreducible, it suffices to show that, for any fixed $j \neq k$ in M, there exist $j_1, \ldots, j_\ell$ in M such that

$$\lambda_{jj_1}\lambda_{j_1j_2}\cdots\lambda_{j_\ell k} > 0. \tag{1.8}$$

Choose x and $\tilde{x}$ in $\mathbb{E}$ such that x_j and $\tilde{x}_k$ are positive. The irreducibility of X ensures that there exist $j_1, \ldots, j_\ell$ in M such that the states

$$x,\ x^1 = T_{jj_1}x, \ldots, x^\ell = T_{j_{\ell-1}j_\ell}x^{\ell-1},\ \tilde{x} = T_{j_\ell k}x^\ell, \tag{1.9}$$

form a feasible path from x to $\tilde{x}$, and so

$$q(x, x^1)q(x^1, x^2)\cdots q(x^\ell, \tilde{x}) > 0. \tag{1.10}$$

This selection of states first chooses j_1 and j_ℓ such that $q(x, x^1)$ and $q(x^\ell, \tilde{x})$ are positive, and then chooses $x^2, \ldots, x^{\ell-1}$ such that $x^1, \ldots, x^\ell$ is a feasible path from x^1 to x^ℓ. Since the ϕ_j's implicit in (1.10) must be positive, then (1.10) yields (1.8).

Now, suppose the routing process is irreducible. Fix any $x \neq \tilde{x}$ in $\mathbb{E}$. Choose $j \neq k$ such that x_j and $\tilde{x}_k$ are positive, and then choose $j_1, \ldots, j_\ell$ in M that satisfy (1.8). Consider the states defined by (1.9). Then (1.8) and the positiveness of the ϕ_j's yield (1.10). This shows that the process X can communicate between any x and $\tilde{x}$, and hence it is irreducible. □

Remark 1.11. *(Routing in Jackson Networks).* Suppose the network is a Jackson network and the routing rates λ_{jk} are reducible with n disjoint recurrent communication classes in M. Then one can view the network as consisting of n subnetworks. The subnetworks would be independent, because the service intensity at each node does not depend on units elsewhere. Consequently, one could analyze the subnetworks as a collection of separate irreducible network processes. Therefore, with no loss in generality, we will assume that the routing rates are irreducible for Jackson networks.

A Whittle network with reducible routing, however, would be a collection of subnetworks that are independent in their routing, but dependent through their service intensities. We discuss such interacting subnetworks and partially open networks as well in Sections 1.13 and 3.1.

1.5 Equilibrium Behavior

We are now ready to characterize invariant measures for Jackson and Whittle processes.

In addition to the notation above, let w_j, $j \in M$, denote a positive invariant measure that satisfies the *routing balance equations* or *traffic equations*

$$w_j \sum_{k \in M} \lambda_{jk} = \sum_{k \in M} w_k \lambda_{kj}, \quad j \in M. \tag{1.11}$$

To simplify some expressions, we adopt the convention that $w_0 = 1$ when the network is open. The existence of such an invariant measure is ensured, because M is a finite set and the routing process does not have transient states. When the network is closed, one may want to normalize w to be a probability distribution. Then it would be an invariant distribution for the routing rates λ_{jk} and also for the routing probabilities $p_{jk} = \lambda_{jk} / \sum_{k'} \lambda_{jk'}$.

When X is a Jackson process, we assume, as mentioned above, that λ_{jk} is irreducible. Hence X is irreducible by Proposition 1.10. When X is a Whittle process, we allow λ_{jk} to be reducible or irreducible—hence X may be reducible or irreducible.

The following results describe the equilibrium behavior of Jackson processes.

Theorem 1.12. *If X is a closed Jackson process with ν units, then it is ergodic and its stationary distribution is*

$$\pi(x) = c \prod_{j=1}^{m} w_j^{x_j} \prod_{n=1}^{x_j} \phi_j(n)^{-1}, \quad x \in \mathbb{E} = \{x : |x| = \nu\}, \tag{1.12}$$

where the w_j's satisfy (1.11). The c is the normalizing constant given by

$$c^{-1} = \sum_{x \in \mathbb{E}} \prod_{j=1}^{m} w_j^{x_j} \prod_{n=1}^{x_j} \phi_j(n)^{-1}.$$

Theorem 1.13. *If X is an open Jackson process with finite capacity ν, then the assertions of Theorem 1.12 with $\mathbb{E} = \{x : |x| \leq \nu\}$ apply to this process.*

Theorem 1.14. *If X is an open Jackson process with unlimited capacity, then it has an invariant measure of the form (1.12) with $\mathbb{E} = \{x : |x| < \infty\}$. Hence, the process is positive recurrent if and only if*

$$c_j^{-1} \equiv \sum_{x_j=0}^{\infty} w_j^{x_j} \prod_{n=1}^{x_j} \phi_j(n)^{-1} < \infty, \quad j = 1, \ldots, m.$$

In this case, its stationary distribution is

$$\pi(x) = \pi_1(x_1)\cdots\pi_m(x_m), \quad x \in \mathbb{E}, \tag{1.13}$$

where

$$\pi_j(x_j) = c_j w_j^{x_j} \prod_{n=1}^{x_j} \phi_j(n)^{-1}, \quad n \geq 0.$$

Recall that the node-dependent intensities $\phi_j(x_j)$ of a Jackson network are balanced by $\Phi(x) = \prod_{j=1}^{m} \prod_{n=1}^{x_j} \phi_j(n)^{-1}$. Consequently, the preceding theorems for Jackson networks are special cases of the following theorem for Whittle networks.

This result describes the equilibrium behavior of a Whittle process; the function Φ that balances its system-dependent service rates is characterized later in Proposition 1.46.

Theorem 1.15. *An invariant measure for the Whittle process X is*

$$\pi(x) = \Phi(x) \prod_{j=1}^{m} w_j^{x_j}, \quad x \in \mathbb{E}, \tag{1.14}$$

where the w_j's satisfy (1.11). The measure π also satisfies the partial balance equations

$$\pi(x) \sum_{k \in M} q(x, T_{jk}x) = \sum_{k \in M} \pi(T_{jk}x) q(T_{jk}x, x), \quad j \in M, x \in \mathbb{E}. \tag{1.15}$$

PROOF. Because the process X has single-unit movements, the balance equations an invariant measure π must satisfy are

$$\pi(x) \sum_j \sum_k q(x, T_{jk}x) = \sum_j \sum_k \pi(T_{jk}x) q(T_{jk}x, x), \quad x \in \mathbb{E}.$$

Since these equations are the sum of (1.15) over j, it follows that any measure satisfying (1.15) is an invariant measure. Therefore, it suffices to show that π given by (1.14) satisfies (1.15).

To this end, fix a $j \in M$ and $x \in \mathbb{E}$. If $x_j = 0$, then both sides of (1.15) are zero since $T_{jk}x \notin \mathbb{E}$ for each k. Now, assume $x_j > 0$. By the definitions of q and w_j, the left side of (1.15) is

$$\pi(x) \sum_k q(x, T_{jk}x) = \pi(x)\phi_j(x) \sum_k \lambda_{jk} = \pi(x)\phi_j(x) w_j^{-1} \sum_k w_k \lambda_{kj}.$$

Next, note that the definition of π and the Φ-balance property yield the identity

$$\pi(x)\phi_j(x) w_j^{-1} w_k = \pi(T_{jk}x)\phi_k(T_{jk}x), \quad k \in M.$$

Using this in the preceding equation, we have

$$\pi(x) \sum_k q(x, T_{jk}x) = \sum_k \pi(T_{jk}x)\phi_k(T_{jk}x)\lambda_{kj} = \sum_{k \in M} \pi(T_{jk}x) q(T_{jk}x, x).$$

Thus, π satisfies the partial balance equations (1.15). □

Here are some observations about the preceding results.

Remark 1.16. *(Partial Balance).* From the law of large numbers for Markov processes, condition (1.15) says that the average number of units departing from node j per unit time when X is in state x equals the average number of units entering node j per unit time that land X in state x. Or, loosely speaking, the equilibrium flow of units out of node j, for any state x, equals the flow into j. Equations (1.15) are called *partial balance equations* because they are only a part of the total balance equations. They are also called *station balance equations* because they say the flow into each station or node equals the flow out of the station. The partial balance equations are also satisfied by Jackson processes described in the theorems above.

Remark 1.17. *(Traffic Equations).* Although the traffic equations (1.11) precede Theorem 1.15, they are also a consequence of the result. Namely, upon substituting the measure π given by (1.14) in the partial balance equations (1.15), the service rate functions cancel and the traffic equations are what is left. In other words, the traffic equations are a necessary and sufficient condition for π to satisfy (1.15).

Remark 1.18. *(Nonuniqueness of the w_j's).* Note that the invariant measure π in the results above is the same for *any* positive solution w to the traffic equations. This follows since the normalization constant c is a function of w. In particular, the w_j's need not sum to one.

Remark 1.19. *(Can Any Measure be an Invariant Measure?).* Any measure on $\mathbb{E}$ can be an invariant measure of a Whittle process. For instance, the process with $q(x, T_{jk}x) = \Psi(x - e_j)/\pi(x)$, where $\lambda_{jk} \equiv 1$ and $w_j \equiv 1$, has invariant measure π. For Jackson processes, however, only product form measures can be invariant measures, since Φ is always a product form.

Remark 1.20. *(Additional Modeling Capabilities).* In Chapter 3, we discuss applications of the results in this chapter to the following types of networks.

- Jackson and Whittle networks with multiple types of units.
- Kelly networks in which units have deterministic routes depending on their type.
- BCMP networks with multiple types of units and processor sharing.
- Networks in which the service time distributions can be general rather than exponentially distributed.
- Networks with an infinite number of nodes and units.

Remark 1.21. *(Other Types of State Spaces).* One can define Jackson and Whittle processes on state spaces other than the three standard spaces we are considering. For instance, one may want to restrict the number of units at the nodes to be below certain levels. Network processes on other spaces, however, may not have invariant measures as those above. The reason is that there may be boundary effects in the spaces that do not have the same balance properties. There are some networks with reversible or locally reversible routing, however, that still have invariant measures as above; see Sections 2.4, 2.5, 2.7, 2.9, 3.5, and 3.6.

Remark 1.22. *(Weak Coupling of Services and Routing).* The transition function q of the Whittle process is a "weak coupling" of two transition functions in the sense that

$$q(x, y) = q_1(x, y)q_2(x, y), \quad x \neq y \in \mathbb{E}, \tag{1.16}$$

where $q_1(x, T_{jk}x) \equiv \lambda_{jk}$ involves only the routing rates, and $q_2(x, T_{jk}x) \equiv \phi_j(x)$ involves only the service rates. Theorem 1.15 with $\phi_j \equiv 1$ says that $\pi_1(x) = \prod_{j=1}^m w_j^{x_j}$ is an invariant measure for q_1. Similarly, Theorem 1.15 with $\lambda_{jk} \equiv 1$ says that $\pi_2(x) = \Phi(x)$ is an invariant measure for q_2. In addition, Theorem 1.15 says that an invariant measure for q is the product $\pi(x) = \pi_1(x)\pi_2(x)$. This product form does not automatically follow by the coupling (1.16). It is due to these additional properties:
(i) q_2 is reversible with respect to π_2 (because the service intensities are Φ-balanced; see Section 1.13).
(ii) $\pi_2(x)q_2(x, T_{jk}x)$ is independent of k for each j and x.
These are strong conditions, which are generally not satisfied for Markov processes.

The preceding remark raises the following question: Are there more general routing and service rates that lead to tractable stationary distributions? Some insight into this issue is given by the following result. Suppose the transition rates of the process X are of the form

$$q(x, y) = \begin{cases} \phi_j(x)\lambda_{jk}(x) & \text{if } y = T_{jk}x \text{ for some } j \neq k \text{ in } M \\ 0 & \text{otherwise,} \end{cases} \tag{1.17}$$

where $\lambda_{jk}(x)$ is a routing rate as a function of the state x.

Proposition 1.23. (State-dependent Routing) *For the network process with transition rates (1.17), assume that the ϕ_j are Φ-balanced and that there is a positive function Λ on $\mathbb{E}$ such that*

$$\Lambda(x)\sum_k \lambda_{jk}(x) = \sum_k \Lambda(T_{jk}x)\lambda_{kj}(T_{jk}x), \quad j \in M,\ x \in \mathbb{E} \text{ with } x_j \geq 1. \tag{1.18}$$

Then an invariant measure of the process is

$$\pi(x) = \Phi(x)\Lambda(x), \quad x \in \mathbb{E}. \tag{1.19}$$

PROOF. The proof is similar to that of Theorem 1.15. The approach is to show that $\pi(x) = \Phi(x)\Lambda(x)$ satisfies the partial balance equations (1.15). The main step is that, for each $x \in \mathbb{E}$ with $x_j \geq 1$,

$$\begin{aligned}\pi(x)\sum_k q(x, T_{jk}x) &= \Phi(x)\phi_j(x)\Lambda(x)\sum_k \lambda_{jk}(x) \\ &= \sum_k \Phi(T_{jk}x)\phi_k(T_{jk}x)\Lambda(T_{jk}x)\lambda_{kj}(T_{jk}x) \\ &= \sum_k \pi(T_{jk}x)q(T_{jk}x, x). \qquad \square\end{aligned}$$

Although this result provides a general framework for state-dependent routing, it does not solve the problem of finding π. This is because obtaining a Λ that satisfies (1.18)—without any more information about $\lambda_{jk}(x)$—is essentially equivalent to finding π. In other words, for general routing or service transition functions in the coupled transition rate (1.16), the problem amounts to finding invariant measures for general transition rates of the form $q(x, T_{jk}x) = q_{jk}(x)$.

1.6 Production–Maintenance Network

Before developing further properties of networks, we give an application in this section of a closed Jackson network. This network is indicative of maintenance networks that arise in industrial and military settings for maintaining expensive equipment to produce goods or services or to perform a mission.

Consider a system shown in Figure 1.2 in which ν machines (subsystems, trucks or electronic equipment) are available for use at some facility or location called node 1. At most s_1 machines can be in use at node 1 at any time for producing goods or services. Therefore, if x_1 machines are present then $\min\{x_1, s_1\}$ of these will be in use. After a machine is put into use, it operates continuously until it fails or degrades to a point that it requires a repair. The total operating time is exponentially distributed with rate μ_1. At the end of this time, the unit is transported to a repair facility. The transportation system (which may involve initial processing and rail or air travel) is called node 2, and the unit's time at this node is exponentially distributed with rate μ_2; there is no queueing for the transportation.

The repair facility consists of nodes 3, 4, 5, which are single-server nodes with respective rates μ_3, μ_4, μ_5. Depending on the nature of the repair, the unit goes to one of these nodes with respective probabilities p_{23}, p_{24}, p_{25}. After its repair, the unit goes to another transportation system, called node 6, for an exponentially distributed time with rate μ_6. And then it enters node 1 to begin another production/repair cycle.

Let X denote the process representing the numbers of machines at the respective nodes. Under the preceding assumptions, X is a closed Jackson process in which each node j is an s_j-server node, where $s_2 = s_6 = \infty$ and $s_j = 1$ for $j = 3, 4, 5$. The rate of each server at node j is μ_j. The routing intensities are the routing probabilities $\lambda_{12} = \lambda_{56} = \lambda_{61} = 1$ and $\lambda_{2k} = p_{2k}$ for $k = 3, 4, 5$; the other λ_{jk}'s are 0. The traffic equations (1.11) for these routing probabilities have a solution $w_j = 1$ for $j = 1, 2, 6$ and $w_j = p_{2j}$ for $j = 3, 4, 5$. Then by Theorem 1.12, the

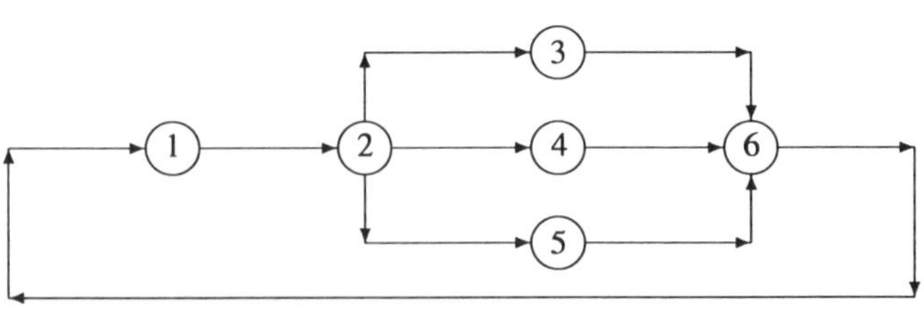

FIGURE 1.2. Production–Maintenance Network

stationary distribution of X is

$$\pi(x) = c\frac{1}{x_2!x_6!}\prod_{n=1}^{x_1}\frac{1}{\min\{n, s_1\}}\prod_{j=1}^{6}(w_j/\mu_j)^{x_j}, \qquad x \in \mathbb{E},$$

where c is the normalization constant.

The quality of this maintenance system is measured by the number of machines in productive use at node 1. Suppose the aim is to find the number of machines ν^* to provision for the network such that the probability of having less than $\bar{x}_1$ machines in use at node 1 is below β (for instance .10). From the stationary distribution above, it follows that the equilibrium probability of having less than $\bar{x}_1$ machines at node 1 (as a function of ν) is

$$\begin{aligned} p(\nu) &= \sum_x \pi(x)1(x_i < \bar{x}_1) \\ &= c\sum_{n=0}^{\bar{x}_1-1}\frac{1}{\mu_1^n n!}\sum_{x_2,\dots,x_6} 1\Big(\sum_{j=2}^{6} x_j = \nu - n\Big)\frac{1}{x_2!x_6!}\prod_{j=2}^{6}(w_j/\mu_j)^{x_j}. \end{aligned}$$

Then the desired provisioning quantity is $\nu^* = \min\{\nu : p(\nu) < \beta\}$. This is obtained by computing $p(\nu)$ for $\nu = \bar{x}_1, \bar{x}_1 + 1 \dots$ until it falls below β.

1.7 Networks with Special Structures

The structure of a network is determined by the communication graph of its Markov routing process. The set of all communication graphs of Markov processes is vast. When the finite node set M is not too large, one can use a standard numerical procedure for finding an invariant measure w_j for the transition probabilities $p_{jk} = \lambda_{jk}/\sum_\ell \lambda_{j\ell}$. In some cases, however, these measures have closed-form expressions. This section describes several elementary examples that are relevant for networks.

Example 1.24. *Nearest Neighbor Travel.* Suppose the routing of units in a closed network is a *simple random walk* on the nodes $1, \dots, m$ in which a unit at node j moves to $j+1$ or $j-1$ with respective probabilities p_j and $1 - p_j$, where $p_1 = 1$ and $p_m = 0$. In this case, it is well known that an invariant measure of the routing rates is $w_1 = 1$ and

$$w_j = \frac{p_0 \cdots p_{j-1}}{(1-p_1)\cdots(1-p_j)}, \qquad j = 2, \dots, m. \qquad \square$$

Example 1.25. *Progress-or-Return-to-Origin Network.* This type of closed network with $m = 5$ nodes has a communication graph shown in Figure 1.3 below. Here the routing rates $\lambda_{12}, \lambda_{j,j+1}$ and λ_{j1} are positive for $1 \le j \le m$, and all other rates are 0. Then the traffic equations are

$$w_1\lambda_{12} = \sum_{j=2}^{m} w_j\lambda_{j1}, \quad w_j(\lambda_{j1} + \lambda_{j,j+1}) = w_{j-1}\lambda_{j-1,j}, \quad 2 \le j \le m.$$

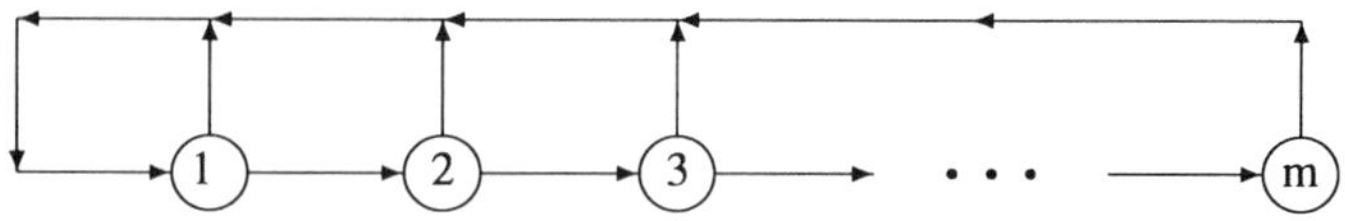

FIGURE 1.3. Progress-or-Return-to-Origin Network

A solution is $w_1 = 1$ and

$$w_j = \prod_{i=1}^{j}(\lambda_{i-1,i}/(\lambda_{i1} + \lambda_{i,i+1})), \quad 2 \le j \le m. \qquad \square$$

When the routing rates λ_{jk} are reversible, then its invariant measure has a closed-form expression given by Theorem 2.8. We will see later in Example 2.25 that reversible routing is a necessary and sufficient condition for a Jackson or Whittle network process to be reversible. For instance, Example 2.10 describes a circular network with reversible routing. Another example is the following special case of Example 2.25.

Example 1.26. *Star-Shaped Network.* The graph of a *star-shaped* or *central-processor* network with $m = 5$ nodes is shown in Figure 1.4. Node 1 is the center node and nodes $2, \dots, m$ are points of the star such that the routing rates λ_{1j} and λ_{j1} are positive. All the other routing rates are 0.

In this case, the traffic equations (1.11) are $w_j\lambda_{j1} = w_1\lambda_{1j}$, $2 \le j \le m$. Then the routing rates are reversible with respect to the invariant measure $w_1 = 1$ and $w_j = \lambda_{1j}/\lambda_{j1}$, $2 \le j \le m$. $\square$

1.8 Properties of Jackson Equilibrium Distributions

In this section, we discuss how one can obtain the normalization constants and marginal distributions for the equilibrium distribution of a Jackson network. For this discussion, we assume that X is an ergodic Jackson process for a network that may be open or closed.

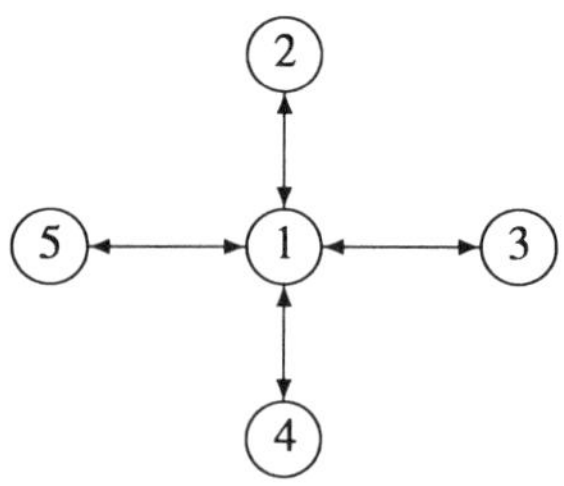

FIGURE 1.4. Star-Shaped Network

First, consider the case in which the network is open with unlimited capacity. By Theorem 1.14, the stationary distribution of X is

$$\pi(x) = \prod_{j=1}^{m} \pi_j(x_j) = \prod_{j=1}^{m} \left[c_j \prod_{j=1}^{m} w_j^{x_j} \prod_{n=1}^{x_j} \phi_j(n)^{-1} \right], \qquad |x| < \infty.$$

Recall that the w_j's satisfy the traffic equations

$$w_j \sum_{k\in M} \lambda_{jk} = \sum_{k\in M} w_k \lambda_{kj}, \qquad j \in M.$$

This distribution π is a product of its marginal distributions π_j. Consequently, if X is stationary, then, for *each fixed* t, its m components $X_t^1, \ldots, X_t^m$ are independent. Of course, X_s^j and X_t^j for $s \neq t$ are dependent, and hence so are X_s and X_t. Note that each marginal distribution π_j is the equilibrium distribution of a birth–death queueing process with transition rates

$$q(n, n') = w_j 1(n' = n + 1) + \phi_j(n) 1(n' = n - 1 \geq 0).$$

In other words, each node j in equilibrium appears to be like a single node in isolation in which units arrive by a Poisson process with intensity w_j, and when n units are present, they are served at the rate $\phi_j(n)$. The actual arrival process of units into node j in equilibrium, however, is generally not a Poisson process.

Next, consider the case in which the Jackson network is closed with ν units. By Theorem 1.12, the stationary distribution of the process X is

$$\pi(x) = c \prod_{j=1}^{m} f_j(x_j), \qquad x \in \mathbb{E}, \tag{1.20}$$

where $f_j(n) = w_j^n \prod_{r=1}^{n} \phi_j(r)^{-1}$ and the w_j's satisfy the traffic equations. Although the distribution π is a product form, it is not a distribution of independent random variables since $|x| = \nu$.

We now show that the normalization constant and marginal distributions are expressible in terms of convolutions of functions. The convolution $f \star g$ of two functions f and g on the nonnegative integers is defined by

$$f \star g(n) = \sum_{i=0}^{n} f(i) g(n - i), \qquad n \geq 0.$$

For a sequence of such functions $f_1, f_2, \ldots$, it follows by induction on m that

$$f_1 \star \ldots \star f_m(n) = \sum_{x:|x|=n} \prod_{j=1}^{m} f_j(x_j), \qquad n \geq 0,\ m \geq 1. \tag{1.21}$$

This property yields the following computational procedure.

Remark 1.27. (*Normalizing Constant for Closed Jackson Network*). The normalizing constant c has the representation

$$c^{-1} = \sum_{x:|x|=\nu} \prod_{j=1}^{m} f_j(x_j) = f_1 \star \ldots \star f_m(\nu). \tag{1.22}$$

One can compute c by the following procedure. For each $\ell = 1, \ldots, \nu$, define $g_\ell(n) = f_1 \star \ldots \star f_\ell(n)$, for $0 \le n \le \nu$. Then compute these convolutions by the recursion

$$g_\ell(n) = f_\ell \star g_{\ell-1}(n), \quad 0 \le n \le \nu,$$

for $\ell = 2, \ldots, m$. The final iteration yields g_m, from which one obtains $c^{-1} = g_m(\nu)$. The number of computations for this procedure is of the order $m\nu$.

Remark 1.28. (*Marginal Distributions of Closed Jackson Networks*). Knowing π, one can obtain the marginal equilibrium distribution of the number of units at a single node or of the numbers in sets of nodes as follows. We call a subset of nodes $J \subset M$ a *sector* of the network. Associated with J, we define $x(J) = \sum_{j \in J} x_j$, and let f_J denote the convolution of the functions $\{f_j,\ j \in J\}$. Now consider any disjoint sectors $J_1, \ldots, J_\ell$ whose union is M. The joint equilibrium distribution of $n_1, \ldots, n_\ell$ units in these sectors is

$$\begin{aligned}\pi_{J_1,\ldots,J_\ell}(n_1, \ldots, n_\ell) &= c \prod_{i=1}^{\ell} \sum_{x_j : j \in J_i} 1(x(J_i) = n_i) \prod_{j \in J_i} f_j(x_j) \\ &= c \prod_{i=1}^{\ell} f_{J_i}(n_i), \quad n_1 + \ldots + n_\ell = \nu.\end{aligned}$$

The last equality follows by property (1.21) for convolutions. From these distributions, one can obtain means, variances, covariances, and other items of interest such as expected costs for the process. In particular, the *marginal* equilibrium distribution of the number of units in the sector J is

$$\pi_J(n) = c f_J(n) f_{J^c}(\nu - n), \quad 0 \le n \le \nu, \tag{1.23}$$

where J^c is the complement of J.

Next, suppose X is an open Jackson process with capacity ν. Here, the normalization constant for its stationary distribution (1.20) (where $\mathbb{E} = \{|x| \le \nu\}$) has the representation

$$c^{-1} = \sum_{x:|x| \le \nu} \prod_{j=1}^{m} f_j(x_j) = \sum_{n=0}^{\nu} f_1 \star \ldots \star f_m(n).$$

Also, the joint equilibrium probability of $n_1, \ldots, n_\ell$ units in the respective sectors $J_1, \ldots, J_\ell$ that partition M is

$$\pi_{J_1,\ldots,J_\ell}(n_1, \ldots, n_\ell) = c \prod_{i=1}^{\ell} f_{J_i}(n_i), \quad n_1 + \ldots + n_\ell \le \nu.$$

Keep in mind that w_j implicit in the functions f_j is the invariant measure of the routing intensities for the "open" network; consequently, $f_1 \star \ldots \star f_m(n)$ is not necessarily the inverse of the normalization constant for the related closed network with n units (although it appears to be).

Another useful observation is that one can interpret this finite-capacity open network process X as a closed network process on $\{0, 1, \ldots, m\}$ with the same routing and service intensities as X, plus the intensity $\phi_0(\cdot) \equiv 1$ for node 0. Then clearly $\pi(x_1, \ldots, x_m) = \bar{\pi}(\nu - |x|, x_1, \ldots, x_m)$, where $\bar{\pi}$ is the stationary distribution of the closed network.

Jackson networks with infinite-server nodes are useful for modeling storage systems or service systems in which the units move independently and there is no queueing. Expressions for their equilibrium distributions are as follows.

Example 1.29. *Jackson Networks with Infinite-Server Nodes.* Suppose X is a Jackson network process, where each node j is an infinite-server node and each server has rate μ_j. The departure intensity is therefore $\phi_j(x_j) = x_j\mu_j$. Then its stationary distribution (1.20) is

$$\pi(x) = \frac{c}{x_1! \cdots x_m!} r_1^{x_1} \cdots r_m^{x_m}, \tag{1.24}$$

where $r_j = w_j/\mu_j$. In case X is an open network with unlimited capacity,

$$\pi(x) = \prod_{j=1}^{m} e^{-r_j} r_j^{x_j}/x_j!,$$

which is a product of Poisson distributions. In case X is a closed network, its distribution (1.24) is the multinomial distribution

$$\pi(x) = \frac{\nu!}{x_1! \cdots x_m!} p_1^{x_1} \cdots p_m^{x_m}, \quad |x| = \nu,$$

where $p_j = r_j/(r_1+\cdots+r_m)$. This follows by applying the multinomial expansion

$$(r_1 + \cdots + r_m)^\nu = \sum_{x:|x|=\nu} \frac{\nu!}{x_1! \cdots x_m!} r_1^{x_1} \cdots r_m^{x_m}$$

to (1.24). In this closed network, each unit is moving independently as a Markov chain whose stationary distribution is $\{p_j\}$. One would therefore anticipate that, in equilibrium, the numbers of units at the nodes have the preceding multinomial distribution. Because of the multinomial form, the number of units in any sector J has a binomial distribution with parameters ν and $\sum_{j\in J} p_j$. Finally, when X is an open network with finite capacity, then

$$\pi(x) = \frac{|x|!}{x_1! \cdots x_m!} p_1^{x_1} \cdots p_m^{x_m}, \quad |x| \le \nu.$$

This is a conditional multinomial distribution given that there are $|x|$ units in the system. □

Example 1.30. *Jackson Networks with Single-Server Nodes.* Suppose X is a Jackson process in which each node j is a single-server node with rate μ_j. Then its stationary distribution (1.20) is

$$\pi(x) = c r_1^{x_1} \cdots r_m^{x_m}, \tag{1.25}$$

where $r_j = w_j/\mu_j$. In case X is an open network with unlimited capacity, then

$$\pi(x) = \prod_{j=1}^{m}(1 - r_j)r_j^{x_j}.$$

This is a product of equilibrium distributions of birth–death processes with birth rates w_j and death rates μ_j. If X is closed or open with finite capacity ν, then the distribution (1.25) has normalization constants given respectively by

$$c^{-1} = f_1 \star \cdots \star f_m(\nu), \quad c^{-1} = \sum_{n=0}^{\nu} f_1 \star \cdots \star f_m(n),$$

where $f_j(n) = r_j^n$. These are very special convolutions that have closed form expressions given in the next section. □

1.9 Convolutions for Single-Server Nodes

For closed or finite-capacity open Jackson networks with single-server nodes, the preceding example showed that their equilibrium distributions have normalization constants that are functions of the convolution $f_1 \star \cdots \star f_m(\nu)$, where $f_j(n) = r_j^n$. Their marginal distributions involve similar convolutions. The following results are closed form expressions for these convolutions.

Proposition 1.31. *If $r_1, \ldots, r_m$ are distinct, then*

$$f_1 \star \cdots \star f_m(\nu) = \sum_{j=1}^{m} r_j^{\nu+m-1} \prod_{k\neq j}(r_j - r_k)^{-1}. \tag{1.26}$$

PROOF. Consider the generating function

$$G(z) = \sum_{\nu=0}^{\infty} f_1 \star \cdots \star f_m(\nu) z^{\nu}.$$

Then we can write

$$f_1 \star \cdots \star f_m(\nu) = G^{(\nu)}(0)/\nu!, \tag{1.27}$$

where $G^{(\nu)}$ is the νth derivative of G. Since the generating function of a convolution of functions is the product of the generating functions of the convolved functions, and $f_j(n) = r_j^n$, we have

$$G(z) = \prod_{j=1}^{m} \sum_{\nu=0}^{\infty} f_j(\nu) z^{\nu} = \prod_{j=1}^{m}(1 - r_j z)^{-1}.$$

Now, since $r_1, \ldots, r_m$ are distinct, the standard partial fraction expansion of this product is

$$G(z) = \sum_{j=1}^{m} \frac{c_j}{(1 - r_j z)},$$

where

$$c_j = \lim_{z \to 1/r_j} (1 - r_j z) G(z) = \prod_{k \neq j} (1 - r_k / r_j).$$

Then clearly $G^{(\nu)}(0) = \nu! \sum_{j=1}^{m} c_j r_j^{\nu}$. This and (1.27) yield (1.26). □

When the parameters r_j are not distinct, $f_1 \star \cdots \star f_m(\nu)$ has the following expression.

Proposition 1.32. *Let $\bar{r}_1, \ldots, \bar{r}_{\bar{m}}$ be the distinct r_j's and let n_ℓ denote the number of r_j's equal to $\bar{r}_\ell$. Then*

$$f_1 \star \cdots \star f_m(\nu) = \sum_{\ell=1}^{\bar{m}} (-1)^{n_\ell - 1} (\bar{r}_\ell)^{\nu + m - n_\ell} h(n_\ell, m, \bar{m}) \tag{1.28}$$

where

$$h(n_\ell, m, \bar{m}) = \sum_{i_1 + \cdots + i_{\bar{m}} = n_\ell - 1} (-1)^{i_\ell} \frac{(m + i_\ell)!}{i_\ell! m!} \times \prod_{u=1, u \neq \ell}^{\bar{m}} \frac{(n_u + i_u - 1)!}{i_u!(n_u - 1)!} (\bar{r}_u)^{i_u} / (\bar{r}_\ell - \bar{r}_i)^{n_u + i_u}.$$

PROOF. Proceeding as in the proof above, we have (1.27), where

$$G(z) = \prod_{j=1}^{m} (1 - r_j z)^{-1} = \prod_{\ell=1}^{\bar{m}} (1 - \bar{r}_\ell z)^{-n_\ell}.$$

Clearly G is analytic in the complex plane, except at $1/\bar{r}_1, \ldots 1/\bar{r}_{\bar{m}}$, which are poles of G of orders $n_1, \ldots, n_{\bar{m}}$, respectively. Then by the Cauchy integral formula for derivatives,

$$G^{(\nu)}(0) = \frac{\nu!}{2\pi i} \int_{|z| = \varepsilon} f(z)\, dz, \tag{1.29}$$

where $f(z) = G(z)/z^{\nu+1}$ and the integral is counterclockwise on the curve $|z| = \varepsilon$, for some $\varepsilon < \min\{1/\bar{r}_1, \ldots 1/\bar{r}_{\bar{m}}\}$.

For a fixed $b > \max\{1/\bar{r}_1, \ldots 1/\bar{r}_{\bar{m}}\}$, the function f is analytic in the region $|z| \leq b$, except at $0, 1/\bar{r}_1, \ldots 1/\bar{r}_{\bar{m}}$, which are poles of f. Then by the residue theorem for complex integrals,

$$\frac{1}{2\pi i} \int_{|z| = b} f(z)\, dz = \mathrm{Res}_{\{z=0\}} f + \sum_{\ell=1}^{\bar{m}} \mathrm{Res}_{\{z = 1/\bar{r}_\ell\}} f = 0, \tag{1.30}$$

where $\mathrm{Res}_{\{z=\zeta\}} f$ denotes the residue of f at ζ. That this integral is zero is a standard result for a function such as f that is a ratio of polynomials in which the denominator has a degree that is at least two more than the numerator. Now,

$$\mathrm{Res}_{\{z=0\}} f = \frac{1}{2\pi i} \int_{|z| = b} f(z)\, dz,$$

and since $1/\bar{r}_\ell$ is a pole of order n_ℓ,

$$\mathrm{Res}_{\{z=1/\bar{r}_\ell\}} f = \frac{1}{(n_\ell - 1)!} \left(\frac{d}{dz}\right)^{n_\ell - 1} [(z - 1/\bar{r}_\ell)^{n_\ell} f(z)] \,|_{z=1/\bar{r}_\ell}.$$

Consequently, (1.27), (1.29), and (1.30) yield

$$f_1 \star \cdots \star f_m(\nu) = -\sum_{\ell=1}^{\bar{m}} \frac{(-1/\bar{r}_\ell)^{n_\ell}}{(n_\ell - 1)!} \left(\frac{d}{dz}\right)^{n_\ell - 1} [z^{-\nu-1} \prod_{k \neq \ell}^{\bar{m}} (1 - \bar{r}_k z)^{-n_k}] \,|_{z=1/\bar{r}_\ell}.$$

This reduces to (1.28) by applying the multinomial derivative formula

$$\left(\frac{d}{dz}\right)^n h_1(z) \cdots h_{\bar{m}}(z) = \sum_{i_1+\cdots+i_{\bar{m}}=n} \frac{n!}{i_1! \cdots i_{\bar{m}}!} \times \left(\frac{d}{dz}\right)^{i_1} h_1(z) \cdots \left(\frac{d}{dz}\right)^{i_{\bar{m}}} h_{\bar{m}}(z).$$

This concludes the proof. □

For a closed or finite-capacity Jackson network, recall that the joint equilibrium distribution of the numbers of units in sectors $J_1, \ldots, J_\ell$ that partition the nodes is given by $\pi_{J_1,\ldots,J_\ell}(n_1, \ldots, n_\ell) = c \prod_{i=1}^{\ell} f_{J_i}(n_i)$, where f_{J_i} is the convolution of the functions f_j for $j \in J_i$. These convolutions can be computed by the formulas in the preceding results. The joint equilibrium distribution is also useful for computing other quantities. For instance, for the closed network in equilibrium, the expected number of units in the sector J is

$$L_J = c \sum_{n=1}^{\nu} n f_J(n) f_{J^c}(\nu - n).$$

The c is the normalization constant that we have already discussed. For convenience, assume $r_1, \ldots, r_m$ are distinct. Then (1.26) and a little algebra yield

$$L_J = c \sum_{j \in J} \sum_{k \in J^c} \Big[r_{jk}^{|J|} r_k^{\nu+m} h_j(J) h_k(J^c) \times [1 - r_{jk}^{\nu}(\nu r_{jk} + \nu + 1)]/(1 - r_{jk})^2 \Big],$$

where $|J|$ is the number of nodes in J and $h_j(J) = \prod_{i \in J, i \neq j} (r_j - r_i)^{-1}$. An analogous formula can be obtained from (1.28) for nondistinct r_j's.

1.10 Throughputs and Expected Sojourn Times

The performance or quality of a network is typically measured by its expected queue lengths, the speeds at which units move through it (throughput rates), expected sojourn times of the network, and expected sojourn times of units at the nodes. In this section, we will describe these quantities for Jackson and Whittle networks.

Consider a Jackson or Whittle network that is represented by an ergodic process X whose equilibrium distribution π is given in Theorems 1.12–1.15. The number of units that move from a node j to a node k in the time interval $(0, t]$ is

$$N_{jk}(t) = \sum_n 1(X_{\tau_n} = T_{jk}X_{\tau_{n-1}}, \tau_n \in (0, t]),$$

where $0 \equiv \tau_0 < \tau_1 < \tau_2 \ldots$ are the transition times of X. More generally, the number of units that move from a sector J to a sector K during $(0, t]$ is

$$N_{JK}(t) = \sum_{j \in J} \sum_{k \in K} N_{jk}(t).$$

The sets J and K may overlap. The average number of such movements per unit time is

$$\rho_{JK} = \lim_{t \to \infty} t^{-1} N_{JK}(t).$$

Also, $\rho_{JK} = EN_{JK}(1)$ when the process X is stationary. The ρ_{JK} is called the *throughput from J to K*. Another performance measure is the *throughput of sector J* defined by $\lambda_J = \rho_{J^cJ}$, which is the average number of units that enter J per unit time. It also equals the average number of units ρ_{JJ^c} that exit J per unit time since the process is ergodic. Note that

$$\rho_{JK} = \sum_{j \in J} \sum_{k \in K} \rho_{jk}, \quad \text{and} \quad \lambda_J = \sum_{j \in J^c} \sum_{k \in J} \rho_{jk}.$$

By the law of large numbers for Markov processes, we have the general expression

$$\rho_{jk} = \sum_{x \in \mathbb{E}} \pi(x) q(x, T_{jk}x) = \lambda_{jk} \sum_{x \in \mathbb{E}} \pi(x) \phi_j(x) 1(x_j \geq 1), \quad j, k \in M. \quad (1.31)$$

This expression simplifies in the following cases.

Proposition 1.33. *Suppose that X is a Jackson process, or that X is a Whittle process with service rates of the form*

$$\phi_j(x) = \Phi(x - e_j)/\Phi(x), \quad x \in \mathbb{E}, \ j \in M.$$

If X is open with unlimited capacity, then

$$\rho_{jk} = w_j \lambda_{jk}, \quad j, k \in M.$$

If X is closed with ν units (or open with capacity ν), then

$$\rho_{jk} = c_\nu c_{\nu-1}^{-1} w_j \lambda_{jk}, \quad j, k \in M.$$

Here c_ν is the normalizing constant for the equilibrium distribution of the closed network with ν units (or the open network with capacity ν).

PROOF. Under the hypothesis, (1.31) is

$$\rho_{jk} = w_j \lambda_{jk} \sum_{x \in \mathbb{E}} \pi(x - e_j) 1(x_j \geq 1). \tag{1.32}$$

Then the first assertion follows since the last sum is 1 for the unlimited capacity network with $|x| < \infty$. Also, the second assertion follows from (1.32), since the summation equals $c_\nu / c_{\nu-1}$, which is clear by the definition of c_ν. □

Another important feature of a network process is its sojourn or waiting times in certain sets of its state space. The following expression for average waiting times follows from Theorem 1.3.

Proposition 1.34. *The average waiting time of X in a set $A \subset \mathbb{E}$ is*

$$W(A) = \pi(A)/\lambda(A),$$

where

$$\lambda(A) = \sum_{x \in A^c} \pi(x) q(x, T_{jk}x) 1(T_{jk}x \in A).$$

The following is a typical example. Here we define $x(J) = \sum_{j \in J} x_j$ and let f_J denote the convolution of the functions $\{f_j : j \in J\}$.

Example 1.35. *Busy Periods and High-Level Exceedances.* Suppose X represents a closed Jackson network with ν units. Consider the length of time that the number of units in a sector J exceeds a level b. This is the sojourn time of X in the set $A = \{x : x(J) > b\}$. By the preceding proposition, the average sojourn time in A is

$$W(A) = \frac{\sum_{n=b+1}^{\nu} f_J(n) f_{J^c}(\nu - n)}{f_J(b) f_{J^c}(\nu - b) \sum_{j \in J} \sum_{k \in J^c} \lambda_{jk}}. \tag{1.33}$$

Indeed, the numerator is $\pi(A)/c$ (recall the marginal distribution (1.23)), and the denominator is $\lambda(A)/c$. This formula for $W(A)$ is especially nice because it does not involve the normalization constant c.

For a high level b, the average *high-level exceedance period* $W(A)$ might be used as a guide in designing a network. For instance, one might select service rates such that $W(A)$ is below a certain value. One could use (1.33) to characterize the set of service rates for which the constraint is satisfied. Another quantity of interest is the average duration of a *busy period for sector J*. This quantity is given by (1.33) with $b = 0$. The approach in this example can be used to obtain average exceedance periods for other types of Markovian networks. □

We now turn to expected queue lengths and sojourn times of units in a sector J of the network. First note that the average number of units in J per unit time is

$$L_J = \lim_{t \to \infty} t^{-1} \int_0^t X_s(J)\, ds = \sum_x x(J) \pi(x) \quad \text{w.p.1},$$

where $X_t(J) = x(J) \equiv \sum_{j \in J} x_j$ when $X_t = x$. This follows by the law of large numbers for Markov processes. For convenience, we assume L_J is finite.

Next, consider the sojourn times (or waiting times) $W_1(J)$, $W_2(J)$, ... of units in J, where $W_i(J)$ is the sojourn time of the ith unit to enter J. There is no restriction on the nodes at which the units enter and leave J; a unit may have multiple visits to the nodes in J before it exits, and units need not exit J in the same order in which they entered. We only assume $J \neq M$ when the network is closed (otherwise all sojourns would be infinite). Then the average waiting time of units in the sector J is

$$W_J = \lim_{n\to\infty} n^{-1} \sum_{i=1}^{n} W_i(J) \quad \text{w.p.1},$$

provided the limit exists.

The existence of these average waiting times is justified by the following Little laws. These results follow immediately from Theorems 5.1 and 5.2 that hold for Markovian systems that are recurrently empty. The emptiness condition is that the network contains a state x with $x(J) = 0$. This is automatically true in this case by the form of the state space.

Theorem 1.36. *The average waiting time W_J exits and $L_J = \lambda_J W_J$.*

When the process X is stationary, W_J is an expected waiting time as follows—its expectation is with respect to a Palm probability of the process, which is defined in Chapter 4.

Theorem 1.37. *Suppose the process X is stationary and let W_J denote the expected sojourn time in J with respect to the Palm probability of the stationary process X conditioned that a unit enters J at time 0. Then the expectation W_J is finite and $L_J = \lambda_J W_J$. Furthermore, $L_J = EX_t(J)$ and $\lambda_J = EN_{J^cJ}(1)$.*

When J is a single node j, we write the preceding Little law as $L_j = \lambda_j W_j$. Since the number of units in J at any time is the sum of the units at the single nodes in J, it follows that $L_J = \sum_{j\in J} L_j$. Similarly, $\lambda_J = \sum_{j\in J} \lambda_j$. Although this additivity is not generally true for waiting times, it is clear that $W_J = \sum_{j\in J} W_j$ if and only if a unit's waiting time in J involves exactly one visit to each node in J.

In an open network, a quantity of interest is the average time a unit spends in J (in possibly multiple visits to J) during its total stay in the network. This and related quantities for closed networks are as follows. The average waiting time of a unit in a sector J while it is in a larger sector $K \supset J$ is defined by

$$W_J^K = \lim_{n\to\infty} n^{-1} \sum_{i=1}^{n} W_i(J)^K \quad \text{w.p.1}.$$

Here $W_i(J)^K$ is the time the ith unit entering K spends in J before exiting K. A unit may have several visits to J while in K, and so $W_i(J)^K$ is the sum of all these waits in J. Note that $W_J^K \neq 0$ since each unit entering K has a positive probability of entering J (if the probability of moving from $K\backslash J$ to J is 0, then the irreducibility of the process ensures that there is a positive probability that a unit may enter J directly from K^c).

The general Little laws that justify the preceding results also apply to yield the following Little laws for W_J^K. Let L_J^K and λ_J^K denote the associated average queue length and arrival rate. Note that $L_J^K = L_J$ and $\lambda_J^K = \lambda_J$ since $J \subset K$. An analogue of Theorem 1.36 is as follows. There is also an obvious analogue of Theorem 1.37.

Theorem 1.38. *The average waiting time* W_J^K *exists and* $L_J = \lambda_K W_J^K$. *Furthermore,* $W_J^K = \lambda_J \lambda_K^{-1} W_J$.

1.11 Algorithms for Performance Parameters

This section contains recursive equations for computing the performance parameters for closed Jackson networks; namely, marginal equilibrium distributions, average queue lengths, waiting times, and throughputs. Recall that an open Jackson network with finite capacity can be interpreted as a closed Jackson process with node set $\{0, 1, \dots, m\}$ and service rate $\phi_0(\cdot) \equiv 1$ for node 0. Therefore, the performance parameters for this open network can also be computed by appropriate modifications of the results below.

Throughout this section, we assume the network process X represents a closed Jackson network. Let $\mathcal{M}$ be a collection of disjoint sectors whose union is $M = \{1, \dots, m\}$. The aim is to obtain the performance parameters L_J, W_J, λ_J, π_J for each sector $J \in \mathcal{M}$. One's choice of $\mathcal{M}$ would depend on the sectors of interest, but $\mathcal{M}$ must be a partition of M for completeness in the computations. For instance, if one were interested in the performance parameters of each node, then $\mathcal{M}$ would be all singleton nodes. If one were interested in the single nodes j, k, and ℓ along with sectors J, J' not containing these nodes, then $\mathcal{M}$ would be $\{\{j\}, \{k\}, \{\ell\}, J, J', J''\}$, where J'' is the sector consisting of the remaining nodes. If one were interested in sectors J, K that are not disjoint, then the procedure below would have to be performed separately for the two partitions $\mathcal{M} = \{J, J^c\}$ and $\mathcal{M} = \{K, K^c\}$.

Due to the structure of the equilibrium distribution of the process X, the performance parameters for this ν-unit network process are expressible as functions of the performance parameters of a $(\nu - 1)$-unit network process with the same routing probabilities and service rate functions. This is the key idea in the following result, which evaluates the performance parameters successively for the network with n units in it, where $n = 1, 2, \dots, \nu$. Here we let $L_J(n)$, $W_J(n)$, $\lambda_J(n)$, $\pi_J(i; n)$ denote the parameters for the n-unit network and each $J \in \mathcal{M}$.

Let f_J denote the convolution of the functions $\{f_j : j \in J\}$, where $f_j(i) = w_j^i \prod_{r=1}^i \phi_j(r)^{-1}$. Define

$$\alpha_J = \sum_{\ell \in J^c} \sum_{j \in J} w_\ell \lambda_{\ell j}, \qquad h_J(i) = \alpha_J^{-1} f_J(i) f_J(i-1)^{-1}.$$

Proposition 1.39. *For each $n = 1, 2, \ldots, \nu$ and $J \in \mathcal{M}$,*

$$W_J(n) = \sum_{i=1}^{n} i h_J(i)\pi_J(i-1; n-1), \tag{1.34}$$

$$\lambda_J(n) = n\alpha_J / \sum_{J' \in \mathcal{M}} \alpha_{J'} W_{J'}(n), \tag{1.35}$$

$$L_J(n) = \lambda_J(n) W_J(n), \tag{1.36}$$

$$\pi_J(i; n) = \lambda_J(n) h_J(i) \pi_J(i-1; n-1), \quad 1 \le i \le n, \tag{1.37}$$

$$\pi_J(0; n) = 1 - \sum_{i=1}^{n} \pi_J(i; n), \tag{1.38}$$

where $L_J(0) = 0$ and $\pi_J(0; 0) = 1$, for $J \in \mathcal{M}$.

Remark 1.40. To use these equations for computations, set $n = 1$ and compute (1.34)–(1.38) for each $J \in \mathcal{M}$. Then repeat the computations for $n = 2, 3, \ldots, \nu$.

PROOF. Consider the equations (1.34)–(1.38) from last to first. Expression (1.38) simply says the probabilities there must sum to 1. To verify (1.37), recall from the preceding section that the marginal distribution for sector J is

$$\pi_J(i; n) = c_n f_J(i) f_{J^c}(n-i)$$

where c_n is the normalizing constant for the n-unit network. This expression and its analogue for the $(n-1)$-unit network yield

$$\pi_J(i; n) = \alpha_J c_n / c_{n-1} h_J(i) \pi_J(i-1; n-1).$$

Also, we know that the throughput for J is $\lambda_J(n) = \alpha_J c_n / c_{n-1}$. Substituting this in the preceding display proves (1.37).

Next observe that (1.36) is the Little law in Theorem 1.36. To prove (1.35), we use the facts that the network is closed, $\mathcal{M}$ is a partition of M, and (1.36) to obtain

$$n = \sum_{J' \in \mathcal{M}} L_{J'}(n) = \sum_{J' \in \mathcal{M}} \lambda_{J'}(n) W_{J'}(n). \tag{1.39}$$

From the throughput expression above, $\lambda_{J'}(n) = \lambda_J(n)\alpha_{J'}/\alpha_J$. Substituting this in (1.39) yields (1.35).

Finally, by (1.36) and the definition of $L_J(n)$, it follows that

$$W_J(n) = \lambda_J(n)^{-1} L_J(n) = \lambda_J(n)^{-1} \sum_{i=1}^{n} i\pi_J(i; n).$$

Applying (1.37) to the last summation yields (1.34). □

Note that the evaluation of the performance parameters L_J, λ_J, and W_J by the preceding result requires the marginal distributions π_J. We now show that the marginal distributions are not needed when the network has only single-server or infinite-server nodes. Here we only consider the performance parameters for single nodes and not sectors, since the sector parameters can be obtained from single node parameters via the relations

$$L_J = \sum_{j \in J} L_j, \quad \lambda_J = \sum_{j \in J} \lambda_j, \quad W_J = L_J/\lambda_J.$$

Corollary 1.41. *Suppose each node j in the closed network is a single-server node with $\phi_j(n) = \mu_j$ and first-in, first-out service, or an infinite-server node with $\phi_j(n) = n\mu_j$ (each server has rate μ_j). Then, for each $n = 1, \dots, \nu$ and $j \in M$,*

$$W_j(n) = \begin{cases} [1 + L_j(n-1)](\mu_j \sum_k \lambda_{jk})^{-1} & \text{if } j \text{ is single-server} \\ (\mu_j \sum_k \lambda_{jk})^{-1} & \text{if } j \text{ is infinite-server,} \end{cases}$$

$$\lambda_j(n) = n\alpha_j / \sum_{j'} \alpha_{j'} W_{j'}(n),$$

$$L_j(n) = \lambda_j(n) W_j(n),$$

where $L_j(0) = 0$ and $\alpha_j = w_j \sum_k \lambda_{jk}$.

Proof. These equations are simply special cases of (1.34)–(1.36) with J representing the single node j. Namely, if node j is a single-server node with $\phi_j(i) = \mu_j$ and $f_j(i) = w_j^i \mu_j^{-i}$, then (1.34) reduces to

$$\begin{aligned} W_j(n) &= (\mu_j \sum_k \lambda_{jk})^{-1} \left(\sum_{i=1}^{n} (i-1)\pi_j(i-1; n-1) + 1 \right) \\ &= (\mu_j \sum_k \lambda_{jk})^{-1} \left(L_j(n-1) + 1 \right). \end{aligned}$$

And if j is an infinite-server node with $\phi_j(i) = i\mu_j$, then $f_j(i) = w_j^i \mu_j^{-i}/i!$ and $h_j(i) = w_j/(\alpha_j \mu_j i)$. Hence (1.34) reduces to $W_j(n) = w_j/(\mu_j \alpha_j)$. □

1.12 Monte Carlo Estimation of Network Parameters

We now discuss Monte Carlo procedures for estimating performance parameters of Jackson and Whittle networks. This approach is an alternative to the computational algorithms above for Jackson networks, and it is especially useful for Whittle networks that do not have such algorithms. We will give a brief overview of two procedures: random sampling and Metropolis Markov chain sampling. The latter has been used in several areas including estimating parameters in Gibbs distributions of Markov random fields and in optimization via simulated annealing.

Suppose that X is an ergodic Jackson or Whittle process that represents a closed or finite-capacity open network. The procedures below are for finite state spaces, but they can be extended to the unlimited capacity open network. We write the stationary distribution of X as

$$\pi(x) = c\eta(x), \quad x \in \mathbb{E},$$

where $\eta(x) = \Phi(x)\prod_{j=1}^m w_j^{x_j}$ and $c = 1/\sum_x \eta(x)$ is the normalization constant. Assume that $\Phi(x)$, w_j and hence $\eta(x)$ are known, but that c is not known. Many performance parameters of the process can be expressed as an expected value of the form

$$\mu = \sum_{x \in \mathbb{E}} g(x)\pi(x),$$

for some function g. We will focus on describing estimators for μ.

A typical example of μ is the throughput on the arc from node j to node k, which according to (1.31) is

$$\rho_{jk} = \sum_{x \in \mathbb{E}} \pi(x) q(x, T_{jk}x).$$

Another family of examples is as follows. Suppose $f(x)$ is a cost rate of being in state x and $h(x, y)$ is the cost or value of a transition from x to y. Then by the law of large numbers for Markov processes, the average cost per unit time is

$$\lim_{t\to\infty} t^{-1}[\int_0^t f(X_s)\,ds + \sum_{s \le t} h(X_{s-}, X_s)]$$
$$= \sum_{x \in \mathbb{E}} [f(x) + \sum_y q(x, y)h(x, y)]\pi(x) \quad \text{w.p.1.}$$

Note that the mean $\mu = \sum_{x\in\mathbb{E}} g(x)\pi(x)$ depends on the unknown normalization constant c of the distribution π. Also, note that c is a special case of μ with $g(x) = 1/\eta(x)$. We will now show how to estimate μ as well as c from data generated by a Markov chain. For this, we use an ergodic Markov chain $Y = \{Y_n : n \ge 0\}$ on $\mathbb{E}$ with a stationary distribution p, which is specified. This is an artificially constructed chain (separate from the network process X) that is to be simulated. We consider two cases in which p is known and $p = \pi$. The latter may seem surprising since the normalizing constant for π is unknown.

Upon observing the values $Y_1, \ldots, Y_n$ of the chain for n steps, the procedure is to use estimators c_n, μ_n defined as follows for c, μ. An estimator μ_n of μ is said to be *consistent* if $\mu_n \to \mu$ w.p.1.

Proposition 1.42. *Suppose Y is an ergodic Markov chain on $\mathbb{E}$ with stationary distribution p. Then a consistent estimator for c^{-1} is*

$$\hat{c}_n^{-1} = n^{-1} \sum_{i=1}^n \eta(Y_i)/p(Y_i).$$

Furthermore, a consistent estimator for $\mu = \sum_{x\in\mathbb{E}} g(x)\pi(x)$ is

$$\hat{\mu}_n = \frac{\sum_{i=1}^n g(Y_i)\eta(Y_i)/p(Y_i)}{\sum_{i=1}^n \eta(Y_i)/p(Y_i)}. \tag{1.40}$$

For the case $p = \pi$, this estimator reduces to $\hat{\mu}_n = n^{-1}\sum_{i=1}^n g(Y_i)$.

PROOF. The first assertion follows since, by the law of large numbers for Markov chains,

$$\hat{c}_n^{-1} \to \sum_x \eta(x) = c^{-1} \quad \text{w.p.1.}$$

This and another application of the law of large numbers yields

$$\hat{\mu}_n = \hat{c}_n n^{-1} \sum_{i=1}^{n} g(Y_i)\eta(Y_i)/p(Y_i) \to c \sum_x g(x)\eta(x) = \mu \quad \text{w.p.1.} \qquad \Box$$

To use the estimator $\hat{\mu}_n$, one would simulate the Markov chain Y_n for a large number of steps n and then take the resulting $\hat{\mu}_n$ as the value of μ. To complete the description of this Monte Carlo procedure, it remains to select a probability law for Y_n that is easy to simulate. Two standard approaches for this are as follows.

Example 1.43. *Random Sampling*. Take Y_n to be independent and identically distributed with distribution p. The challenge is to choose an efficient distribution p. It is natural to choose p such that the variance of $\eta(Y_i)/p(Y_i)$ is small and its shape is consistent with that of η. Typical choices to use in the estimator (1.40) are

$$p(x) = b \prod_{j \in M} r_j^{x_j}, \quad \text{and} \quad p(x) = b \prod_{j \in M} r_j^{x_j}/x_j!. \qquad \Box$$

Example 1.44. *Metropolis Method of Sampling*. We now describe a method of choosing Markov transition probabilities for the Markov chain Y, whose stationary distribution is $p = \pi$. Surprising, this is possible even though the normalization constant c for π is unkown. A general candidate for the transition probabilities is

$$P(x, y) = \gamma(x, y)/\eta(x), \quad x, y \in \mathbb{E}, \tag{1.41}$$

where γ satisfies $\sum_y \gamma(x, y) = \eta(x)$ and $\gamma(x, y) = \gamma(y, x)$ for each $x, y \in \mathbb{E}$. By Theorem 1.5, these transition probabilities are reversible and π is their stationary distribution. The reversibility property is not especially important. However, reversible chains do have a fast geometric rate of convergence to their stationary distribution.

The *Metropolis Markov chain* is a special case in which the transition probabilities (1.41) have the following special form. Let

$$\mathbb{E}(x) = \{y \in \mathbb{E} : y = T_{jk}x, \text{ for some } j \neq k \text{ in } M\}, \quad x \in \mathbb{E}.$$

This is the set of all states that can be reached by subtracting one unit from some coordinate j and adding one unit to some coordinate $k \neq j$ (the j or k may be 0 if the network is open). For each $x \in \mathbb{E}$ and $y \in \mathbb{E}(x)$, define

$$P(x, y) = \begin{cases} 1/|\mathbb{E}(x)| & \text{if } \eta(y) \geq \eta(x) \\ \eta(y)/[\eta(x)|\mathbb{E}(x)|] & \text{if } \eta(y) < \eta(x). \end{cases}$$

Also, let $P(x, y) = 0$ for $y \notin \mathbb{E}(x) \cup \{x\}$ and let

$$P(x, x) = 1 - \sum_{y \in \mathbb{E}(x)} P(x, y).$$

Computations of these probabilities involve computing

$$\eta(y)/\eta(x) = w_k w_j^{-1} \phi_j(x)/\phi_k(T_{jk}x), \quad \text{when } y = T_{jk}x.$$

This expression follows because

$$\pi(x) = c\eta(x) = c\Phi(x) \prod_{j=1}^{m} w_j^{x_j}.$$

Clearly, the transition probabilities $P(x, y)$ defined above are of the form (1.41), and hence the resulting Markov chain has the stationary distribution π. In this case, $\hat{\mu}_n = n^{-1} \sum_{i=1}^{n} g(Y_i)$ is the appropriate estimator for μ.

To simulate a Markov chain with these transition probabilities, one generates a transition from a state x to the next state y as follows:
(1) Randomly select a $y = T_{jk}x \in \mathbb{E}(x)$ with probability $1/|\mathbb{E}(x)|$ to be a candidate for the next state of the chain. (This amounts to selecting a pair $j \neq k$.)
(2) Accept $y = T_{jk}x$ as the next state with probability $\min\{1, \eta(y)/\eta(x)\}$. If y is not accepted, then take the next state to be the current state x.
This procedure is easy to implement because each transition involves changing only one or two coordinates of the state x. □

1.13 Properties of Whittle Networks

In a Whittle network, the service rates at a node may depend on the numbers of units at the other nodes, while in a Jackson network, a node's service rate depends only on the number of units at that node. This section gives insight into the added richness of routing and Φ-balanced service rates in Whittle networks. We explain the meaning of Φ-balance and give a few examples of sector-dependent services. Expanding on the ideas here, Section 3.1 shows how networks with multiclass customers can be modeled by Whittle networks.

Throughout this section, assume that X is a Whittle process with routing rates λ_{jk} and service rates ϕ_j, which are Φ-balanced. We begin with some observations about reducible routing and subnetworks. Suppose that $M_1, \ldots, M_n$ denote subsets of M such that each node $j \neq 0$ is in exactly one of the subsets and, when the network is open, the outside node 0 may be in several of the sets. Assume the routing rates λ_{jk} are irreducible on each M_i. Consider the network process X as the vector process $X_t = (X_t^1, \ldots, X_t^n)$ on $E = E_1 \times \cdots \times E_n$, where $X_t^i = (X_t(j) : j \in M_i \backslash \{0\})$ is the restriction of X to the subnetwork $M_i \backslash \{0\}$, and its state space E_i may be closed or open.

The process X^i is irreducible on E_i since the routing rates are irreducible on M_i; this is justified by the argument used in proving Proposition 1.10. Then the network process X is irreducible and invariant measures for it are given by Theorem 1.15. The X is called a *mixed* process if some of the X^i's operate like open networks and the others operate like closed networks. Although the routings of the X^i's are separate, these processes are dependent via the system-dependent service rates.

In some cases, it is also natural that each $\phi_j(x)$ depends only on x restricted to the nodes in the subnetwork M_i that contains j. In this case, the X^i's may still be dependent.

The preceding observations show that reducible routing is a viable option for Whittle networks. In contrast, we saw that reducible routing is not of interest for Jackson networks.

We now explain the meaning of Φ-balance and give several characterizations of this property. Recall that the service intensities ϕ_j are Φ-balanced if Φ is a positive function on $\mathbb{E}$ such that, for each $x \in \mathbb{E}$ and $j, k \in M$ with $T_{jk}x \in \mathbb{E}$,

$$\Phi(x)\phi_j(x) = \Phi(T_{jk}x)\phi_k(T_{jk}x).$$

To understand the motivation for this assumption, consider the transition rate function

$$\tilde{q}(x, T_{jk}x) \equiv \phi_j(x),$$

which is the same as q given by (1.6) with $\lambda_{jk} = 1$. By Definition 1.4 of reversibility, it follows that ϕ_j are Φ-balanced if and only if $\tilde{q}$ is reversible with respect to Φ. The importance of this reversibility was discussed in the remark on weak coupling following Theorem 1.15.

The preceding observation and the canonical form of reversible transition rates in Theorem 1.5 yield the following result.

Proposition 1.45. *The ϕ_j are Φ-balanced if and only if each ϕ_j is of the form*

$$\phi_j(x) = \Psi(x - e_j)/\Phi(x), \qquad x \in \mathbb{E},$$

for some nonnegative function Ψ defined on $\{x - e_j : x \in \mathbb{E},\ j \in M\}$.

The preceding canonical form for Φ-balanced service intensities ϕ_j is useful when Φ is known. How about when Φ is unknown? Can one construct Φ as a function of the ϕ_j's? The next result gives such a construction. It also characterizes the Φ-balance property directly in terms of the ϕ_j's. Here we say that $x^0, \ldots, x^n \in \mathbb{E}$ is a *direct path from x^0 to x^n* if $x^i = x^{i-1} - e_{j_i} + e_{j_i'}$ for some j_i, j_i' in M such that $n = |x^0 - x^n|$.

Proposition 1.46. *The service intensities ϕ_j are Φ-balanced if and only if, for each $j, k, \ell \in M$ and $x \in \mathbb{E}$ with $T_{j\ell}x, T_{k\ell}x \in \mathbb{E}$,*

$$\phi_j(x)\phi_k(T_{j\ell}x)\phi_\ell(T_{k\ell}x) = \phi_k(x)\phi_j(T_{k\ell}x)\phi_\ell(T_{j\ell}x). \tag{1.42}$$

In this case, ϕ_j are Φ-balanced by

$$\Phi(x) = \prod_{i=1}^{n} \phi_{j_i}(x^{i-1})/\phi_{j_i'}(x^i), \qquad x \in \mathbb{E}, \tag{1.43}$$

for any direct path $x^0, \ldots, x^n$ from a fixed reference state x^0 to $x^n = x$, where $x^i = x^{i-1} - e_{j_i} + e_{j_i'}$ for some j_i, j_i' in M such that $n = |x^0 - x|$.

Proof. We pointed out above that ϕ_j are Φ-balanced if and only if the rates $\tilde{q}(x, T_{jk}x) = \phi_j(x)$ are reversible with respect to Φ. Then the assertions follow from Theorem 2.8 and Proposition 2.21 in the next chapter. Expression (1.42) is a special case of the Kolmogorov reversibility criterion, and (1.43) is a special case of the canonical distribution for reversible processes. □

Condition (1.42) is easy to verify for specific ϕ_j's and, when it holds, one can easily construct Φ by (1.43). This condition, which involves three-step paths, has a simpler version involving only two-step paths when the service intensities satisfy another mild condition; see Exercise 13.

A useful class of service intensities are *compound service intensities* of the form $\phi_j(x) = \prod_{i=1}^n \phi_j^i(x)$, where ϕ_j^i are Φ_i-balanced for each i. In this case, ϕ_j are Φ-balanced, where $\Phi(x) = \prod_{i=1}^n \Phi_i(x)$. Such compound intensities are natural when there are several sources contributing to the departure intensity. A large class of compound service intensities is as follows.

Example 1.47. *Sector-dependent Service Rates.* Associated with the Whittle network process X we are studying, let S denote the collection of all subsets (or sectors) of $\{1, \ldots, m\}$. Let $S_j \subset S$ denote the family of sectors that contain node j. For each sector $J \in S$ there is a nonnegative function $\phi_J(n)$ defined on the nonnegative integers n that is 0 only if $n = 0$. Think of $\phi_J(x(J))$ as a "departure intensity," which depends on the number of units $x(J) \equiv \sum_{j \in J} x_j$ in J. The ϕ_J affects the departures at each node j in J. Specifically, we assume these sector intensities are compounded such that the departure intensity at each node $j \neq 0$ is

$$\phi_j(x) = \prod_{J \in S_j} \phi_J(x(J)), \quad x \in E. \tag{1.44}$$

Typically, there will be sectors J with $\phi_J(\cdot) \equiv 1$; they do not affect *any* node and hence are not relevant. In addition, if the network is open, we assume the intensity ϕ_0 is a positive function of the form $\phi_0(|x|)$. We call these ϕ_j's *sector-dependent departure intensities.* □

Invariant measures of Whittle network processes with sector-dependent departure intensities are as follows.

Theorem 1.48. *The sector-dependent departure intensities described above are balanced by the function*

$$\Phi(x) = \prod_{i=1}^{|x|} \phi_0(i-1) \prod_{J \in S} \prod_{n=1}^{x(J)} \phi_J(n)^{-1}, \quad x \in E,$$

where $\phi_0 \equiv 1$ if the network is closed. Hence, the Whittle network process X with these sector-dependent departure intensities has an invariant measure

$$\pi(x) = \Phi(x) \prod_{j=1}^m w_j^{x_j}, \quad x \in E. \tag{1.45}$$

PROOF. The first assertion follows by Proposition 1.45, since an easy check shows that $\phi_j(x) = \phi_0(|x - e_i|)\Phi(x - e_i)/\Phi(x)$. The second assertion of the theorem follows by Theorem 1.15. □

There are many types of sector-dependent departure rates based on interacting subpopulations. The trick is to identify relevant sectors J and their compounding intensities ϕ_J to model the dependency at hand. Here are some illustrations.

Example 1.49. *Treelike Networks with Load Balancing*. Suppose the Whittle network process X represents an open network, and the communication graph of the routing intensities λ_{jk} is a tree with one root node and each unit moves up some branch beginning from the root node. Assume that its nonzero transition rates are

$$q(x, T_{jk}x) = \begin{cases} \lambda_{0k}\phi_0(|x|) & \text{if } j = 0 \\ \lambda_{jk}\phi_j(x_j)\phi_{B_j}(x(B_j)) & \text{if } j \neq 0. \end{cases}$$

The ϕ_j is a "local service intensity" and $\phi_{B_j}(x(B_j))$ is a "load-balancing intensity" that is a function of the number of units in the branch B_j that contains node j. The departure intensities are clearly sector dependent with the relevant sectors being branches and single nodes. Then the invariant measure (1.45) for the process is

$$\pi(x) = \prod_{j=1}^{m} w_j^{x_j} \prod_{i=1}^{|x|} \phi_0(i-1) \prod_{n=1}^{x_j} \phi_j(n)^{-1} \prod_{n'=1}^{x(B_j)} \phi_{B_j}(n')^{-1}, \qquad x \in E.$$

Another natural dependency would be that departures at a node depend on the number of units "immediately above" the node on the branch, a special case being route-to-the-shortest-queue. Unfortunately, these dependencies (which are well known to be intractable) cannot be modeled by sector-dependent rates defined here. □

Example 1.50. *Manufacturing Networks with Work Centers*. A manufacturing facility commonly consists of work centers that contain several work stations whose processing rates may depend on the congestion in the work center and the overall congestion in the facility as well. As an elementary example, consider the Whittle network process X where the m nodes represent work stations and they are partitioned into work centers. Assume the arrival intensity from outside into node k is $\lambda_{0k}\phi_0(|x|)$, and assume the departure intensity at node j is the compound intensity $\phi_j(x_j)\phi_{C_j}(x(C_j))\phi_{M_0}(|x|)$. Here $\phi_j(x_j)$ is the work-station intensity, $\phi_{C_j}(x(C_j))$ is the work-center intensity depending on the number of units $x(C_j)$ in the work center C_j containing station j, and $\phi_{M_0}(|x|)$ is the network intensity depending on the total number of units in $M_0 \equiv \{1, \ldots, m\}$. These intensities are sector dependent, where the relevant sectors are single nodes, work center node sets, and M_0. Then the process has an invariant measure given by (1.45), where

$$\Phi(x) = \prod_{i=1}^{|x|} \phi_0(i-1) \prod_{n=1}^{x_j} \phi_j(n)^{-1} \prod_{n'=1}^{x(C_j)} \phi_{C_j}(n')^{-1} \prod_{n''=1}^{|x|} \phi_{M_0}(n'')^{-1}.$$

□

1.14 Exercises

1. Suppose X is a stochastic process on a countable state space $\mathbb{E}$ such that the sequence of states it visits is a Markov chain with transition probabilities $\bar{p}(x, y)$ and, whenever the process is in state x, the time to the next transition is exponentially distributed with rate $\lambda(x)$. The probability $\bar{p}(x, x)$ of a transition from state x back to itself may be positive. Show that X is a Markov process with transition rates $q(x, y) = \lambda(x)\bar{p}(x, y)$. To do this, note that each sojourn time of X in a state x (including possible feedbacks) is equal in distribution to $W \equiv \sum_{i=1}^{N+1} W_i$, where $W_1, W_2, \ldots$ are independent and exponentially distributed with rate $\lambda(x)$, and N is independent of these variables. The N represents the number of feedbacks to x until another state is selected, and $N = 0$ when $\bar{p}(x, x) = 0$. Show that W is exponentially distributed with rate $q(x) \equiv \lambda(x)(1 - \bar{p}(x, x))$.
2. Consider a tandem network as in Section 1.2 that consists of only two nodes. Assume the network process is stationary. Find the percentage of time that node 2 contains more units than node 1. This is the percentage of time X is in the set of states $A = \{x : x_1 < x_2\}$. Show that the average waiting time in this set is $W(A) = 1/(\lambda(1 - \rho_2))$.
3. Give an expression for an invariant measure w_j for routing intensities λ_{jk}, whose communication graph determines the following types of networks. (a) Closed cyclic network: the nodes form a circle that each unit traverses clockwise. (b) A closed treelike network with one root node (node 1) and the units move up the tree from the root to the leaves and a unit departing from a leaf node returns to node 1. (c) An open treelike network with one root node (node 1) and the units move up the tree from the root to the leaves and a unit departing from a leaf node exits the network. (d) An open feedforward network: The nodes can be labeled such that node 1 has no predecessors ($\lambda_{j1} = 0$ for $j \neq 0$), node m has no successors ($\lambda_{mj} = 0$ for $j \neq m$), and $\lambda_{jk} = 0$ if $j < k$, for each $1 < j < m$.
4. An open *in-tree network* is a treelike network with one root node, where all units move from the leaves to the root. The communication graph of such a network is shown in Figure 1.5. Such networks arise when routes of customers merge near the end of their network sojourns. Find a formula for an invariant measure for the routing rates λ_{jk} of the in-tree as in Figure 1.5. Extend this formula for a general in-tree network.
5. Give an expression for the equilibrium distribution of a Jackson process in which each node j is an s_j-server node ($1 \leq s_j \leq \infty$), where each server works at the rate μ_j.
6. For an open ergodic Jackson network with unlimited capacity, show that if $w_j < \liminf_{n\to\infty} \phi_j(n)$, then the average queue length L_j at node j is finite.

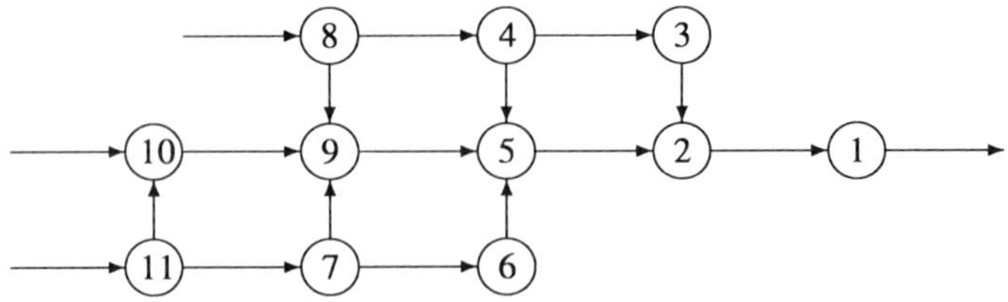

FIGURE 1.5. In-Tree Network

7. Consider an ergodic Jackson or Whittle process, and let $r = (r_1, \ldots, r_\ell)$ denote distinct nodes that form a route in the network. Find conditions on the routing and service intensities under which the average travel time of units on this route is $W_{r_1} + \ldots + W_{r_\ell}$, where W_j is the average sojourn time in node j.
8. Suppose X is an ergodic open or closed Whittle process with service rates of the form $\phi_j(x) = \Phi(x - e_j)/\Phi(x)$. Consider the event that a transition consists of a unit moving on any one of the arcs $1 \to 2$, $2 \to 1$, $3 \to 4$, or $6 \to 5$ ($j \to k$ means the unit moves from node j to node k). Let $N(t)$ denote the number of times in the time interval $(0, t]$ that this transition event occurs. Assuming X is stationary, give an expression for $EN(1)$.
9. *Open Jackson Networks with Population-dependent Entries.* Suppose X is an open Jackson network process with the added generality that, instead of $\phi_0(\cdot) \equiv 1$, the rate ϕ_0 is a positive function $\phi_0(|x|)$ of the total population $|x|$ in the network. This implies that the arrivals into node k from outside form a system-dependent Poisson process with intensity $\lambda_{0k}\phi_0(|x|)$. Show that X is a Whittle process and that an invariant measure for it is

$$\pi(x) = \prod_{i=0}^{|x|-1} \phi_0(i) \prod_{j=1}^{m} \prod_{n=1}^{x_j} \phi_j(n)^{-1}, \quad x \in \mathbb{E}.$$

For the unlimited-capacity case, give a necessary and sufficient condition for the process to be positive recurrent and describe its normalization constant. Consider the case in which X has a finite capacity ν and $\phi_0(|x|) = \psi(\nu - |x|)$, where $\psi(r)$ is the intensity when there is room for r more units in the network. Show that the stationary distribution for X is

$$\pi(x) = c \prod_{i=1}^{\nu-|x|} \psi(i) \prod_{j=1}^{m} \prod_{n=1}^{x_j} \phi_j(n)^{-1}, \quad x \in \mathbb{E}.$$

10. Prove the convolution formula (1.21).
11. Prove expression (1.26) using a direct induction argument.
12. Proposition 1.39 contains an algorithm for computing performance parameters for a closed Jackson process. One might think this result automatically applies to an open Jackson process with finite capacity. The proof, however, contains a key equation that requires a closed network, and the equation is not valid for an open network. Specify this key equation.

13. *Two-Step Criterion for Φ-balance.* Prove the following results for system-dependent service intensities $\phi_j(x)$.
(a) Suppose the network is open and ϕ_0 has the form $\phi_0(|x|)$. Then ϕ_j are Φ-balanced if and only if, for each $j, k \in M\backslash\{0\}$ and $x \in \mathbb{E}$ with $x - e_j, x - e_k \in \mathbb{E}$,

$$\phi_j(x)\phi_k(x - e_j) = \phi_k(x)\phi_j(x - e_k).$$

In this case,

$$\Phi(x) = \prod_{i=1}^{n} \phi_0(i-1)/\phi_{k_i}(x^i), \quad x \in \mathbb{E},$$

for any direct path $x^0, \ldots, x^n$ from $x^0 = 0$ to $x^n = x$, where $n = |x|$ and $x^i = x^{i-1} + e_{k_i}$.
(b) Suppose the network is closed and there is a node ℓ such that ϕ_ℓ has the form $\phi_\ell(x_\ell, |x|)$. Then ϕ_j are Φ-balanced if and only if, for each $j, k \in M\backslash\{\ell\}$ and $x \in \mathbb{E}$ with $T_{j\ell}x, T_{k\ell}x \in \mathbb{E}$,

$$\phi_j(x)\phi_k(T_{j\ell}x) = \phi_k(x)\phi_j(T_{k\ell}x).$$

In this case,

$$\Phi(x) = \prod_{i=1}^{\nu} \phi_\ell(\nu - i + 1, i - 1)/\phi_{k_i}(x^i), \quad x \in \mathbb{E},$$

for any direct path $x^0, \ldots, x^\nu$ from $x^0 = \nu e_\ell$ to $x^\nu = x$, where $x^i = x^{i-1} - e_\ell + e_{k_i}$.
These results can be proved directly by induction. Another approach is to use the property that Φ-balance is equivalent to $\tilde{q}(x, T_{jk}x) = \phi_j(x)$ being reversible with respect to Φ. Then show that the Kolmogorov criterion (in the next chapter) for reversibility, for paths of any length, is implied by each criterion above, which is a Kolmogorov criterion for paths of length two.

14. *Networks with Single-Server and Infinite-Server Nodes.* Suppose X is a closed Jackson network process in which each node j in a certain sector J is a single-server node with rate μ_j and each node $k \in J^c$ is an infinite-server node with rate μ_k for each of its servers. Show that the stationary distribution of X is

$$\pi(x) = c \prod_{j \in J} r_j^{x_j} \prod_{k \in J^c} r_k^{x_k}/x_k!,$$

where $r_j = w_j/\mu_j$ and the normalization constant is given by

$$c^{-1} = \sum_{\ell=1}^{\bar{m}} (\bar{r}_\ell)^{\nu+|J|-n_\ell} H_\ell(J) \sum_{n=0}^{\nu} (r/\bar{r}_\ell)^n/n!. \tag{1.46}$$

Here $|J|$ is the number of nodes in J and $r = \sum_{k \in J^c} r_k$. Also, as in Proposition 1.32, the $\bar{r}_1, \ldots, \bar{r}_{\bar{m}}$ denote the distinct r_j's in J and $H_\ell(J)$ is the term in brackets in (1.28). In particular, if the r_j's in J are distinct, then

$$c^{-1} = \sum_{j \in J} (r_j)^{\nu+|J|-1} \prod_{k \in J, k \neq j} (r_j - r_k)^{-1} \sum_{n=0}^{\nu} (r/r_j)^n / n!.$$

Hint: Use the expression

$$c^{-1} = \sum_{n=0}^{\nu} \left[\sum_{x(J)=\nu-n} \prod_{j \in J} r_j^{x_j} \right] \left[\sum_{x^{J^c}=n} \prod_{k \in J^c} r_k^{x_k} / x_k! \right].$$

15. *Networks with $\cdot/M/s$ Nodes or Limited Queue Dependency.* Suppose X is a closed Jackson network process in which each node has *limited queue dependency* in the sense that its departure rate function is constant when the queue length is above a certain level. Let $\bar{n}_j$ be such that $\phi_j(n) = \phi_j(\bar{n}_j)$, for $n \geq \bar{n}_j$. An example is an s_j-server node with departure rate $\phi_j(n) = \mu_j \min\{n, s_j\}$, which equals the constant $\mu_j s_j$ for $n \geq s_j$. Then the stationary distribution of X is $\pi(x) = c \prod_j f_j(x_j)$, where $f_j(n) = \prod_{n=1}^{x_j} w_j \phi_j(n)^{-1}$. By an obvious evaluation, the generating function of the convolution $f_1 \star \cdots \star f_m$ (which is the product of their generating functions) is

$$G(z) = \prod_{j=1}^{m} \left[b_j + \sum_{\nu=0}^{\bar{n}_j - 1} (b_j - \phi_j(\nu)/w_j) f_j^{-1}(\nu) z^{\nu} \right] / (b_j - z),$$

where $b_j = w_j / \phi_j(\bar{n}_j)$. Then proceeding as in Proposition 1.32, obtain a closed form expression for $c^{-1} = G^{(\nu)}(0)/\nu!$.

1.15 Bibliographical Notes

Further background on Markov processes can be obtained in standard texts on stochastic processes such as Asmussen (1987), Cinlar (1975), Kulkarni (1995), Resnick (1992), and Ross (1983). Jackson networks originated in Jackson (1957), with additional extensions by Gordon and Newell (1967) and Whittle (1968). The theory of these and related stochastic networks was advanced considerably by Kelly (1979). In (1975) he identified system-dependent service rates at nodes for modeling more intricate dependencies in networks. This theme was developed further by Whittle (1986b). Other monographs on stochastic networks are Disney and Kiessler (1987), Gelenbe and Pujolle (1987), Walrand (1988), and van Dijk (1993). Applications of stochastic networks in manufacturing and computer systems are covered in the respective texts by Buzacott and Shanthikumar (1993) and Haverkork (1998).

The convolution formulas in Proposition 1.31 came from Harrison (1985) and Bertozzi and McKenna (1993). For references on Little laws, see Chapter 5. The example on busy periods and high-level exceedances is based on Daduna (1988/89). The algorithms for performance parameters were distilled from Buzen (1973) and

Reiser and Lavenberg (1980). Other approaches are in Gerasimov (1995) (via integral representations) and in Choudhury et al. (1995) (via numerical inversion of generating functions). Fishman (1996) gives a survey of Monte Carlo simulation techniques; see Ross et al. (1994) for more details on the independent sampling approach for Jackson networks. The characterizations of Φ-balance and sector-dependent service rates are from Serfozo (1993). Examples of other system dependencies are in Towsley (1980) and Hordijk and van Dijk (1984).

2

Reversible Processes

An ergodic Markov process is reversible if, in equilibrium, the expected number of transitions per unit time from one state to another is equal to the expected number of the transitions in the reverse order. This is also equivalent to a time-reversibility property that, at any instant, the future of the process is stochastically indistinguishable from viewing the process in reverse time. A remarkable feature of such a process is that its equilibrium distribution is readily obtainable as a certain product of ratios of its transition rates. A classic example is a birth–death queueing process. This chapter describes a wide class of reversible Markov network processes with batch or multiple-unit movements as well as single-unit movements. Examples include multivariate birth–death processes with single and batch increments and reversible Jackson and Whittle processes. The last two sections cover partition-reversible processes, which are generalizations of reversible processes. Invariant measures for such processes are obtainable by solving balance equations separately on subsets that partition the state space.

2.1 Reversibility

Reversible Markov processes are very tractable because their transition rates and equilibrium distributions have canonical forms. We will describe these and other fundamental properties of reversibility in the first four sections. The rest of the chapter covers reversible network processes.

Unless specified otherwise, we assume that $\{X_t : t \geq 0\}$ is a Markov jump process on a countable state space $\mathbb{E}$, and its transition rates are denoted by $q(x, y)$.

Recall from Definition 1.4 that the process X is *reversible* if there is a positive measure π on $\mathbb{E}$ that satisfies the *detailed balance equations*

$$\pi(x)q(x, y) = \pi(y)q(y, x), \quad x, y \in \mathbb{E}. \tag{2.1}$$

We also say that *q is reversible with respect to π*. The measure π is necessarily an invariant measure of q (or of X) since it satisfies the total balance equations, which equal the sum of (2.1) over y.

Equation (2.1) implies that, for an ergodic process, the average number of transitions of the process from state x to state y is equal to the average number of transitions in the reverse direction from y to x. These average numbers of transitions are also expected numbers of transitions per unit time when the process is stationary. Note that if q is reversible, then it has the *two-way communication property* that, for each $x \neq y \in \mathbb{E}$, the $q(x, y)$ and $q(y, x)$ are both positive or both equal to 0. This yields the simple but useful criterion that a process is "not reversible" if a transition from some x to y is possible, but a transition in the reverse direction is not possible.

Recall from Theorem 1.5 that q is reversible if and only if it is of the form

$$q(x, y) = \gamma(x, y)/\pi(x), \quad x \neq y \in \mathbb{E}, \tag{2.2}$$

for some positive function π on $\mathbb{E}$ and some nonnegative function γ on $\mathbb{E} \times \mathbb{E}$ such that $\gamma(x, y) = \gamma(y, x)$, $x, y \in \mathbb{E}$. In this case, q is reversible with respect to π.

In addition to its use for verifying reversibility, the canonical representation (2.2) is useful for constructing reversible processes or modifying processes to be reversible—we will see examples shortly. Another observation is that any positive distribution π on $\mathbb{E}$ is the stationary distribution of a reversible Markov process with transition rates (2.2).

Note that the process X need not be stationary. Another approach to reversibility is to define it in terms of time reversals (discussed in the next section) and then show its equivalence to the preceding algebraic definition. This alternate approach requires the Markov process to be stationary, which is not needed for many results.

The definition of reversibility also applies to discrete-time Markov chains, in which case $q(x, y)$ are its one-step transition probabilities. All the results below hold for Markov chains with $q(x, y)$ interpreted as transition probabilities and the time parameter is discrete instead of continuous. Note that the definition applies to any nonnegative rates or probabilities as an algebraic property, not necessarily associated with a stochastic process. Recall that the sequence of states visited by X is a Markov chain with transition probabilities $p(x, y) = q(x, y)/q(x)$, where $q(x) = \sum_y q(x, y)$. Clearly, q is reversible with respect to π if and only if p is reversible with respect to $\pi(x)q(x)$.

A quintessential reversible process is the following classical birth–death process.

Example 2.1. *Birth–Death Process.* Suppose the process X represents the number of units in a service system (or in any population) and its state space $\mathbb{E}$ is the set of nonnegative integers. Assume that whenever there are x units in the system, the time to the next arrival (birth) is exponentially distributed with rate $\lambda(x)$, and the time to the next departure (death) is exponentially distributed with rate $\mu(x)$. Then the transition rates for the process are

$$q(x, y) = \begin{cases} \lambda(x) & \text{if } y = x + 1 \\ \mu(x) & \text{if } y = x - 1 \geq 0 \\ 0 & \text{otherwise.} \end{cases}$$

This process is the classical *birth–death process*. Its detailed balance equations for $y = x + 1$ and $y = x - 1$ are respectively

$$\pi(x)\lambda(x) = \pi(x+1)\mu(x+1), \quad x \geq 0,$$

$$\pi(x)\mu(x) = \pi(x-1)\lambda(x-1), \quad x \geq 1.$$

But these two equations are the same. The second one yields

$$\pi(x) = \pi(x-1)\lambda(x-1)/\mu(x), \quad x \geq 1.$$

By a backward iteration of this equation, it follows that it has a solution

$$\pi(x) = \pi(0) \prod_{n=1}^{x} \lambda(n-1)/\mu(n), \quad x \geq 1. \tag{2.3}$$

Thus, the process is reversible with invariant measure π. Furthermore, π is its stationary distribution and

$$\pi(0)^{-1} = 1 + \sum_{x=1}^{\infty} \prod_{n=1}^{x} \lambda(n-1)/\mu(n),$$

provided this sum is finite. □

The next result is a sufficient condition for reversibility. The *communication graph* of the rate function q is an undirected graph whose set of vertices is the state space $\mathbb{E}$ and there is an edge linking a pair x, y if either $q(x, y)$ or $q(y, x)$ is not 0. The graph is connected when X is irreducible (which we have assumed).

Theorem 2.2. *If the process X is ergodic and its communication graph is a tree, then X is reversible.*

Proof. Let π denote the stationary distribution of X. Recall that

$$\pi q(A, B) = \sum_{x \in A} \sum_{y \in B} \pi(x) q(x, y)$$

is the average rate of flow from A to B, and we noted in (1.2) that $\pi q(A, A^c) = \pi q(A^c, A)$ for any A.

Now, suppose x, y are vertices in the communication graph that are linked by an edge. Let A_x be the set of states in $\mathbb{E}$ reachable from x if the edge were deleted. Since the graph is a tree, it follows by the definition of A_x and the observation above that

$$\pi(x)q(x, y) = \pi q(A_x, A_x^c) = \pi q(A_x^c, A_x) = \pi(y)q(y, x).$$

Thus, the detailed balance equations are satisfied and hence X is reversible. □

Note that the communication graph of the classical birth–death process in Example 2.1 is a tree, but there are many reversible processes whose communication graphs are not trees.

2.2 Time Reversal

Let us see how reversibility is related to the behavior of a process in reverse time. For fixed $\tau > 0$, consider the process

$$X_t^\tau = X_{\tau - t}, \quad 0 \le t \le \tau.$$

This process X^τ on the time set $[0, \tau]$ is the *time reversal of X at* τ. It represents the evolution of X in reverse time beginning at τ. If one thinks of a sample path of X as a video tape, where X_t is the picture at time t, then one would see the corresponding sample path of X^τ by viewing the tape in reverse beginning at time τ.

Proposition 2.3. *The process X^τ is a Markov jump process with (time-dependent) transition probabilities*

$$P\{X_t^\tau = y | X_s^\tau = x\} = \frac{P\{X_{\tau-t} = y\}}{P\{X_{\tau-s} = x\}} P\{X_{t-s} = x | X_0 = y\}, \quad 0 \le s \le t \le \tau. \tag{2.4}$$

If X is stationary with distribution π, then X^τ is also a stationary Markov process with distribution π, and its transition rates are

$$q_\tau(x, y) = \pi(x)^{-1}\pi(y)q(y, x), \quad x \ne y \in \mathbb{E}. \tag{2.5}$$

PROOF. Consider the probability

$$P\{X_t^\tau = y | X_s^\tau = x, A\} = \frac{P\{X_{\tau-t} = y, X_{\tau-s} = x, A\}}{P\{X_{\tau-s} = x, A\}},$$

for any $0 < s \le t \le \tau$ and event A generated by $\{X_r^\tau : 0 \le r < s\}$. To prove the first assertion, it suffices to show that this fraction equals the right side of (2.4). But this equality follows since the denominator equals

$$P\{X_{\tau-s} = x\}P\{A | X_{\tau-s} = x\}$$

and the numerator equals

$$P\{X_{\tau-t} = y\}P\{X_{\tau-s} = x | X_{\tau-t} = y\}P\{A | X_{\tau-s} = x\},$$

because X is Markovian and A is generated by $\{X_r : \tau < r < \tau + s\}$.

Now, suppose X is stationary with distribution π. By the first assertion, X^τ is a Markov process, where the transition probabilities (2.4) now reduce to

$$\pi(x)^{-1}\pi(y)P\{X_t = x | X_0 = y\}.$$

Dividing this by t and letting $t \to 0$ proves (2.5). Also, X^τ is stationary and its distribution is π since, for each t, the X_t^τ has the same distribution as $X_{\tau-t}$, which is π. □

The evolution of X backward in time is equal in distribution to its evolution forward in time if X^τ and X are equal in distribution on $[0, \tau]$ for each $\tau > 0$. That is, for each $t_1 < \ldots < t_n \le \tau$,

$$(X_{\tau-t_1}, \ldots, X_{\tau-t_n}) \overset{\mathcal{D}}{=} (X_{t_1}, \ldots, X_{t_n}). \tag{2.6}$$

This property is related to reversibility as follows.

Theorem 2.4. *The Markov process X is stationary and reversible if and only if (2.6) holds.*

PROOF. Suppose X is stationary and reversible and its distribution is π. Then $X_0^\tau \overset{\mathcal{D}}{=} X_0$. Also, by Proposition 2.3, X^τ is a stationary Markov process with transition function q_τ given by (2.5). This expression and the reversibility of q implies $q_\tau = q$. Since X^τ and X are Markov processes, they are equal in distribution on $[0, \tau]$ for each $\tau > 0$, which is equivalent to (2.6).

Conversely, suppose (2.6) holds. Then, in particular, $X_0 \overset{\mathcal{D}}{=} X_\tau$ for each τ, and since X is Markovian, it is therefore stationary. Then by Proposition 2.3, X^τ is a stationary Markov process with transition rates (2.5). Also, (2.6) implies $X^\tau \overset{\mathcal{D}}{=} X$, and hence $q_\tau = q$. This and (2.5) establish that q is reversible. □

Statement (2.6) is sometimes used to define reversibility of X. It says that the distribution of X is invariant under the compound operation of reflecting the time scale about 0 and then shifting it by any amount τ. Also, a sufficient, but not necessary condition for X to be reversible is that

$$(X_{t_1}, \ldots, X_{t_n}) \overset{\mathcal{D}}{=} (X_{t_n}, \ldots, X_{t_1}), \quad t_1 < \ldots < t_n.$$

It is often natural to consider X as a process $\{X_t : t \in \mathbb{R}\}$ whose time set is the entire real line $\mathbb{R}$. In this case, X^τ, for each τ, is defined on the time interval $(-\infty, \tau)$. Furthermore, if X is stationary, then it is reversible if and only if its distribution is invariant under a reflection of the time axis about 0. That is, $X^0 \overset{\mathcal{D}}{=} X$ or, equivalently,

$$(X_{t_1}, \ldots, X_{t_n}) \overset{\mathcal{D}}{=} (X_{-t_1}, \ldots X_{-t_n}), \quad t_1 < \ldots < t_n.$$

We end this section with more insights into time-reversal processes. This material does not involve the notion of reversibility. The following result relates the balance equations for a Markov process to the time reversal of the process.

Theorem 2.5. *Suppose the Markov process X is ergodic and there exists a positive distribution π on $\mathbb{E}$ and a transition function $\bar{q}$ on $\mathbb{E}$ such that*

$$\bar{q}(x, y) = \pi(x)^{-1}\pi(y)q(y, x), \quad x, y \in \mathbb{E}, \tag{2.7}$$

$$\sum_y q(x, y) = \sum_y \bar{q}(x, y), \quad x \in \mathbb{E}. \tag{2.8}$$

Then π is the stationary distribution of X. Also, $\bar{q}$ is the transition function of a time reversal of X when X is stationary.

PROOF. Under the assumptions,

$$\pi(x) \sum_y q(x, y) = \pi(x) \sum_y \bar{q}(x, y) = \sum_y \pi(y) q(y, x).$$

This proves the first assertion. The second assertion follows by Proposition 2.3. □

The preceding result may be used to verify that a conjectured distribution π satisfies the balance equations for a Markov process X, and, at the same time, obtain the time reversal transition rate $\bar{q}$. Namely, define a transition function $\bar{q}$ by (2.7), where π is a conjectured stationary distribution of X. Alternatively, one could conjecture the form of $\bar{q}$ for a time reversal of a stationary version of X, and this would define π by (2.7). In either case, if $\bar{q}$ satisfies (2.8), then π is the stationary distribution of X.

Remark 2.6. Note that verifying (2.8) is the same as verifying directly by substitution that π satisfies the balance equations (Exercise 3 is an example). Consequentially, the preceding result may not be as useful as it appears. However, the act of conjecturing what $\bar{q}$ is for the time reversal of X and using (2.7) might give insight into candidates for π.

The time reversal of a process need not be the same type of process as the original one. This property is easy to check, however, if one knows the stationary distribution of the process.

Example 2.7. *Time Reversal of Jackson and Whittle Processes.* Suppose X is a stationary Jackson or Whittle process. Then by Proposition 2.3, its time reversal $\bar{X}_t$ is an ergodic, stationary Markov process with the same stationary distribution π as X and its transition rates are

$$\begin{aligned} \bar{q}(x, T_{jk}x) &= \pi(T_{jk}x)\pi(x)^{-1} q(T_{jk}x, x) \\ &= \bar{\lambda}_{jk}\phi_j(x), \end{aligned}$$

where $\bar{\lambda}_{jk} = w_k w_j^{-1} \lambda_{jk}$. The w_j's that satisfy the traffic equations for λ_{jk} also satisfy the traffic equations for $\bar{\lambda}_{jk}$, since the latter is the time reversal of the former. Thus, $\bar{X}$ is the same type of network process as X; the service rates are the same, but the routing rates are the reversal of the original routing rates. Note that this result does not say anything about the reversibility of the process X. □

2.3 Invariant Measures

We now present the canonical form of invariant measures for reversible Markov processes. This is linked to Kolmogorov's criterion for characterizing a reversible transition function.

Recall that the Markov process X has the two-way communication property that either $q(x, y)$ and $q(y, x)$ are both positive or both equal to 0, for each $x \neq y \in \mathbb{E}$. Throughout this section, we will assume (at no loss in generality) that X has this property. We say that a sequence of states $x^0, x^1, \ldots, x^n$ in $\mathbb{E}$ is a *path* if $q(x^{i-1}, x^i) > 0$, $i = 1, \ldots, n$. We also use the ratio of rates

$$\rho(x, y) \equiv q(x, y)/q(y, x), \qquad \text{for a path } x, y.$$

The most remarkable feature of a reversible Markov process is that an invariant measure for it is automatically given by expression (2.9) below, which is a product of ratios of the transition rates. In proving this canonical form, we also establish Kolmogorov's criterion, which is a condition equivalent to reversibility, but which only involves q. Statement (iii) is a "ratio form" of Kolmogorov's criterion, which is often easier to exploit.

Theorem 2.8. *The following statements are equivalent.*
(i) *The transition function q is reversible.*
(ii) (Kolmogorov Criterion) *For each n and $x^0, x^1, \ldots, x^n$ in $\mathbb{E}$ with $x^n = x^0$,*

$$\prod_{i=1}^{n} q(x^{i-1}, x^i) = \prod_{i=1}^{n} q(x^i, x^{i-1}).$$

(iii) *For each path $x^0, x^1, \ldots, x^n$ in $\mathbb{E}$, the product $\prod_{i=1}^{n} \rho(x^{i-1}, x^i)$ depends on $x^0, \ldots, x^n$ and n only through x^0, x^n.*
If q is reversible, then an invariant measure for it is

$$\pi(x) = \prod_{i=1}^{n} \rho(x^{i-1}, x^i), \qquad x \in \mathbb{E}\backslash\{x^0\}, \tag{2.9}$$

and $\pi(x^0) = 1$, where $x^0, x^1, \ldots, x^n = x$ is any path and x^0 is an arbitrary state viewed as an origin.

Remark 2.9. One can construct the π in (2.9) by the following recursion. Set $\mathbb{E}_0 = \{x^0\}$ and

$$\mathbb{E}_{n+1} = \{x \in \mathbb{E}\backslash\mathbb{E}_n : q(y, x) > 0 \text{ for some } y \in \mathbb{E}_n\}.$$

Then let $\pi(x^0) \equiv 1$ and for each $n \geq 1$, define

$$\pi(x) = \rho(y, x)\pi(y), \qquad \text{for } x \in \mathbb{E}_{n+1}\backslash\mathbb{E}_n \text{ and any } y \in \mathbb{E}_n \text{ with } q(y, x) > 0.$$

PROOF. (i) $\Rightarrow$ (ii). If q is reversible with respect to π, then, for each $x^0, x^1, \ldots, x^n = x^0$ in $\mathbb{E}$,

$$\prod_{i=1}^{n} \pi(x^{i-1})q(x^{i-1}, x^i) = \prod_{i=1}^{n} \pi(x^i)q(x^i, x^{i-1}).$$

Cancelling the π's yields (ii).
(ii) $\Rightarrow$ (iii). To prove (iii), it suffices to show

$$\prod_{i=1}^{n} \rho(x^{i-1}, x^i) = \prod_{i=1}^{\ell} \rho(\tilde{x}^{i-1}, \tilde{x}^i), \tag{2.10}$$

where $x^0, \ldots, x^n$ and $\tilde{x}^0, \ldots, \tilde{x}^\ell$ are two paths with $\tilde{x}^0 = x^0$ and $\tilde{x}^\ell = x^n$. Since $x^0, \ldots, x^n, \tilde{x}^{\ell-1}, \ldots, \tilde{x}^1, \tilde{x}^0$ is a path from x^0 to itself, (ii) implies

$$\prod_{i=1}^{n} q(x^{i-1}, x^i) \prod_{i=1}^{\ell} q(\tilde{x}^i, \tilde{x}^{i-1}) = \prod_{i=1}^{\ell} q(\tilde{x}^{i-1}, \tilde{x}^i) \prod_{i=1}^{n} q(x^i, x^{i-1}).$$

These quantities are positive, by the definition of a path. Then, dividing both sides of this equation by the second and fourth products yields (2.10).
(iii) $\Rightarrow$ (i). Suppose (iii) holds, and let π be defined by (2.9). We will show that q is reversible with respect to π. For a fixed x, let $x^0, \ldots, x^n = x$ be a path. Choose any y such that $q(x, y) > 0$. Then using (2.9),

$$\begin{aligned}
\pi(x)q(x, y) &= \prod_{i=1}^{n} \rho(x^{i-1}, x^i) q(x, y) \\
&= q(y, x) \prod_{i=1}^{n} \rho(x^{i-1}, x^i) q(x, y)/q(y, x) = q(y, x)\pi(y).
\end{aligned}$$

These detailed balance equations also hold trivially for x, y with $q(x, y) = q(y, x) = 0$. Thus q is reversible with respect to π. □

To verify the Kolmogorov criterion, or its ratio analogue (iii), one may not have to consider all possible sequences or paths in $\mathbb{E}$. In many instances, certain structural properties of q and $\mathbb{E}$ lead to simpler versions of the Kolmogorov criterion. In particular, for some processes on vector state spaces such as network processes discussed shortly, only a small family of paths generated by the basis vectors need be considered. The following is another special case that is illustrated in the example below.
Fact. *The Kolmogorov criterion holds for all paths, if it holds for paths consisting of distinct states, aside from the same beginning and end states.*
This is because any path can be partitioned into subpaths of distinct states.

Example 2.10. *Circular Birth–Death Process.* Suppose the process X has state space $\mathbb{E} = \{0, 1, \ldots, n\}$, and it moves as follows: From any state j, it may move to $j+1$ or $j-1$, where $n+1 = 0$ and $-1 = n$. Its transition rates are

$$q(x, y) = \begin{cases} \lambda(x) & \text{if } y = x+1 \le n \text{ or } (x, y) = (n, 0) \\ \mu(x) & \text{if } y = x-1 \ge 0 \text{ or } (x, y) = (0, n) \\ 0 & \text{otherwise.} \end{cases}$$

This circular birth–death process may not be reversible as its classical counterpart is. Let us see the type of birth–death rates under which it is reversible. Note that its communication graph is a circle. Consequently, a path of distinct states from any state back to itself consists of all the states. In this case, the Kolmogorov criterion for reversibility is

$$\lambda(0)\cdots\lambda(n) = \mu(0)\cdots\mu(n). \tag{2.11}$$

In other words, the process is reversible if and only if (2.11) holds. In this case, the stationary distribution is

$$\pi(x) = \pi(0)\prod_{k=1}^{x}\lambda(k-1)/\mu(k), \quad 1 \le x \le n,$$

where $\pi(0)^{-1} = \sum_{x=0}^{n}\prod_{k=1}^{x}\lambda(k-1)/\mu(k)$.

This example readily extends to the context in which the communication graph of the process X is a tree with leaf-to-root connections. Specifically, assume there is a single root node 0 for the branches, and there is two-way communication between adjacent nodes that form the branches. In addition there is two-way communication between the ends of the branches (the leaves of the tree) and node 0. This graph is a collection of circular graphs connected at node 0, where the unique set of nodes leading from a leaf to the root 0 forms a circular graph. Note that a path of distinct states (or nodes) from a state back to itself consists of the circular graph that goes through that state. Consequently, the Kolmogorov criterion for reversibility of X is equivalent to the identity (2.11) for each circular graph of the tree (node n would be a leaf of the branch). Some of these equations may be dependent since a node may be on several circular graphs. □

The following is another example where the Kolmogorov ratio criterion simplifies considerably.

Example 2.11. *McCabe's Library.* Consider an infinite number of books or items labeled 0, 1, ... that are placed in a row on an infinite (virtual) bookshelf. The successive book selections by users are independent, and each user selects book b with probability p_b. When a book at location 0 is selected, it is returned to that location. Otherwise, a book selected from location $j \ge 1$ is returned to location $j-1$, and the book there is placed in location j. This switching is done before the next book is selected. The state of the library (called McCabe's library) at any selection is $x = (x_0, x_1, \ldots)$, where x_j denotes the book at location j. Whenever the library is in state x and the book x_j at location j is selected, then the new state is the vector x with the entries x_j and x_{j-1} interchanged if $j \ge 1$, and the state remains the same if $j = 0$. Without loss of generality, assume the initial state of the library is in the set $\mathbb{E}$ of all permutations of the books $(0, 1, \ldots)$ obtained by a finite number of book selections. Then the states of the library at successive book selections is a Markov chain $\{X_n : n \ge 0\}$ on $\mathbb{E}$. Its transition probabilities are $P(x, y) = p_{x_j}$ if y is obtained from x after selecting the book at location j for some $j \ge 0$ and $P(x, y) = 0$ otherwise.

We will show that the Markov chain X_n is reversible and has an invariant measure

$$\pi(x) = \prod_{j=0}^{n(x)} p_{x_j}^{(j-x_j)}, \quad x \in \mathbb{E}, \tag{2.12}$$

where $n(x) = \min\{n : x_j = j,\ j > n\}$. This $n(x)$ is finite since x is obtained from $(0, 1, \ldots)$ by a finite number of book selections. Note that if the book collection were finite, then the resulting Markov chain would have the same invariant measure as above (Exercise 4), where $\mathbb{E}$ is the finite set of all permutations of the books.

To establish reversibility, we use the Kolmogorov ratio criterion. For any path $x^0, \ldots, x^n$ of distinct states,

$$\prod_{i=1}^{n} P(x^{i-1}, x^i)/P(x^i, x^{i-1}) = \prod_{i=1}^{n} p_{b_i}/p_{\tilde{b}_i},$$

where b_i is the book selection that yields x^i from x^{i-1} and $\tilde{b}_i$ is the book selection that yields x^{i-1} from x^i. This product simplifies because of the following properties. To move from x^0 to x^n, each book x_j^n with $(x_j^n < x_j^0)$ has to be selected at least $(x_j^0 - x_j^n)$ times. And after the $(x_j^0 - x_j^n)$th one, each subsequent b_i book selection has to be compensated by the associated book $\tilde{b}_i$. Similarly, to move in reverse from x^n to x^0, each book x_j^0 with $(x_j^0 < x_j^n)$ has to be selected at least $(x_j^n - x_j^0)$ times. And after the $(x_j^n - x_j^0)$th one, each subsequent $\tilde{b}_i$ selection has to be compensated by the associated b_i. Consequently,

$$\prod_{i=1}^{n} P(x^{i-1}, x^i)/P(x^i, x^{i-1}) = \prod_{j=0}^{\infty} p_{x_j^n}^{(x_j^0 - x_j^n)}.$$

This quantity does not depend on the interior states $x^1, \ldots, x^{n-1}$ of the path. Then by Theorem 2.8, the Markov chain X_n is reversible and an invariant measure for it is the preceding product evaluated at $x^n = x$, where x^0 is fixed. In other words, for $x^0 = (0, 1, \ldots)$ this invariant measure is (2.12) as asserted.

Note that the state of the library can also be represented by the vector $z = (z_0, z_1, \ldots)$, where z_b denotes the location of book b. The z is the inverse of the corresponding state x in that $z_{x_j} = j$ and $x_{z_b} = b$. Because of this one-to-one correspondence between x's and z's, the successive values of this shelf variable Z_n also form a reversible Markov chain on the state space $\mathbb{E}$. Since $\{Z_n = z\} = \{X_{nz_b} = b,\ b \geq 0\}$, it follows that an invariant measure for Z_n is

$$\pi_Z(z) = \prod_{b=0}^{n(z)} p_b^{(z_b - b)}, \quad z \in \mathbb{E},$$

where $n(z) = \min\{n : z_b = b,\ b > n\}$. This measure is (2.12) with the variable x_j changed to b and $j = z_{x_j} = z_b$. □

2.4 Construction of Reversible Processes

This section covers elementary results that are handy for identifying or constructing reversible processes. The focus is on reversible processes restricted to subsets of their state spaces, independent reversible processes, and compounding of reversible transition rates.

Our first observation is that a reversible process restricted to any subspace is also reversible. Suppose that X is a Markov process on $\mathbb{E}$ with transition rates $q(x, y)$. Fix a subset $\tilde{\mathbb{E}} \subset \mathbb{E}$, and let $\tilde{X}$ be a Markov process on $\tilde{\mathbb{E}}$ whose transition rates are the rates $q(x, y)$, with x, y restricted to $\tilde{\mathbb{E}}$. The process $\tilde{X}$ is the *restriction of X to $\tilde{\mathbb{E}}$*. This restriction $\tilde{X}$ can be viewed as the process X with its transitions from states inside $\tilde{\mathbb{E}}$ to states outside of $\tilde{\mathbb{E}}$ "blocked" or suppressed. The following result is an immediate consequence of the definition of reversibility.

Proposition 2.12. *If X is reversible with respect to π, then its restriction $\tilde{X}$ to $\tilde{\mathbb{E}}$ is also reversible with respect to π restricted to $\tilde{\mathbb{E}}$. If in addition, X is ergodic and $\tilde{X}$ is irreducible, then $\tilde{X}$ is ergodic and its stationary distribution is*

$$\tilde{\pi}(x) = \pi(x) / \sum_{y \in \tilde{\mathbb{E}}} \pi(y), \quad x \in \tilde{\mathbb{E}}.$$

This is the conditional stationary distribution of X being in state x given that it is in $\tilde{\mathbb{E}}$.

Suppose one is considering a new Markov process and recognizes that it is a restriction of a known reversible process whose stationary distribution is known. Then by the preceding result, one automatically knows the stationary distribution of the new process. Restrictions of reversible processes are also of interest when studying the effect of changing the operation of a reversible process by blocking certain transitions. Here is a typical example.

Example 2.13. *Truncated Birth–Death Process.* Suppose the process X represents the classical birth–death process in Example 2.1 with birth and death rates $\lambda(x)$ and $\mu(x)$. Consider a truncated variation of this process in which the system can only accomodate at most ν units; arrivals are blocked or lost from the system when ν units are present. Also, assume the system does not serve customers whenever the number of customers is below a prescribed lower limit $\nu_0 < \nu$. For instance, the servers may be assigned to other duties such as maintenance when the queue is below ν_0. For simplicity, assume the number of units in the system at time 0 is in the set $\tilde{\mathbb{E}} = \{\nu_0, \nu_0 + 1, \ldots, \nu\}$. Otherwise, the queue length will eventually reach this set and stay there. Under these assumptions, the number of units $\tilde{X}_t$ in the system at time t is a process that is a restriction of X to $\tilde{\mathbb{E}}$. Thus, by Proposition 2.12, the process $\tilde{X}$ is reversible and its stationary distribution is

$$\tilde{\pi}(x) = \tilde{\pi}(\nu_0) \prod_{n=\nu_0+1}^{x} \lambda(n-1)/\mu(n), \quad x \in \tilde{\mathbb{E}},$$

where $\tilde{\pi}(\nu_0)^{-1} = \sum_{x=\nu_0}^{\nu} \prod_{n=\nu_0+1}^{x} \lambda(n-1)/\mu(n)$. □

The next result says that a juxtaposition of independent reversible processes is also reversible. This simple property follows immediately from the definition of reversibility. There are several interesting dependencies that can be modeled by restricting such multivariate independent processes to smaller subspaces. Examples are in the next section.

Proposition 2.14. *Suppose* $X_t = (X_t^1, \dots, X_t^m), t \geq 0$, *where* $X^1, \dots, X^m$ *are independent, irreducible, reversible Markov processes on* $\mathbb{E}_1, \dots, \mathbb{E}_m$, *respectively, and* X^j *has transition rate function* q_j *and invariant measure* π_j. *Then* X *is an irreducible Markov process on* $\mathbb{E} = \mathbb{E}_1 \times \cdots \times \mathbb{E}_m$ *with transition rates*

$$q(x, y) = \begin{cases} q_j(x_j, y_j) & \text{if } y_k = x_k \text{ for } k \neq j \text{ for some } j \in \{1, \dots, m\} \\ 0 & \text{otherwise,} \end{cases}$$

which is reversible with respect to $\pi(x) = \pi_1(x_1) \cdots \pi_m(x_m)$, $x \in \mathbb{E}$.

The next observation is that a transition function is reversible if it is a compounding of reversible transition functions.

Proposition 2.15. *Suppose* $q(x, y) = q_1(x, y)q_2(x, y)$, $x, y \in \mathbb{E}$, *where* q_1 *and* q_2 *are irreducible transition functions on* $\mathbb{E}$. *If* q_1 *and* q_2 *are reversible with respect to* π_1 *and* π_2, *respectively, then* q *is reversible with respect to* $\pi(x) = \pi_1(x)\pi_2(x)$. *Furthermore, if* q *and* q_1 *are reversible with respect to* π *and* π_1, *respectively, then* q_2 *is reversible with respect to* $\pi_2(x) = \pi(x)/\pi_1(x)$.

This result follows immediately from the definition of reversibility. It readily extends to multiple compounds $q(x, y) = q_1(x, y) \cdots q_n(x, y)$ as follows. If any n of the transition functions $q, q_1, \dots, q_n$ are reversible, then the other one is also reversible and their invariant measures are related by $\pi(x) = \pi_1(x) \cdots \pi_n(x)$.

2.5 More Birth–Death Processes

Classical birth–death processes have natural extensions to multivariate processes, including processes with multiple births and deaths. A few examples are as follows.

The first example is indicative of a wide class of multivariate birth–death processes constructed by a coupling together of several one-dimensional birth–death processes.

Example 2.16. *Multiple Birth–Death Processes with Population Constraints.* Consider m populations that operate like independent irreducible birth–death processes, but the vector of the respective population sizes $x = (x_1, \dots, x_m)$ is constrained to be in a subset $\tilde{\mathbb{E}}$ of $\mathbb{E} = \{x : x_j = 0, 1, \dots; 1 \leq j \leq m\}$. For instance, if the total number of units in the populations is constrained to not exceed ν, then $\tilde{\mathbb{E}} = \{x : 0 \leq |x| \leq \nu\}$. To model the m population sizes, consider the process $X_t = (X_t^1, \dots, X_t^m)$, $t \geq 0$, where $X^1, \dots, X^m$ are independent irreducible birth–death processes on the nonnegative integers. An invariant measure for X^j is $\pi_j(n) = \prod_{k=1}^{x} \lambda_j(k-1)/\mu_j(k)$, $n \geq 1$, where $\lambda_j(\cdot)$ and $\mu_j(\cdot)$ are the birth and death rates. Now the sizes of the m populations can be represented by the process $\tilde{X}$ that is the restriction of X to the subset $\tilde{\mathbb{E}}$. By Proposition 2.14, X is reversible with respect to a measure that is the product of the invariant measures

π_j. Then by Proposition 2.12, $\tilde{X}$ is reversible with respect to

$$\tilde{\pi}(x) = \prod_{j=1}^{m} \prod_{n=1}^{x_j} \lambda_j(n-1)/\mu_j(n), \quad x \in \tilde{\mathbb{E}}.$$

The process $\tilde{X}$ is a special multivariate birth–death process where the populations operate independently subject to the constraint of the restricted subspace—hence the populations are dependent. There are many applications of this model in which $\tilde{\mathbb{E}}$ is the set of population vectors x that satisfy linear constraints such as $a_j \leq x_j \leq b_j$, $\sum_{j=1}^m r_j x_j \leq r$, or

$$\sum_{j=1}^{m} r_{ij} x_j \leq r_i, \quad i = 1, \dots I.$$

For instance, the last constraint applies when each unit of population j requires r_{ij} units of a resource i and there are only r_i units of the resource available. Typical resources are space, computer memory, manufacturing tools, and money. Other common constraints are that $f_i(x) \leq 0$, $i = 1, \dots I$, where f_i are nonlinear functions. □

The preceding example, which is a rich source of applications, could be covered in an elementary course that introduces reversibility along with Propositions 2.12 and 2.14. A special case of this example is as follows.

Example 2.17. *Communication Network with Capacity Constraints and Blocking.* Consider a communication network that services m types of units. The units arrive to the network according to independent Poisson processes with respective rates $\lambda_1, \dots, \lambda_m$. For its communication across the network, each type j unit requires the simultaneous use of a_{ij} channels on link i for each i in the set I of links of the network. Some of the a_{ij}'s may be 0. If these quantities of channels are available, they are assigned to the unit, and the unit holds the channels for a time that is exponentially distributed with rate μ_j. At the end of this time, the unit releases the channels and exits the network. The total number of channels available on link $i \in I$ is b_i. If a unit arrives and its required channel quantities are not available, then it cannot enter the network (it is blocked or lost).

Let $X_t = (X_t^1, \dots, X_t^m)$ denote the numbers of the m types of units in the network at time t. Think of the m populations as nodes of a "virtual network," not to be confused with the underlying communication network. When X is in state x, the number of channels in use on link i is $\sum_j a_{ij} x_j$. Then the state space of X is $\mathbb{E} = \{x : 0 \leq \sum_j a_{ij} x_j \leq b_i, i \in I\}$. Note that if the state of the process is x, then a type j unit can enter the network provided $x \in \mathbb{E}_j = \{x : \sum_k a_{ik} x_k \leq b_i - a_{ij}, i \in I\}$. Under these assumptions, X is a constrained multivariate birth–death process as described in the preceding example. It has single-unit movements and its transition rates are $q(x, x + e_j) = \lambda_j$, if $x \in \mathbb{E}_j$ and $q(x, x - e_j) = \mu_j$, if

$x_j \geq 1$. Then its stationary distribution is

$$\pi(x) = c\prod_{j=1}^{m}(\lambda_j/\mu_j)^{x_j}, \quad x \in \mathbb{E},$$

where c is the normalization constant.

The quality of the network is usually assessed in terms of blocked or lost customers. The probability that a type j arrival is blocked in equilibrium is $\pi(\mathbb{E}_j^c)$. Ideally, the channel capacities b_i would be sized such that the blocking probability $\pi(\mathbb{E}_j^c)$ would be less than some small amount such as .01. The π also provides insight into which links cause the blocking. For instance, the probability that a type j is blocked because of the load on link i is full is $\sum_{x\in\mathbb{E}_j^c}\pi(x)1(\sum_k a_{ik}x_k > b_i)$.

What is the average number of type j units blocked per unit time? To determine this, consider $\tau_j(t) = \int_0^t 1(X_s \in \mathbb{E}_j^c)\,ds$, which is the amount of time in $[0, t]$ that type j units are blocked. Now, the number of type j units blocked in $[0, t]$ can be expressed as $N_j(\tau_j(t))$, where $N_j(t)$ is a Poisson process with rate λ_j and N_j is independent of X. Thus, by the strong law of large numbers for $N_j(t)$ and for $\tau_j(t)$, the number of type j units blocked per unit time is

$$\lim_{t\to\infty} t^{-1}N_j(\tau_j(t)) = \lim_{t\to\infty} \tau_j(t)^{-1}N_j(\tau_j(t))\tau_j(t)/t = \lambda_j\pi(\mathbb{E}_j^c) \quad \text{w.p.1.}$$

A related process for assessing loads on the network links is $Y_t = (Y_t^i : i \in I)$ where $Y_t^i = \sum_j a_{ij}X_t^j$ is the number of channels on link i that are in use at time t. Although this process Y is not Markovian, its stationary distribution, as a function of π, is

$$\pi_Y(y) = \sum_{x\in\mathbb{E}}\pi(x)1(\sum_j a_{ij}x_j = y_i, i \in I).$$

This distribution can be used to determine various performance parameters such as the percent of time that link i is idle or the stationary probability that link i has more channels in use than link k. Another parameter of interest is the average number of channels in use on link i, which is $\sum_j a_{ij}\sum_{x\in\mathbb{E}} x_j\pi(x)$. □

Note that any Markov jump process on the nonnegative integers that has only unit increments is a classical birth–death process, and hence it is reversible. What about processes on the integers whose increments may be more than one unit? A necessary and sufficient condition for such a process to be reversible is given in the following example.

Example 2.18. *Batch Birth–Death Process.* Consider a generalization of the classical birth–death queueing process in which there are batch arrivals and batch departures, whose size a is in a set A of allowable increments. Let X_t denote the number of units in the system at time t. The state space $\mathbb{E}$ of X is the set spanned by the elements of A. For simplicity, assume the least common divisor of the elements in A is 1, and so $\mathbb{E}$ is the set of nonnegative integers. Assume the transition rates of this process are

$$q(x,y) = \begin{cases} \lambda_a(x) & \text{if } y = x + a,\ a \in A \\ \mu_a(x) & \text{if } y = x - a \geq 0,\ a \in A \\ 0 & \text{otherwise.} \end{cases}$$

The $\lambda_a(x)$ and $\mu_a(x)$ are the positive rates for births and deaths of size a when the process is in state x. From a result we will prove shortly, Theorem 2.22, it follows that the process X is reversible if and only if the birth and death rates satisfy

$$\lambda_a(x)/\mu_a(x+a) = \prod_{n=x}^{x+a-1} \lambda_1(n)/\mu_1(n+1), \quad x \in \mathbb{E},\ a \in A. \tag{2.13}$$

Assuming this is true, then an invariant measure for X is

$$\pi(x) = \pi(0) \prod_{n=1}^{x} \lambda_1(n-1)/\mu_1(n),$$

which is the same as (2.3) for the classical birth–death process with unit increments. This equality of invariant measures is somewhat surprising. However, we will discuss shortly how it can be explained in terms of the Kolmogorov criteria for reversibility. Another insight is that one can view the batch increment process as a "random time transformation" of the unit increment case, where time is stopped at each batch arrival or departure and the stopped batch process has the same ergodic behavior as the unstopped unit-increment process. □

2.6 Reversible Network Processes

We now characterize invariant measures for reversible network processes. The focus will be on a general network process whose increments are selected by reversible intensities and whose departure–arrival intensities are also reversible. Examples include reversible Jackson and Whittle processes and multi-dimensional batch birth–death processes.

Assume that $\{X_t : t \geq 0\}$ is an m-node Markov network process whose state $x = (x_1, \ldots, x_m)$ represents the number of units at the respective nodes. The network may be open or closed, and its state space $\mathbb{E}$ is any set of m-dimensional vectors with nonnegative integer entries. Assume also that the process is irreducible on $\mathbb{E}$. We envision the units moving in batches or one at a time in the node set M, where $M = \{0, 1, \ldots, m\}$ if the network is open, and $M = \{1, \ldots, m\}$ if the network is closed. A typical transition will be from x to $x - d + a$, where a, d are vectors in a prescribed set A of allowable increments of X. We adopt the convention that either a_j or d_j equals 0 for each $j = 1, \ldots, m$. This means that a and d represent the net numbers of arrivals and departures from the respective nodes. These batch arrivals and departures are sometimes called concurrent or synchronous movements of several units. With no loss in generality, assume A

contains $e_0 = 0$ and the unit vectors $e_1, \ldots, e_m$. This is not a restriction since one can choose any basis that spans the sets A and $\mathbb{E}$ to represent their vectors; the form of the basis is not important here.

We assume that whenever the network process X is in state x, the time to the next transition to state $x - d + a$ is exponentially distributed with rate $\lambda_{da}\phi_{da}(x)$. In other words, the transition rates of X are

$$q(x, y) = \begin{cases} \lambda_{da}\phi_{da}(x) & \text{if } y = x - d + a \in \mathbb{E} \text{ for some } a, d \in A \\ 0 & \text{otherwise.} \end{cases}$$

These rates are analogous to those for Jackson or Whittle processes. Think of λ_{da} as the relative *increment-selection* or routing intensities. They are independent of the state x and $\lambda_{da} = 0$ if $a_j d_j > 0$ for some j. This ensures that d and a are net increments. Also, think of $\phi_{da}(x)$ as the relative *departure–arrival* intensity at which the vector d is deleted from the network and the vector a is added to the network. Assume these intensities are positive except that $\phi_{da}(x) = 0$ if $d_j > x_j$ for some j. We write $\lambda_{jk} = \lambda_{e_j e_k}$ and $\phi_{jk} = \phi_{e_j e_k}$, for $j, k \in M$.

Our aim is to derive a Kolmogorov-type criterion for the process X to be reversible and to give an expression for its invariant measures. We will exploit the fact that the transition rate function q is a compounding (or weak coupling) of increment-selection and departure–arrival intensities. Specifically, we can write

$$q(x, y) = q_\lambda(x, y) q_\phi(x, y), \quad x, y \in \mathbb{E}, \tag{2.14}$$

where q_λ and q_ϕ are transition rate functions defined (excluding the zero entries) by

$$q_\lambda(x, x - d + a) = \lambda_{da}, \quad q_\phi(x, x - d + a) = \phi_{da}(x).$$

These two rate functions define irreducible Markov jump processes on $\mathbb{E}$. The next results give necessary and sufficient conditions for their reversibility.

Proposition 2.19. *The q_λ is reversible if and only if λ_{jk} is reversible with respect to some w_j, $j \in M$; and λ_{da} is reversible with respect to $w(a) = \prod_{j=1}^m w_j^{a_j}$, $a \in A$, where $w_0 = 1$ if the network is open. In this case, an invariant measure of q_λ is*

$$\pi(x) = \prod_{j=1}^{m} w_j^{x_j}, \quad x \in \mathbb{E}.$$

PROOF. Suppose q_λ is reversible. Consider any distinct $a^0, \ldots, a^n$ in A that form a path for the rates λ_{da}. Fix $x^0 \in \mathbb{E}$ and define $x^i = x^{i-1} - a^{i-1} + a^i$, for $1 \le i \le n$. Then $x^0, \ldots, x^n$ is a path in $\mathbb{E}$ for q_λ. Since q_λ is reversible, the Kolmogorov ratio criterion says that

$$\prod_{i=1}^{n} \lambda_{a^{i-1}a^i}/\lambda_{a^i a^{i-1}} = \prod_{i=1}^{n} q_\lambda(x^{i-1}, x^i)/q_\lambda(x^i, x^{i-1})$$

depends only on x^0 and $x^n = x^0 + \sum_{i=1}^n (a^i - a^{i-1}) = x^0 + a^n - a^0$. In other words, this product of λ ratios does not depend on $a^1, \ldots, a^{n-1}$. Hence λ_{da} is reversible on A. Now, λ_{da} is also reversible on any subset of A, and so λ_{jk} is reversible with respect to some w_j, $j \in M$. Since q_λ is reversible with respect to π, for each $a, d \in A$ there is an $x \geq d$ in $\mathbb{E}$ such that

$$w(d)\lambda_{da} = \pi(x)\lambda_{da}/\pi(x-d) = \pi(x-d+a)\lambda_{ad}/\pi(x-d) = w(a)\lambda_{ad}.$$

Consequently, λ_{da} is reversible with respect to $w(a)$.

Conversely, suppose λ_{jk} is reversible with respect to w_j and λ_{da} is reversible with respect to $w(a) = \prod_{j=1}^m w_j^{a_j}$. Then q_λ is reversible with respect to $\pi(x) = \prod_{j=1}^m w_j^{x_j}$, since, for any x, a, and d such that $d \leq x$,

$$\pi(x)\lambda_{da} = \pi(x-d)w(d)\lambda_{da} = \pi(x-d)w(a)\lambda_{ad} = \pi(x-d+a)\lambda_{ad}. \quad \square$$

To describe the reversibility of $q_\phi(x, y) = \phi_{da}(x)$, we will use the following notion.

Definition 2.20. The intensities ϕ_{da} are Φ*-balanced departure–arrival intensities* if Φ is a positive function on $\mathbb{E}$ such that, for $x \in \mathbb{E}$ and $a, d \in A$ with $x-d+a \in \mathbb{E}$,

$$\Phi(x)\phi_{da}(x) = \Phi(x-d+a)\phi_{ad}(x-d+a).$$

This condition is the same as saying that q_ϕ is reversible with respect to Φ. By Theorem 1.5, the ϕ_{da} are Φ-balanced if and only if, for any $x \in \mathbb{E}$ and $d, a, \in A$ such that $x - d + a \in \mathbb{E}$,

$$\phi_{da}(x) = g(x, x-d+a)/\Phi(x), \tag{2.15}$$

for some function g that satisfies $g(x, y) = g(y, x)$ for each $x, y \in \mathbb{E}$.

A more useful characterization is the following special Kolmogorov criterion. Here we say that $x^0, \ldots, x^n \in \mathbb{E}$ is a *direct path from* x^0 *to* x^n if $x^i = x^{i-1} - e_{j_i} + e_{k_i}$ for some j_i, k_i in M such that $n = |x^0 - x^n|$.

Proposition 2.21. *The ϕ_{da} are Φ-balanced departure–arrival intensities if and only if for each $j, k, \ell \in M$, and $x \in \mathbb{E}$ with $T_{j\ell}x, T_{k\ell}x \in \mathbb{E}$,*

$$\phi_{j\ell}(x)\phi_{kj}(T_{j\ell}x)\phi_{\ell k}(T_{k\ell}x) = \phi_{k\ell}(x)\phi_{jk}(T_{k\ell}x)\phi_{\ell j}(T_{j\ell}x), \tag{2.16}$$

and, for each $d, a \in A$ with $x - d + a \in \mathbb{E}$, and any direct path $x^0, \ldots, x^n$ from $x^0 = x$ to $x^n = x - d + a$,

$$\phi_{da}(x)/\phi_{ad}(x-d+a) = \prod_{i=1}^n \phi_{j_i k_i}(x^{i-1})/\phi_{k_i j_i}(x^i). \tag{2.17}$$

In this case,

$$\Phi(x) = \prod_{i=1}^n \phi_{j_i k_i}(x^{i-1})/\phi_{k_i j_i}(x^i), \quad x \in \mathbb{E}, \tag{2.18}$$

for any direct path $x^0, \ldots, x^n$ from a fixed reference state x^0 to $x^n = x$.

PROOF. Since the Φ-balance of the ϕ_{da}'s is equivalent to the reversibility of q_ϕ, it suffices to show that q_ϕ is reversible if and only if (2.16) and (2.17) hold. And if q_ϕ is reversible, then (2.18) is an invariant measure for it. To prove these assertions, first note that transition rates under q_ϕ are positive for any unit increment. Then it follows by Theorem 2.8 that q_ϕ is reversible if and only if the Kolmogorov ratio criterion holds for only *direct paths*. But this criterion for direct paths is clearly equivalent to (2.16) and (2.17). This proves that q_ϕ is reversible if and only if (2.16) and (2.17) hold. Theorem 2.8 also justifies that if q_ϕ is reversible, then (2.18) is an invariant measure for it. □

We are now ready to consider the reversibility and invariant measures of the process X with transition rates $q(x, x-d+a) = \lambda_{da}\phi_{da}(x)$.

Theorem 2.22. *Suppose the following conditions hold:*
(a) λ_{jk} is reversible with respect to w_j, where $w_0 = 1$ if the network is open; and λ_{da} is reversible with respect to $w(a) = \prod_{j=1}^m w_j^{a_j}$, $a \in A$.
(b) ϕ_{da} are Φ-balanced departure–arrival intensities.
Then the network process X is reversible with respect to

$$\pi(x) = \Phi(x) \prod_{i=1}^{m} w_j^{x_j}, \quad x \in \mathbb{E}, \tag{2.19}$$

where Φ is given by (2.18). Conversely, if X is reversible, then (a) is equivalent to (b).

PROOF. Consider the compound transition rate

$$q(x, x-d+a) = \lambda_{da}\phi_{da}(x) = q_\lambda(x, y)q_\phi(x, y).$$

From the definition of reversibility and Proposition 2.15, it follows that if any two of the q, q_λ, and q_ϕ are reversible, then so is the third. In this case, invariant measures for the three rates are related by $\pi(x) = \pi_\lambda(x)\tilde{\pi}(x)$. This observation and Propositions 2.19 and 2.18 prove the assertions of the theorem. □

2.7 Examples of Reversible Networks

Let us explore some examples of the network process X discussed in the preceding section. First note that an important subclass of departure–arrival intensities are separable ones of the form

$$\phi_{da}(x) = \phi_d(x)\psi_a(x).$$

The ϕ_{da} are $\Phi\Psi$-balanced departure–arrival intensities if ϕ_d are Φ-balanced departure intensities and ψ_a are Ψ-balanced arrival intensities as follows.

Definition 2.23. The ϕ_d are *Φ-balanced departure intensities* if Φ is a positive function on $\mathbb{E}$ such that, for each $x \in \mathbb{E}$ and $a, d \in A$ with $x - d + a \in \mathbb{E}$,

$$\Phi(x)\phi_d(x) = \Phi(x-d+a)\phi_a(x-d+a). \tag{2.20}$$

The ψ_a are *Ψ-balanced arrival intensities* if Ψ is a positive function on $\mathbb{E}$ such that, for each $x \in \mathbb{E}$ and $a, d \in A$ with $x - d + a \in \mathbb{E}$,

$$\Psi(x)\psi_a(x) = \Psi(x-d+a)\psi_d(x-d+a).$$

It is clear that ϕ_d are Φ-balanced departure intensities if and only if

$$\phi_d(x) = \Psi(x-d)/\Phi(x), \quad j \in M, \ x \in \mathbb{E},$$

for some nonnegative function Ψ defined on $\{x-d : x \in \mathbb{E}, d \in A\}$. This follows by setting $\Psi(x-d) = \Phi(x-d+a)\phi_a(x-d+a)$ for a fixed a. Similarly, ψ_a are Ψ-balanced arrival intensities if and only if

$$\phi_a(x) = \Psi(x+a)/\Phi(x), \quad j \in M, \ x \in \mathbb{E},$$

for some nonnegative function Ψ defined on $\{x+a : x \in \mathbb{E}, a \in A\}$. These representations are special cases of the canonical form of reversible transition rates in Theorem 1.5.

The following are some illustrations of separable departure–arrival intensities.

Example 2.24. *Networks with Single-Unit Movements and Independent Nodes.* Consider the special case in which the process X has unit increments and its transition rates are

$$q(x,y) = \begin{cases} \lambda_{jk}\phi_j(x_j)\psi_k(x_k) & \text{if } y = x - e_j + e_k \text{ for some } j,k \in M \\ 0 & \text{otherwise.} \end{cases}$$

In addition to the usual departure intensity $\phi_j(x_j)$, there is a pull or attraction intensity $\psi_k(x_k)$ at each node k that affects where the departure from j goes next. The departure–arrival rates $\phi_j(x_j)\psi_k(x_k)$ are clearly Φ-balanced, where

$$\Phi(x) = \prod_{j=1}^{m}\prod_{i=1}^{x_j} \psi_j(i-1)/\phi_j(k), \quad x \in \mathbb{E}.$$

This follows by the criterion (2.15) since

$$\phi_j(x_j)\psi_k(x_k) = \Phi(x - e_j + e_k)/\Phi(x).$$

Then by Theorem 2.22, the process X is reversible if and only if the rates λ_{jk} on M are reversible with respect to $w_0, \ldots, w_m$, where $w_0 = 1$ if the network is open. In this case, an invariant measure for X is $\pi(x) = \Phi(x)\prod_{j=1}^{m} w_j^{x_j}$, $x \in \mathbb{E}$. □

Example 2.25. *Reversible Jackson and Whittle Processes.* Suppose X is a Jackson or Whittle process. This is a special case of the process in the preceding example with $\psi_j(\cdot) = 1$. Therefore, X is reversible if and only if λ_{jk} is reversible. As an illustration, suppose the network is a closed starlike network with the following routing. The communication graph of λ_{jk} is a star whose center consists of the single node 1, and M_i is a collection of subsets of M, called points of the star, whose union is M and whose intersection is the center node 1. This network is a generalization of the one discussed in Example 1.26. Also, $\lambda_{jk} = 0$ if j and k are not in the same point set. This means that in order for a unit to travel from one point of the star to another, it must go through the center node 1 (the central processor). Assume that λ_{jk} restricted to each M_i is reversible with respect to some w_j^i, $j \in M_i$. Then clearly λ_{jk} on M is reversible with respect to $w_j = w_j^i$, for $j \in M_i$. Thus it follows that X is reversible. □

If a network process is reversible, then we know that its restriction to any subspace is also reversible. The next two examples describe restrictions that arise when (1) nodes have finite capacities resulting in blocked transitions; or (2) units require resources for services, and transitions are blocked when the resources are not available.

Example 2.26. *Reversible Network Processes with Communication Blocking.* Consider a reversible network process with single-unit movements that has a known invariant measure (e.g., a Jackson or Whittle process). Now, suppose X is the process with the added restriction that the number of units at each node j cannot exceed a prescribed bound b_j, which may be infinite. That is, the routing and services are the same, but transitions from x to $T_{jk}x$ are not allowed when $x_k = b_k$. This is called *communication blocking*. The standard interpretation is that when $x_k = b_k$, any unit at j that is potentially scheduled to enter k cannot begin its service at j until there is a departure at k. Another equivalent interpretation is that services at j continue, but a departing unit from j scheduled to enter k must return to j for another service as if it were a new arrival. These interpretations are equivalent because the time to a departure is exponentially distributed. Another type of blocking, called *manufacturing blocking*, assumes that when $x_k = b_k$, the services at the other nodes continue, but a job at j attempting to enter k will remain at j until a space at k becomes available, at which time it immediately enters k.

Under the preceding communication-blocking assumption, the process X is the original network process restricted to the state space $\tilde{\mathbb{E}} = \{x \in \mathbb{E} : x \leq b\}$. Hence X is also reversible, and its invariant measures are those of the original process restricted to $\tilde{\mathbb{E}}$. Similar blockings can be defined for networks with batch movements. □

Example 2.27. *Reversible Networks with Resource Constraints.* Consider a reversible network process with single-unit movements that has a known invariant measure (e.g., a Jackson or Whittle process). Now, suppose X is the process with the added restriction that the units require certain sets of resources for their processing as follows. The network contains quantities b_i, $i \in I$, of resources that the units may use. Each unit entering node j requires the prescribed quantities a_{ij}, $i \in I$, of the resources for its processing at that node. If these quantities are available, they are assigned instantaneously to the unit which holds the resources throughout its stay at the node, without sharing them with other units. Upon leaving j, the unit releases the resources so that they can be used again. If the resources are not available for a unit attempting to enter node j, the unit is blocked from being served and its service can begin when the resources become available. Whenever the network is in state x, the quantities of resources held by the units is Ax, where A is the matrix with entries a_{ij}. Under these assumptions, a transition from x to $T_{jk}x$ is blocked if $AT_{jk}x \leq b$ is violated. Then the process X is the original reversible network process restricted to the state space $\tilde{\mathbb{E}} = \{x \in \mathbb{E} : Ax \leq b\}$. Consequently, X is also reversible and its invariant measures are the original invariant measures restricted to $\tilde{\mathbb{E}}$. Note that this example with A equal to the m-dimensional identity matrix is the same as the communication blocking example above. □

The next example, which is a generalization of the classical birth–death process, describes a variety of population models including many service systems with queueing.

Example 2.28. *Multivariate Batch Birth–Death Processes.* Consider the case in which the process X has transition rates

$$q(x, y) = \begin{cases} \psi_a(x) & \text{if } y = x + a \text{ for some } a \in A \\ \phi_a(x) & \text{if } y = x - a \text{ for some } d \in A \\ 0 & \text{otherwise.} \end{cases}$$

Think of the process as representing the sizes of m populations or queues in which a batch arrival $a = (a_1, \ldots, a_m)$ increases the population j by the amount a_j, $1 \leq j \leq m$, and a departure of a decreases the populations similarly. There is no routing among the populations; a unit departing from a population exits the system. The populations are dependent because the arrival and departure rate functions $\psi_a(x)$ and $\phi_a(x)$ may depend on the system state x. We assume that ϕ_d are Φ-balanced departure intensities and that ψ_a are Ψ-balanced arrival intensities.

Note that the transition rates can be written as

$$q(x, x - d + a) = \phi_d(x)\psi_a(x)1(\, a = 0 \text{ or } d = 0).$$

Now, the routing intensity function (the indicator function) is automatically reversible with respect to $w(a) \equiv 1$. Also, the departure–arrival intensities are $\Phi\Psi$-balanced. Then by Theorem 2.22, the process X is reversible with respect to $\pi(x) = \Phi(x)\Psi(x)$, $x \in \mathbb{E}$. We call such a process a *multivariate batch birth–death process.*

Consider the special case in which X represents the size of a single population ($m = 1$) and there are no assumptions on the departure or arrival intensities. Then it follows by Proposition 2.21 and Theorem 2.22 that X is reversible if and only if

$$\phi_a(x)/\psi_a(x + a) = \prod_{n=x}^{x+a-1} \phi_1(n)/\psi_1(n + 1), \quad x \in \mathbb{E},\ a \in A. \tag{2.21}$$

In this case, an invariant measure for X is

$$\pi(x) = \prod_{n=1}^{x} \psi_1(n - 1)/\phi_1(n), \quad x \in \mathbb{E}.$$

The key observation for this result is that, according to Proposition 2.21, the condition (2.21) is necessary and sufficient for $\phi_d\psi_a$ to be Φ-balanced departure–arrival intensities, where $\Phi = \pi$. This one-dimensional batch birth–death process was mentioned in Example 2.18. □

2.8 Partition-Reversible Processes

In this and the next section we study a generalization of reversibility called partition-reversibility. A Markov process is partition-reversible if its average flows rates are balanced in a certain way over sets that partition the state space. This property is a "macro" version of the detailed balance property of reversible processes. A key feature of a partition-reversible process is that its stationary distribution is obtainable by solving the balance equation separately on the sets of the partition.

Throughout this section, we assume $\{X_t : t \geq 0\}$ is an ergodic Markov jump process with a countable state space $\mathbb{E}$, transition rates $q(x, y)$, and stationary distribution π. Here is an example of what lies ahead.

Example 2.29. Suppose X takes values in the set of integers. Assume that in order for it to move between the positive and negative integers it must pass through 0 and, it can enter 0 only from states 1 or -1. The communication graph of the process is therefore a star with center set $\mathbb{E}_0 = \{0\}$ and point sets $\mathbb{E}_1 = \{1, 2, \ldots\}$ and $\mathbb{E}_2 = \{\ldots, -2, -1\}$. Assume the restrictions of the process to the sets $\mathbb{E}_i$ and to $\mathbb{E}_0 \cup \mathbb{E}_i$ are ergodic, and let π_i and π_{0i} denote their respective stationary distributions. Under these minimal assumptions, the stationary distribution of the process has the form

$$\pi(x) = \pi(0)(\pi_{0i}(0)^{-1} - 1)\pi_i(x) = \pi(0)\pi_{0i}(0)^{-1}\pi_{0i}(x), \quad x \in \mathbb{E}_i,\ i = 1, 2,$$

where $\pi(0)^{-1} = \pi_{01}(0)^{-1} + \pi_{02}(0)^{-1} - 1$. The first equality says that π is a "collage" or pasting together of π_1 and π_2. Similarly, the second equality says that π is a collage of π_{01} and π_{02}. □

We now develop this theme for the general ergodic Markov process X. Suppose there is a partition $\{\mathbb{E}_i : i \in I\}$ of the state space $\mathbb{E}$ such that q restricted to $\mathbb{E}_i$ defines an ergodic Markov process on $\mathbb{E}_i$, and let π_i denote its stationary distribution. Let L denote the set of pairs of indices (i, j) such that the process X can jump (in one transition) from $\mathbb{E}_i$ to $\mathbb{E}_j$ or vice versa; L consists of the "links" in the partition. For each such pair, assume that q restricted to $\mathbb{E}_i \cup \mathbb{E}_j$ defines an ergodic Markov process, and let π_{ij} denote its stationary distribution.

We say that the distribution π is a *collage* of $\{\pi_i : i \in I\}$ if π is a multiple of π_i on $\mathbb{E}_i$, for each $i \in I$. The aim is to characterize this property in terms of how the process moves between pairs of sets $\mathbb{E}_i$ and $\mathbb{E}_j$.

Definition 2.30. The process X (or q) is *reversible over the partition* $\{\mathbb{E}_i : i \in I\}$ if, for each $(i, j) \in L$,

$$\pi(x) \sum_{y \in \mathbb{E}_j} q(x, y) = \sum_{y \in \mathbb{E}_j} \pi(y) q(y, x), \quad x \in \mathbb{E}_i \cup \mathbb{E}_j. \tag{2.22}$$

That is, the average number of jumps per unit time from x to $\mathbb{E}_j$ equals the average for the reverse jumps from $\mathbb{E}_j$ to x. Being symmetric in i and j, this equation also holds with $\mathbb{E}_i$ replaced by $\mathbb{E}_j$.

The following is a characterization of partition-reversibility in terms of the "local" distributions π_i and π_{ij}. The stationary distribution of a partition-reversible process is the collage (2.24), or its relative (2.25). Condition (c) is a convenient criterion for establishing partition-reversibility.

Theorem 2.31. *The following statements are equivalent.*
(a) *The process X is reversible over the partition $\{\mathbb{E}_i : i \in I\}$.*
(b) *The distribution π is a collage of $\{\pi_i : i \in I\}$, and π on $\mathbb{E}_i \cup \mathbb{E}_j$ is a multiple of π_{ij}, for each $(i, j) \in L$.*
(c) *For each $(i, j) \in L$, the distribution π_{ij} balances q on $\mathbb{E}_i$, and the matrix*

$$r_{ij} = \begin{cases} \pi_{ij}(\mathbb{E}_j) & \text{if } (i, j) \in L \\ 0 & \text{otherwise,} \end{cases} \tag{2.23}$$

is reversible.
(d) *For each $(i, j) \in L$, the distribution π_{ij} balances q on $\mathbb{E}_i$ and $\pi_{ij}(\mathbb{E}_i) = p_i/(p_i + p_j)$ for some positive probability measure p_i, $i \in I$.*
If these statements hold, then

$$\pi(x) = p_i \pi_i(x), \quad x \in \mathbb{E}_i,\ i \in I, \tag{2.24}$$

where p_i, $i \in I$ is the stationary distribution of r_{ij}; furthermore, $p_i = \pi(\mathbb{E}_i)$ and

$$\pi(x) = (p_i + p_j)\pi_{ij}(x), \quad x \in \mathbb{E}_i \cup \mathbb{E}_j,\ (i, j) \in L. \tag{2.25}$$

PROOF. For convenience, let

$$\pi q(A, B) = \sum_{x \in A} \sum_{y \in B} \pi(x) q(x, y).$$

Then the balance equations that determine π are $\pi q(x, \mathbb{E}) = \pi q(\mathbb{E}, x)$, $x \in \mathbb{E}$.

(a) $\Leftrightarrow$ (b). Clearly (a) is equivalent to the conditions

$$\pi q(x, \mathbb{E}_i) = \pi q(\mathbb{E}_i, x), \quad x \in \mathbb{E}_i, i \in I,$$
$$\pi q(x, \mathbb{E}_i) + \pi q(x, \mathbb{E}_j) = \pi q(\mathbb{E}_i, x) + \pi q(\mathbb{E}_j, x), \quad x \in \mathbb{E}_i \cup \mathbb{E}_j,\ (i, j) \in L.$$

The latter uses (2.22) with $\mathbb{E}_i$ replaced by $\mathbb{E}_j$. Now, these equations say that π balances q on $\mathbb{E}_i$, $i \in I$, and π balances q on $\mathbb{E}_i \cup \mathbb{E}_j$, $(i, j) \in L$. But this statement is equivalent to (b) by the uniqueness property of invariant measures. Thus statements (a) and (b) are equivalent.

(c) $\Leftrightarrow$ (d). If (c) holds, then, by the definition of reversibility, there exists a positive probability measure p_i on I such that

$$p_i \pi_{ij}(\mathbb{E}_j) = p_j \pi_{ij}(\mathbb{E}_i), \quad (i, j) \in L.$$

This and $\pi_{ij}(\mathbb{E}_i) + \pi_{ij}(\mathbb{E}_j) = 1$ yield $\pi_{ij}(\mathbb{E}_i) = p_i/(p_i + p_j)$. Thus (c) implies (d). Conversely, if (d) holds, then $r_{ij} = p_i/(p_i + p_j)$ is reversible since this is the canonical form of reversible rates; recall Theorem 1.5. Hence (d) implies (c).

(b) $\Leftrightarrow$ (d). Note that in statement (d) the condition that π_{ij} balances q on $\mathbb{E}_i$ is equivalent (since π_{ij} is a multiple of π_i) to

$$\pi_{ij}(x) = \pi_{ij}(\mathbb{E}_i)\pi_i(x), \quad x \in \mathbb{E}_i.$$

This relation also holds with $\mathbb{E}_i$ replaced by $\mathbb{E}_j$ since $\mathbb{E}_i \cup \mathbb{E}_j$ is symmetric in i and j. From these observations, it follows that (d) is equivalent to the following:

(d′) There is a probability measure p_i on I such, for each $(i, j) \in L$,

$$\pi_{ij}(x) = \begin{cases} \dfrac{p_i}{p_i + p_j}\pi_i(x) & \text{if } x \in \mathbb{E}_i \\ \dfrac{p_j}{p_i + p_j}\pi_j(x) & \text{if } x \in \mathbb{E}_j. \end{cases} \tag{2.26}$$

We will complete the proof by showing that (b) and (d′) are equivalent. Suppose (d′) holds. Then (2.26) implies that π_{ij} balances q on $\mathbb{E}_i$. Consider the distribution π on $\mathbb{E}$ defined by

$$\pi(x) = p_i\pi_i(x) = (p_i + p_j)\pi_{ij}(x), \quad x \in \mathbb{E}_i,\ i \in I. \tag{2.27}$$

For any $x \in \mathbb{E}$ and i such that $x \in \mathbb{E}_i$, the property that π_i balances q on $\mathbb{E}_i$ and the assumption that π_{ij} balances q on $\mathbb{E}_j$ yield

$$\begin{aligned} \pi q(x, \mathbb{E}) &= p_i\pi_i(x)q(x, \mathbb{E}_i) + \sum_{j \neq i}(p_i + p_j)\pi_{ij}(x)q(x, \mathbb{E}_j)1((i, j) \in L) \\ &= p_i \sum_{y \in \mathbb{E}_i} \pi_i(y)q(y, x) + \sum_{j \neq i}(p_i + p_j) \sum_{y \in \mathbb{E}_j} \pi_{ij}(y)q(y, x)1((i, j) \in L) \\ &= \pi q(\mathbb{E}, x). \end{aligned}$$

Hence π defined by (2.27) is the stationary distribution of the process X, and its structure implies statement (b).

Now suppose (b) holds. Then the stationary distribution π of X satisfies (2.27) with $p_i = \pi(\mathbb{E}_i)$. The second equality in (2.27) says that π_{ij} is given by (2.25), and so (d′) is true.

The last sentence of the theorem was justified in proving that (b) is equivalent to (d′) and by the reversibility of r_{ij}. □

Note that when checking condition (c) for partition-reversibility, one can take advantage of the theory of reversibility to determine whether the transition rates r_{ij} defined by (2.23) are reversible. The obvious benefit for a partition-reversible process is that the problem of obtaining its stationary distribution reduces to finding several stationary distributions on smaller subspaces, either by analytical means or simulations or by a combination of both. Partition-reversibility is also a natural framework for analyzing Markov processes in random environments, Markov-modulated processes, or controlled Markov processes. Here the environment or control parameters (possibly dependent on the parent process) determine the appropriate partition of the state space. Examples are in the next section.

Theorem 2.31 and the other results here also apply to a discrete-time Markov chain with transition probabilities $P(x, y)$ by viewing these probabilities as transition rates for a continuous-time process. In this setting, $\pi_{ij}(x) = \bar{\pi}_{ij}(x)/P(x, \mathbb{E}_i \cup \mathbb{E}_j)$, where $\bar{\pi}_{ij}$ is the stationary distribution of the Markov chain matrix $P(x, y)$ restricted to $\mathbb{E}_i \cup \mathbb{E}_j$, which is $\bar{P}_{ij}(x, y) = P(x, y)/P(x, \mathbb{E}_i \cup \mathbb{E}_j)$.

Since reversibility and partition-reversibility are defined in terms of average numbers of transitions per unit time, these notions readily extend to non-Markovian processes in continuous or discrete time. To see this, suppose X is a stochastic process that takes jumps in its countable state space $\mathbb{E}$ at the times $0 = T_0 < T_1 < \ldots$, where $T_n \to \infty$ as $n \to \infty$ w.p.1. The average number of jumps per unit time that X makes from A to B is

$$\lambda(A, B) = \lim_{n\to\infty} T_n^{-1} \sum_{i=1}^{n} 1(X_{T_i} \in A, X_{T_{i+1}} \in B).$$

The process X is *reversible* if $\lambda(x, y) = \lambda(y, x)$ for each x, y in $\mathbb{E}$. Similarly, X is *reversible over the partition* $\{\mathbb{E}_i : i \in I\}$ if

$$\lambda(x, \mathbb{E}_i) = \lambda(\mathbb{E}_i, x), \quad \text{for each } x \in \mathbb{E}_i \cup \mathbb{E}_j, (i, j) \in L. \tag{2.28}$$

2.9 Examples of Partition-Reversible Processes

We now discuss special cases of partition-reversible processes whose communication structure between sets of the partition form circles, trees, or stars.

We will use the notation of the previous section. We say that $\{\mathbb{E}_1, \ldots, \mathbb{E}_\ell\}$ is a *circular partition* if whenever X is in $\mathbb{E}_i$, it can make a transition only into $\mathbb{E}_{i-1} \cup \mathbb{E}_i \cup \mathbb{E}_{i+1}$, where $\mathbb{E}_{\ell+1} = \mathbb{E}_1$ and $\mathbb{E}_0 = \mathbb{E}_\ell$. The following is a characterization of circular partition-reversible processes.

Corollary 2.32. *If* $\{\mathbb{E}_1, \ldots, \mathbb{E}_\ell\}$ *is a circular partition, then X is reversible over this partition if and only if each* $\pi_{i,i+1}$ *balances q on* $\mathbb{E}_i$ *and*

$$\alpha_1\alpha_2 \ldots \alpha_\ell = (1 - \alpha_1)(1 - \alpha_2) \cdots (1 - \alpha_\ell), \tag{2.29}$$

where $\alpha_i = \pi_{i,i+1}(\mathbb{E}_{i+1})$. *In this case,* $\pi(x) = p_i\pi_i(x)$, $x \in \mathbb{E}_i$, $i = 1, \ldots, \ell$, *where*

$$p_i = p_1 \prod_{n=2}^{i} \alpha_{n-1}/(1 - \alpha_n), \quad 2 \le i \le \ell,$$

and $p_1^{-1} = 1 + \sum_{i=2}^{\ell} \prod_{n=2}^{i} \alpha_{n-1}/(1 - \alpha_n)$.

Proof. Consider the rates defined by (2.23). The communication graph of these rates is circular because the partition is circular. Then the rates are reversible by Example 2.10 if and only if (2.29) holds. This result and Theorem 2.31 (c) prove the assertion. □

We now discuss treelike and starlike partition-reversible processes. We say the partition $\{\mathbb{E}_i : i \in I\}$ of the process is a *tree* if it has a single root set $\mathbb{E}_0$ and, whenever X is in some set $\mathbb{E}_i$, its one-step transitions can be back into $\mathbb{E}_i$ or into one of its neighboring sets (its single predecessor or its possibly multiple successors in the tree). That is, X can move up and down each branch, and it can move from branch to branch only via $\mathbb{E}_0$. This partition is a *star* if each branch consists of $\mathbb{E}_0$ and some $\mathbb{E}_i$; the $\mathbb{E}_i$'s are points of the star with center $\mathbb{E}_0$.

When the partition for X is a tree, the communication graph of the rates r_{ij} defined by (2.23) is a tree. Then these rates are reversible by Proposition 2.2 and their stationary distribution is

$$p_j = p_0 r_{0i_1} \cdots r_{i_n j} / r_{i_1 0} \cdots r_{j i_n}, \quad j \neq 0, \tag{2.30}$$

where $0, i_1, \ldots, i_n, j$ is the unique subbranch from 0 to j in the tree.

Corollary 2.33. *Suppose the partition of the process X is a tree. Then X is partition-reversible if and only if, for each predecessor-successor pair $\mathbb{E}_i$ and $\mathbb{E}_j$, the distribution π_{ij} balances q on $\mathbb{E}_i$. In this case, the stationary distribution π of X has the form (2.24) or (2.25) with p_i given by (2.30). If the partition is a star, then X is partition-reversible if and only if each π_{0j} balances q on $\mathbb{E}_0$ (which is automatically true when the center set $\mathbb{E}_0$ is a singleton). In this case,*

$$\pi(x) = p_0 \pi_0(x), \quad x \in \mathbb{E}_0,$$

and, for $x \in \mathbb{E}_i$ and $i \neq 0$,

$$\pi(x) = p_0 (\pi_{0i}(\mathbb{E}_0)^{-1} - 1) \pi_i(x) = p_0 \pi_{0i}(\mathbb{E}_0)^{-1} \pi_{0i}(x),$$

where $p_0 = \pi(\mathbb{E}_0) = [1 + \sum_{i \neq 0} (\pi_{0i}(\mathbb{E}_0)^{-1} - 1)]^{-1}$.

PROOF. Since the rates r_{ij} are reversible, the first assertion is a consequence of Theorem 2.31 (part (c) and (2.25)). The other assertions follow immediately from the first one and $p_i = p_0 r_{0i} / r_{i0}$, $i \neq 0$. □

Example 2.34. *A Multiclass Service System with Blocking.* Consider a service system that operates as follows. The system serves m classes of customers that arrive according to m independent Poisson processes with respective rates $\lambda_1, \ldots, \lambda_m$. The system can serve only one class of customer at any time. While it is serving customers of class i, any arrivals of other classes of customers cannot enter the system and are turned away, but new type i arrivals are permissible. Also, the number of these type i customers in the system behaves as an ergodic Markov process with transition rates $q_i(x, y)$. Here $q_i(x, x + 1) = \lambda_i$, but the transition rates for departures are left unspecified. We assume the stationary distribution of q_i can be obtained either analytically or by a simulation. Assume the system starts empty—thereafter it can contain, at most, one class of customer.

We represent the system as an m-dimensional queueing process X with states of the form $x = (x_1, \ldots, x_m)$, where x_i is a nonnegative integer and, at most, one of the x_i's is positive. The state space $\mathbb{E}$ is a star with center $\mathbb{E}_0 = \{0\}$ and point sets

$$\mathbb{E}_i = \{(x_1, \ldots, x_m) : x_i > 0, x_l = 0, l \neq i\}, \quad i = 1, \ldots, m.$$

That is, the process X cannot transfer from a state in $\mathbb{E}_i$ to a state in $\mathbb{E}_j$, $j \neq i$, unless it passes through 0. Under the preceding assumptions, it follows that the transition rates of X are

$$q(x, y) = \begin{cases} q_i(x, y) & \text{if } x, y \in \mathbb{E}_0 \cup \mathbb{E}_i, \text{ and } y = e_i \text{ if } x = 0; \ i = 1, \ldots, m \\ 0 & \text{otherwise.} \end{cases}$$

Since the state space of X is a star and $\mathbb{E}_0$ is the single state 0, its stationary distribution π is given by Corollary 2.33. In this case, $\pi_{0i}(x) = \tilde{\pi}_i(x_i)$ where $\tilde{\pi}_i$ is the stationary distribution of q_i. Therefore,

$$\pi(x) = p_0 \tilde{\pi}_i(0)^{-1} \tilde{\pi}_i(x_i), \quad x \in \mathbb{E}_i, i \neq 0,$$

where $\pi(0) = p_0 = [1 + \sum_{j \neq 0} (\tilde{\pi}_j(0)^{-1} - 1)]^{-1}$. □

Example 2.35. *Service System with State-dependent Service Rates.* Consider a service system in which customers arrive at a single server according to a Poisson process with rate λ. The service times are independent exponentially distributed with rate depending on the number of customers present. When there is one customer present, the service rate is μ and remains at this value until the number of customers reaches the level M. At that instance, the service rate takes a higher value μ' and remains there until a departure leaves m customers behind ($m < M$)—then the rate returns to μ. Assume $\lambda < \mu'$, which is necessary and sufficient for stability, as the analysis below shows.

Under these assumptions, the system is described by a Markov process X with states denoted by i (or i') when there are i customers in the system and μ (or μ') is in use. Then the state space is a star with center

$$\mathbb{E}_0 = \{i, i' : m \leq i \leq M - 1\} \cup \{M'\},$$

and point sets

$$\mathbb{E}_1 = \{i : 0 \leq i < m\}, \quad \mathbb{E}_2 = \{i' : i > M\}.$$

Since q is a birth–death process on each of $\mathbb{E}_1$ and $\mathbb{E}_2$, it follows that

$$\pi_1(i) = (1-\rho)\rho^i/(1-\rho^m), \quad 0 \leq i < m$$
$$\pi_2(i') = (1-\rho')\rho'^{(M-i')}, \quad i' \geq M,$$

where $\rho = \lambda/\mu$, $\rho' = \lambda/\mu'$. And solving the balance equations on $\mathbb{E}_0$ yields

$$\pi_0(i) = a\rho^{i-m}(1-\rho^{M-i})/(1-\rho), \quad m \leq i < M$$
$$\pi_0(i') = a\rho^{M-m-1}\rho'(1-\rho'^{(i-m)})/(1-\rho'), \quad m < i \leq M,$$

where a is the normalization constant.

We now establish that the stationary distribution of X is the collage of π_0, π_1, and π_2. By Corollary 2.33, it suffices to show that each π_{0i} balances the transition function on $\mathbb{E}_0$. Since the process X restricted to $\mathbb{E}_1 \cup \{m\}$ is a truncated birth–death process,

$$\lambda \pi_{01}(m-1) = \mu \pi_{01}(m). \tag{2.31}$$

And because $\mathbb{E}_0$ and $\mathbb{E}_1$ communicate only via states $m-1$ and m, the preceding equation implies that π_{01} balances q on $\mathbb{E}_0$.

Now the balance equations for π_{02} on $\mathbb{E}_0 \cup \mathbb{E}_2$ are

$$\lambda \pi_{02}(m) = \mu \pi_{02}(m+1) + \mu' \pi_{02}((m+1)')$$
$$(\lambda + \mu)\pi_{02}(i+1) = \lambda \pi_{02}(i) + \mu \pi_{02}(i+2), \quad m \leq i \leq M-2$$
$$(\lambda + \mu)\pi_{02}(M-1) = \lambda \pi_{02}(M-2)$$

and

$$(\lambda + \mu')\pi_{02}((m+1)') = \mu'\pi_{02}((m+2)')$$
$$(\lambda + \mu')\pi_{02}(i') = \lambda\pi_{02}((i+1)') + \mu'\pi_{02}((i+1)'), \quad m+2 \le i < M$$
$$(\lambda + \mu')\pi_{02}(M') = \lambda\pi_{02}(M-1) + \lambda\pi_{02}((M-1)') + \mu'\pi_{02}((M+1)')$$
$$(\lambda + \mu')\pi_{02}((i+1)') = \lambda\pi_{02}((i-1)') + \mu'\pi_{02}((i+1)'), \quad i > M.$$

Since $\mathbb{E}_0$ and $\mathbb{E}_2$ communicate only via states M' and $M+1$, the π_{02} balances q on $\mathbb{E}_0$ and on $\mathbb{E}_2$ if and only if

$$\lambda\pi_{02}(M') = \mu'\pi_{02}((M+1)'). \tag{2.32}$$

To show this, note that balance equations above yield

$$\lambda\pi_{02}(M-1) = \mu'\pi_{02}((m+1)'),$$

$$\lambda\pi_{02}((M-1)') + \mu'\pi_{02}((m+1)') = \mu'\pi_{02}(M').$$

These equations imply

$$\mu'\pi_{02}(M') = \lambda\pi_{02}(M-1) + \lambda\pi_{02}((M-1)').$$

But this is equivalent, by the balance equation for the state M', to (2.32).

In summary, the stationary distribution π of X is a collage of π_0, π_1, and π_2 as in (2.24), where (2.26), (2.31), and (2.32) yield

$$p_1 = p_0\mu\pi_0(m)/(\lambda\pi_1(m-1)), \quad p_2 = p_0\lambda\pi_0(M')/(\mu'\pi_2((M+1)')),$$

and p_0 is determined by $p_0 + p_1 + p_2 = 1$. □

2.10 Exercises

1. Consider a Markov jump process whose transition rates are

$$q(x, y) = \tilde{q}(f(x), f(y)), \quad x, y \in \mathbb{E},$$

where $\tilde{q}$ is an ergodic transition rate on $\tilde{\mathbb{E}}$ and f is a function from $\mathbb{E}$ to $\tilde{\mathbb{E}}$. Show that if $\tilde{q}$ is reversible with respect to $\tilde{\pi}$ on $\tilde{\mathbb{E}}$, then q is reversible with respect to $\pi(x) = \tilde{\pi}(f(x))$.

2. *Networks with Environmental Influences.* Suppose X is a Markov jump process that represents the state of an m-node network that is subject to environmental influences such as the status of machines or quantities of resources available for services. The state of the process x is in a countable set $\mathbb{E}$ of vectors, matrices, or functions that contains all the pertinent information about the network and environment. When the system is in state x, the numbers of units at the respective nodes are given by the function $n(x) = (n_1(x), \ldots, n_m(x))$. Assume the transition rates of the process are

$$q(x, y) = q_1(n(x), n(y))q_2(x, y), \quad x, y \in \mathbb{E}.$$

This is a "compounding" of a population rate q_1 and an environment rate q_2. Suppose q_1 is the rate for a reversible, ergodic Whittle or Jackson process. Prove that q is reversible if and only if q_2 is reversible. In this case, an invariant measure of q is $\pi(x) = \pi_1(n(x))\pi_2(x)$, where π_i is an invariant measure for q_i.

3. Use Theorem 2.5 on time reversals to prove Theorem 1.15 that an invariant measure for the Whittle process X is $\pi(x) = \Phi(x)\prod_{j=1}^m w_j^{x_j}$. Is this approach simpler than the direct-substitution proof of Theorem 1.15?
4. *Finite McCabe Library*. Show that an invariant measure for a McCabe library with n books is given by (2.12) where $\mathbb{E}$ is the finite set of all permutations of the n books.
5. *M/M/1 Queue with Variable Waiting Space*. Consider an M/M/1 queueing system with arrival rate λ and service rate μ in which the allowable number in the system varies randomly over time. Specifically, the number of customers in the system at time t, denoted by X_t, cannot exceed a value Y_t. The Y operates like an irreducible reversible Markov process with transition rates $q_Y(y, y')$ and stationary distribution $\pi_Y(y)$, but it is constrained by the inequality $X_t \le Y_t$. That is, whenever $X_t = Y_t$, the arrivals for X are turned away; also, transitions of Y_t below X_t are not allowed. More precisely, assume that (X_t, Y_t) is an irreducible Markov process on the space $\mathbb{E} = \{(x, y) : x \le y\}$ and its transition rates are

$$q((x, y), (x', y')) = \begin{cases} q_X(x, x') & \text{if } y' = y \text{ and } x' \le y \\ q_Y(y, y') & \text{if } x' = x \text{ and } y' \ge x \\ 0 & \text{otherwise.} \end{cases}$$

Here $q_X(x, x')$ is the transition rate function for the unrestricted M/M/1 queueing process on the nonnegative integers. Show that the process (X, Y) is reversible with respect to $\pi(x, y) = (\lambda/\mu)^x \pi_Y(y)$.
6. *Networks with Variable Waiting Spaces*. Consider an m-node open network process $X_t = (X_t^1, \ldots, X_t^m)$ that represents the numbers of units at the nodes at time t. Suppose the waiting spaces at the nodes vary such that $Y_t = (Y_t^1, \ldots, Y_t^m)$ is the maximum numbers of units allowed at the nodes at time t. Suppose $\{(X_t, Y_t) : t \ge 0\}$ is an irreducible Markov process on $\mathbb{E} = \{(x, y) \in \mathbb{E}_X \times \mathbb{E}_Y : x \le y\}$, where $\mathbb{E}_X = \{x : |x| < \infty\} = \mathbb{E}_Y$. Assume that its transition rates are

$$q((x, y), (x', y')) = \begin{cases} \lambda_{jk}\phi_j(x_j) & \text{if } x' = T_{jk}x,\ y' = y \\ & \text{and } x_k < y_k \text{ for some } j, k \in M \\ q_Y(y, y') & \text{if } x' = x \text{ and } y' \ge x'. \\ 0 & \text{otherwise.} \end{cases}$$

The X is an open Jackson process whose node populations are restricted by the process Y with transition rates q_Y. Assume that the routing rates λ_{jk} are reversible with respect to w_j and that q_Y is reversible with respect to π_Y. Show that the process (X, Y) is reversible with respect to $\pi(x, y) = \pi_Y(y)\prod_{j=1}^m w_j \prod_{n=1}^{x_j} \phi_{(}n)^{-1}$. Describe similar results for a Whittle process and any network process that is reversible when it is unrestricted. Take care in defining the state spaces.

2.11 Bibliographical Notes

Kolmogorov (1936) was the founder of reversible Markov processes. A related article is Hostinsky and Potocek (1935). Kingman (1969) was the first to model reversible stochastic networks. Reversibility of Markov processes and networks was developed further by Kelly (1979) and Whittle (1986b). The McCabe library and related library models in computer science are discussed in Letac (1974) and Suomela (1979). The material on batch movements in birth–death processes and reversible networks is from Serfozo (1993), and partition-reversible processes are introduced in Alexopoulos et al. (1999).

3

Miscellaneous Networks

This chapter deals with several applications and variations of the network models developed in the preceding chapters. We first show how to use Whittle processes to model networks with multiple types of units, where the routings and services may depend on a customer's type. This includes Kelly networks with deterministic routes for units, and BCMP networks with Cox and general service times depending on a unit's type. We also discuss several forms of blocking in networks, and bottlenecks in closed Jackson networks. The chapter ends with a discussion of partial balance equations in modeling networks.

3.1 Networks with Multiple Types of Units

Chapter 1 covered Jackson and Whittle networks in which the routing and departure intensities are the same for each unit. We will now show that the results for these networks with homogeneous units also apply to networks with multiple types of units, where the routing and services may depend on a unit's type in a certain way. The only difference is that we now keep track of the number of units of each type at a node.

Consider an m-node network in which each unit carries an attribute or class label from a finite set. A unit's class is a distinguishing characteristic that determines its routing or service rates. The class label may be permanent, or temporary and subject to change as the unit moves. Examples of permanent labels are:

• The size of a unit when it is a batch of subunits such as data packets, orders to be filled, or capacity of a circuit.

• The type of part or tool in a manufacturing network.
• The origin or destination of a unit.
• The general direction in which a unit moves through the network (e.g., north to south).
Examples of temporary labels are:
• The status of a part as it is being produced.
• The number of nodes a unit has visited.
• The number of times a unit has been fed back to the node where it resides.
• The phase of service that a unit is undergoing, when it has a phase-type distribution.

We make the following assumptions about the network, which are consistent with those for Jackson and Whittle networks. The evolution of the network over time is represented by a Markov process $\{X_t : t \geq 0\}$ whose state is a vector $x = (x_{\alpha j} : \alpha j \in M,\ j \neq 0)$, where $x_{\alpha j}$ is the number of α-units at node j. The number of units at node j is $x_j \equiv \sum_\alpha x_{\alpha j}$. We now envision that each unit moves in the set M of all pairs αj, where α is a class label and j is a node number, possibly 0, when the network is open. We denote the state space by $\mathbb{E}$. The network may be closed ($\sum_{\alpha j} x_{\alpha j} = \nu$), or open with finite or unlimited capacity. In addition, the network may be a mixture of these three types: The network may consist of several subprocesses that operate like closed or open networks.

Whenever the process is in a state x, a typical transition consists of an α-unit departing from node j and moving instantaneously into a node k and entering there as a β-unit. We denote the new state by $T_{\alpha j,\beta k}x \equiv x - e_{\alpha j} + e_{\beta k}$, where $e_{\alpha j}$ denotes the unit vector with a 1 in component αj and 0 elsewhere, and $e_{\alpha 0} \equiv 0$. It is allowable that $k = j$ or $\alpha = \beta$, provided $\alpha j \neq \beta k$.

We assume that the transition rates of the process X are of the form

$$q(x, y) \equiv \begin{cases} \lambda_{\alpha j,\beta k}\phi_{\alpha j}(x) & \text{if } y = T_{\alpha j,\beta k}x \in \mathbb{E} \text{ for some } \alpha j \neq \beta k \text{ in } M \\ 0 & \text{otherwise.} \end{cases}$$

The $\phi_{\alpha j}(\cdot)$ are *service rate functions* or intensities and $\lambda_{\alpha j,\beta k}$ are *routing rates* or intensities. The service rates $\phi_{\alpha j}$ are *Φ-balanced* in that Φ is a positive function on $\mathbb{E}$ such that, for each αj and x with $\beta k \neq \alpha j$ and $T_{\alpha j,\beta k}x \in \mathbb{E}$,

$$\Phi(x)\phi_{\alpha j}(x) = \Phi(T_{\alpha j,\beta k}x)\phi_{\beta k}(T_{\alpha j,\beta k}x).$$

The routing rates $\lambda_{\alpha j,\beta k}$ may be reducible, but they do not contain transient states. We let $w_{\alpha j}$ be positive numbers that satisfy the *traffic equations*

$$w_{\alpha j} \sum_{\beta k \in M} \lambda_{\alpha j,\beta k} = \sum_{\beta k \in M} w_{\beta k}\lambda_{\beta k,\alpha j}, \quad \alpha j \in M. \tag{3.1}$$

And $w_{\alpha 0} = 1$ when the network is open.

We call the process X with these properties a *multiclass Whittle network process*. We call X a *multiclass Jackson network process* if each service intensity $\phi_{\alpha j}(x)$ is a function $\phi_{\alpha j}(x_{\alpha j})$ only of $x_{\alpha j}$ and $\phi_0(\cdot) \equiv 1$ when the network is open. In this case, the service intensities are Φ-balanced by

$$\Phi(x) = \prod_{\alpha j \in M} \prod_{n=1}^{x_j} \phi_{\alpha j}(n)^{-1}.$$

The following result describes the equilibrium behavior of multiclass Whittle and Jackson networks. This result is just a restatement of Theorem 1.15 with the single subscripts j replaced by double subscripts αj.

Theorem 3.1. *For the multiclass Whittle network process X, an invariant measure is*

$$\pi(x) = \Phi(x) \prod_{\alpha j \in M} w_{\alpha j}^{x_{\alpha j}}, \quad x \in \mathbb{E}.$$

It is clear that the basic theory of Whittle and Jackson networks for homogeneous units in Chapter 1 also applies to their multiclass analogues—one just replaces all the single-node subscripts j in Chapter 1 with a double subscript αj. We already saw this in the traffic equation above. An important difference, however, is that multiclass labels can be exploited for modeling additional features or dependencies in networks.

To obtain an invariant measure for a multiclass network, one proceeds as in a network with homogeneous units by evaluating the function Φ and determining the $w_{\alpha j}$'s that satisfy the traffic equations. The characterizations of the function Φ in Section 1.13 for homogeneous units readily extend to the present context with multiclass units. One result is that the $\phi_{\alpha j}$ are Φ-balanced if and only if each $\phi_{\alpha j}$ is of the form

$$\phi_{\alpha j}(x) = \Psi(x - e_{\alpha j})/\Phi(x), \quad x \in \mathbb{E}, \tag{3.2}$$

for some nonnegative function Ψ defined on $\{x - e_{\alpha j} : x \in \mathbb{E},\ \alpha j \in M\}$.

The following examples give more insights on service rates; also see Exercise 1.

Example 3.2. *Service Rates Proportional to Local Populations.* A natural processor-sharing service discipline is one with service intensities

$$\phi_{\alpha j}(x) = \frac{x_{\alpha j}}{x_j} \phi_j(x_1, \ldots, x_m).$$

Think of $\phi_j(x_1, \ldots, x_m)$ as the total service capacity at node j, and the amount of this allocated to α-units is the proportion $x_{\alpha j}/x_j$ of those units present. Another interpretation is that $x_{\alpha j}/x_j$ is the probability of an α-unit departing when the intensity is $\phi_j(x_1, \ldots, x_m)$. Suppose ϕ_j are $\tilde{\Phi}$-balanced. Then the service rates $\phi_{\alpha j}$ are balanced (Exercise 2) by the function

$$\Phi(x) = \tilde{\Phi}(x_1, \ldots, x_m) \prod_{j=1}^{m} x_j! \prod_{\alpha} \frac{1}{x_{\alpha j}!}. \qquad \square$$

Example 3.3. *Sector-dependent Service Rates.* The sector-dependent service rates in Example 1.47 also apply as follows to the multiclass network we are studying. Let S denote the collection of all subsets (or sectors) of M. For each sector $J \in S$, there is a "departure intensity" $\phi_J(x(J))$, which is a function of the number of units $x(J) \equiv \sum_{\alpha j \in J} x_{\alpha j}$ in J. This intensity is 0 only if $x(J) = 0$. Assume these sector intensities are compounded such that the departure intensity for $\alpha j \neq \alpha 0$ is

$$\phi_{\alpha j}(x) = \prod_{J \in \mathcal{S}_{\alpha j}} \phi_J(x(J)), \quad x \in \mathbb{E},$$

where $\mathcal{S}_{\alpha j}$ is the collection of subsets that contain αj. Also, in case the network is open, we assume the intensity ϕ_0 is a positive function of the form $\phi_0(|x|)$.

These sector-dependent service rates are balanced (Exercise 2) by the function

$$\Phi(x) = \prod_{k=0}^{|x|-1} \phi_0(k) \prod_{J \in \mathcal{S}} \prod_{n=1}^{x(J)} \phi_J(n)^{-1}, \quad x \in \mathbb{E}, \tag{3.3}$$

where $\phi_0 \equiv 1$ if the network is closed. □

We now turn to properties of the routing rates $\lambda_{\alpha j,\beta k}$. The structure of these rates may be such that the multiclass network contains one or more families of permanent and transient units. This is illustrated in the next three examples.

Example 3.4. *Multichain Routing*. Suppose the routing rates $\lambda_{\alpha j,\beta k}$ are reducible (with no transient states), and let M_i, $i \in I$, denote the disjoint subsets of M upon which the rates irreducible. Then the solution of the traffic equations has the natural partition $\{w_{\alpha j}\} \equiv \{\{w^i_{\alpha j}\} : i \in I\}$, where $\{w^i_{\alpha j}\}$ satisfies the traffic equations on M_i. Similarly, the network process is the partition $X_t = (X^i_t : i \in I\}$, where X^i is the subprocess on M_i. Now, if M_i contains $\alpha 0$ for some α, then the subprocess X^i on M_i operates as an open network. Otherwise, X^i on M_i operates as a closed network with $\nu_i \equiv \sum_{\alpha j \in M_i} x_{\alpha j}$ units permanently in M_i. The number of units in each open subprocess could be limited or unlimited.

Example 3.5. *Permanent Class Labels*. Suppose each unit in the network process X carries a label that does not change. In this case, $\lambda_{\alpha j,\beta k} = 0$ if $\alpha \neq \beta$. Consequently, each α-unit would be routed in the network via the rates $\lambda_{\alpha j,\alpha k}$. Then for each α, the $w_{\alpha j}$ would be a solution to the traffic equations

$$w_{\alpha j} \sum_{k \in M} \lambda_{\alpha j,\alpha k} = \sum_{k \in M} w_{\alpha k} \lambda_{\alpha k,\alpha j}, \quad j \in M.$$

In this setting, there may be several classes of permanent and transient units. □

Example 3.6. *Class Changes Separate from Routing*. Suppose class changes of units in the network process X are independent of their routing, and the routing rates are of the form

$$\lambda_{\alpha j,\beta k} = \tilde{\lambda}_{\alpha\beta} \lambda_{jk}.$$

Interpret $\tilde{\lambda}_{\alpha\beta}$ as the intensity of an α-unit changing to a β-unit and λ_{jk} as the intensity of a unit at j moving into k. If $\tilde{w}_\alpha$ and w_j are respective solutions to their "traffic equations," then it is clear that $w_{\alpha j} = \tilde{w}_\alpha w_j$ is a solution to the traffic equations for $\lambda_{\alpha j,\beta k}$. □

Chapter 1 showed how routing rates determine throughputs in networks. Analogous results apply to multiclass networks as follows. Let $\rho_{\alpha j,\beta k}$ denote the average number of units that move from αj to βk per unit time; this is the *throughput from j to k of α to β class changes*. We know by the law of large numbers for Markov processes that

$$\rho_{\alpha j,\beta k} = \sum_{x \in \mathbb{E}} \pi(x) q(x, T_{\alpha j,\beta k} x) = \lambda_{\alpha j,\beta k} \sum_{x \in \mathbb{E}} \pi(x) \phi_{\alpha j}(x) 1(x_{\alpha j} \geq 1). \tag{3.4}$$

This expression has the following tractable forms. Here, we will assume that the service rates of the multiclass network are of the form

$$\phi_{\alpha j}(x) = \Phi(x - e_{\alpha j})/\Phi(x), \quad x \in \mathbb{E}, \ j \in M. \tag{3.5}$$

This is true, in particular, for the Kelly and BCMP networks we discuss in the next two sections.

Proposition 3.7. *Suppose (3.5) holds, and let M' denote an irreducible class in M for the routing rates as described in Example 3.4. If the subnetwork on M' is open with unlimited capacity, then*

$$\rho_{\alpha j,\beta k} = w_{\alpha j} \lambda_{\alpha j,\beta k}, \quad \alpha j, \ \beta k \in M'.$$

If the subnetwork on M' is closed with ν units (or open with capacity ν), then

$$\rho_{\alpha j,\beta k} = c_\nu c_{\nu-1}^{-1} w_{\alpha j} \lambda_{\alpha j,\beta k}, \quad \alpha j, \ \beta k \in M'.$$

Here c_ν is the normalizing constant for the equilibrium distribution of the closed network with ν units (or the open network with capacity ν).

In the multiclass network we are studying, a *sector* J is a subset of an irreducible routing subset M' of M. The throughput from a sector J to a sector K in M' is

$$\rho_{JK} = \sum_{\alpha j \in J} \sum_{\beta k \in K} \rho_{\alpha j,\beta k}.$$

Also, the *throughput of sector J* is

$$\lambda_J = \rho_{J^c J} = \sum_{\alpha j \in J^c} \sum_{\beta k \in J} \rho_{\alpha j,\beta k}.$$

We now turn to expected sojourn times of units in the sector J. The average sojourn time (or waiting time) of units in J is

$$W_J = \lim_{n \to \infty} n^{-1} \sum_{i=1}^{n} W_i(J) \quad \text{w.p.1},$$

provided the limit exists, where $W_i(J)$ is the waiting time of the ith unit to enter J. We assume $J \neq M'$ when the subnetwork is closed (otherwise all sojourns would be infinite). In addition, assume that the average number of units in J per unit time given by

$$L_J = \sum_x \sum_{\alpha j \in J} x_{\alpha j} \pi(x)$$

is finite. The following result is the analogue of Theorems 1.36 and 1.37.

Theorem 3.8. *The average waiting time W_J exits, and $L_J = \lambda_J W_J$. If the process X is stationary, then $L_J = \lambda_J W_J$, where these terms are expected values: L_J is the expected number of units in J at any time instant, λ_J is the expected number of units entering J per unit time, and W_J is the expected sojourn time in J with respect to the Palm probability of the stationary process X conditioned that a unit enters J at time* 0.

The procedures in Sections 1.11 and 1.12 for computing expected throughputs and waiting times are also valid for multiclass networks.

3.2 Kelly Networks: Route-dependent Services

In this section, we discuss networks in which units are divided into classes depending on their routes through the network, and a unit's service times at the nodes depend on its route. These are multiclass networks introduced by Kelly in 1975.

Consider an open m-node network in which the routing of units is as follows. A typical route of a unit is a finite sequence $r = (r_1, \dots, r_\ell)$ of nodes inside the network, where r_s is the node the unit visits at *stage* s of its route, $1 \le s \le \ell$; the length $\ell \equiv \ell(r)$ is route dependent. Upon leaving the last node r_ℓ, the unit exits the network. A node may appear more than once on a route, and the set of all relevant routes, for simplicity, is finite. Units that traverse a route r arrive to the network according to a Poisson process with rate $\lambda(r)$, and these arrival processes are independent for all the routes. Then the total arrival stream to the network is a Poisson process with rate $\sum_r \lambda(r)$.

The preceding description applies to several scenarios. One is that a deterministic route r is an attribute of a unit and that all units that traverse a given route are in the same class. A second scenario is that each unit carries a permanent class label that determines its route. A third possibility is that deterministic routes are obtained by random routes as follows. The units arrive to the network by a Poisson process with rate λ, and each unit independently selects or is assigned a route r with probability $p(r)$. In this case, $\lambda(r) = p(r)\lambda$. For instance, a route may be selected by Markov probabilities p_{jk} such that $p_{0r_1} p_{r_1 r_2} \cdots p_{r_{\ell-1} r_\ell}$ is the probability of the route $r = (r_1, \dots, r_\ell)$. Combinations of the preceding scenarios yield further possibilities.

To formulate the network as a multiclass Whittle network, we assign a class label to each unit to denote its routing status at any time in the network. Namely, if a unit is traversing route r and is at stage s in this route, we call it a rs-unit. Let M denote the set of all route-stage labels rs, including the outside node 0 as well.

We represent the state of the network by the vector $x = (x_{rs} : rs \in M\backslash\{0\})$, where x_{rs} denotes the number of rs-units in the network at node r_s. The node at which a unit resides is specified by the label rs, and so we need not specify the unit's location separately as we did above by the class-node label αj. Assume that whenever the network is in state x, the time to the next departure of an rs-unit from its current node location r_s is exponentially distributed with rate $\phi_{rs}(x)$. The departing unit goes immediately to its next node r_{s+1} and becomes an $r(s+1)$-unit. In case $s = \ell$, the $r_{\ell+1} = 0$, which means that the route is complete and the unit exits the network. Assume that ϕ_{rs} are Φ-balanced departure intensities.

Let $\{X_t : t \geq 0\}$ denote the stochastic process representing the network. Under the preceding assumptions, X is a Markov process with transition rates

$$q(x, y) = \begin{cases} \lambda(r) & \text{if } y = x + e_{r1} \in \mathbb{E} \text{ for some } r \\ \phi_{rs}(x) & \text{if } y = x - e_{rs} + e_{r(s+1)} \in \mathbb{E} \text{ for some } rs \in M \\ 0 & \text{otherwise.} \end{cases}$$

Note that this is a multiclass Whittle network process, where the class label for a unit describes the route it is taking and where it is on the route.

Using the notation in the preceding section, the routing rates for the units are $\lambda_{0,r1} = \lambda(r)$ and $\lambda_{rs,r(s+1)} = 1$ for $rs \neq 0$. Then the traffic equations for these rates are $w_0 = 1$ and, for each route r,

$$w_{r1} = \lambda(r), \qquad w_{rs} = w_{r(s-1)}, \qquad s = 2, \ldots, \ell.$$

A solution to these equations is $w_{rs} = \lambda(r)$ for each $rs \neq 0$. Consequently, Theorem 3.1 yields the following result.

Corollary 3.9. *An invariant measure for the network process X with transition rates described above is*

$$\pi(x) = \Phi(x) \prod_r \lambda(r)^{x_r}, \qquad x \in \mathbb{E},$$

where $x_r \equiv \sum_{s=1}^{\ell} x_{rs}$.

Since the network model we are discussing is a special case of that in the preceding section, all the results there also apply. For instance, suppose the network is such that each node is a processor-sharing node as discussed in Example 3.2 with service rates

$$\phi_{rs}(x) = \frac{x_{rs}}{x_j} \mu_j(x_j),$$

where $j = r_s$ and $\mu_j(x_j)$ is the departure intensity for node j when it contains $x_j = \sum_{r's'} x_{r's'} 1(r'_{s'} = j)$ units. In this case, the Φ in the preceding result is

$$\Phi(x) = \prod_{j=1}^{m} x_j! \prod_{n=1}^{x_j} \mu_j(n)^{-1} \prod_{rs} \frac{1}{x_{rs}!}.$$

The network process we are discussing is for an open network with unlimited capacity. The following extension covers the finite-capacity case.

Example 3.10. *System-dependent Arrival Rates.* Suppose the arrivals from outside are dependent on the network such that $q(x, x + e_{r1}) = \lambda(r)\phi_{r0}(|x|)$. This would allow for a finite capacity network or subnetworks by assuming $\phi_{r0}(n) = 0$ for

$n = \nu_r$. For this more general arrival rate, the invariant measure from Theorem 3.1 would be

$$\pi(x) = \Phi(x) \prod_r \lambda(r)^{x_r} \prod_{n=1}^{|x|} \phi_{r0}(n-1), \quad x \in \mathbb{E}. \qquad \square$$

3.3 BCMP Networks: Class-Node Service Dependencies

This section describes multiclass BCMP networks, which were introduced by Baskett, Chandy, Muntz, and Palacios in 1975. The distinguishing feature of such a network is that a unit's service rate at a node is a compounding of two intensities—one intensity is a function of the total number of units at the node, and the other intensity is a function of the number of units in the same class as the one being served.

We begin with a general framework for modeling networks with class-node service dependencies. Consider a multiclass Whittle network as described in Section 3.1. Assume that the service rates for each node $j \neq 0$ are of the form

$$\phi_{\alpha j}(x) = g_{\alpha j}(x_j) h_{\alpha j}(x_{\alpha j}), \tag{3.6}$$

where $g_{\alpha j}$ and $h_{\alpha j}$ are functions on the nonnegative integers. These are sector-dependent service rates, where $g_{\alpha j}(x_j)$ is the node intensity and $h_{\alpha j}(x_{\alpha j})$ is the class intensity. In case the network is open, assume, for each α and x, that

$$\phi_{\alpha 0}(x) = g_0(|x|) h_{\alpha 0}(|x_\alpha|), \tag{3.7}$$

where $|x_\alpha| \equiv \sum_{j=1}^{m} x_{\alpha j}$ is the number of α-units in the network.

By Example 3.3, these service rates are balanced by the function

$$\Phi(x) = \prod_{j=0}^{m} f_j(x),$$

where $f_0(x) \equiv 1$ if the network is closed,

$$f_0(x) = \prod_{n=0}^{|x|-1} g_0(n) \prod_\alpha \prod_{n'=0}^{|x_\alpha|-1} h_{\alpha 0}(n'), \quad \text{if the network is open, and}$$

$$f_j(x) = \prod_\alpha \prod_{n=1}^{x_j} g_{\alpha j}(n)^{-1} \prod_{n'=1}^{x_{\alpha j}} h_{\alpha j}(n')^{-1}, \quad j \neq 0.$$

The routing rates $\lambda_{\alpha j, \beta k}$ are the same as those in Section 3.1, and $w_{\alpha j}$ is a solution to the traffic equations (3.1). Then the following result is an immediate consequence of Theorem 3.1.

Corollary 3.11. *Under the preceding assumptions, an invariant measure for the network process is*

$$\pi(x) = \Phi(x) \prod_{\alpha j \in M} w_{\alpha j}^{x_{\alpha j}}, \qquad x \in \mathbb{E}.$$

The main example of this result is as follows.

Example 3.12. *BCMP Networks*. The network described above is a *BCMP network* if each of its nodes is one of the following four types.

• *First-Come, First-Served* node with service rates $\phi_{\alpha j}(x) = \mu_j(x_j)$. Each unit (as in a Jackson network) has exponential service time with the same load-dependent service rate $\mu_j(x_j)$.

• *Processor-Sharing* node with service rates $\phi_{\alpha j}(x) = x_{\alpha j} x_j^{-1} \mu_{\alpha j}(x_j)$. The $\mu_{\alpha j}(x_{\alpha j})$ is a customer-load-dependent service rate, which is apportioned equally among the $x_{\alpha j}$ α-units at the node.

• *Last-Come, First-Served with Preemption* node with service rates as in the preceding PS case.

• *Infinite-Server* node with service rates $\phi_{\alpha j}(x) = x_{\alpha j} \mu_{\alpha j}(x_{\alpha j})$.

Also, in case the network is open, the arrival rates from outside are $\lambda_{\alpha 0} \mu_0(|x|)$.

An invariant measure for this BCMP network is given by Corollary 3.11 with

$$f_0(x) = \prod_{n=0}^{|x|-1} \mu_0(n), \qquad \text{if the network is open,}$$

and the other f_j's are as follows for the four types of nodes:

$$\begin{aligned} f_j(x) \quad &= x_j! \prod_{\alpha} \frac{1}{x_{\alpha j}!} \prod_{n=1}^{x_j} \mu_j(n)^{-1}, && \text{FCFS node} \\ &= x_j! \prod_{\alpha} \frac{1}{x_{\alpha j}!} \prod_{n=1}^{x_j} \mu_{\alpha j}(n)^{-1}, && \text{PS or LCFSPR node} \\ &= \prod_{\alpha} \frac{1}{x_{\alpha j}!} \prod_{n=1}^{x_{\alpha j}} \mu_{\alpha j}(n)^{-1}, && \text{IS node.} \end{aligned}$$

This BCMP network can be extended to model service times with nonexponential distributions. This is explained in the next section. □

3.4 Networks with Cox and General Service Times

Although the Markov network processes we have been studying have exponential times between transitions, the processes can model general service times at the nodes. We will show this for the BCMP networks discussed in the preceding section.

We begin with a few comments on service times. An *Erlang* service time with parameters n and μ is the sum of n independent exponential random variables with rate μ. Its density is

$$f(t|\mu, n) = \mu(\mu t)^{n-1} e^{-\mu t}/(n-1)!, \quad t \geq 0.$$

Think of this Erlang service as consisting of n independent identical exponential phases performed in series. A generalization of this is a *hypo-exponential* service time consisting of a series of n independent exponential phases with respective rates $\mu_1, \ldots, \mu_n$, which may be different.

A versatile generalization of Erlang and hypo-exponential random variables is a Cox random variable defined as follows. Consider a series of n exponential phases with rates $\mu_1, \ldots, \mu_n$ as shown in Figure 3.1. In this system, a service begins by performing phase 1 (an exponential phase with rate μ_1). Upon completing phase 1, the service enters phase 2 with probability p_1 or the service terminates with probability $1 - p_1$. If phase 2 is entered, then upon completing this exponential phase, the service enters phase 3 with probability p_3 or terminates with probability $1 - p_3$. These phases are continued until the service terminates prior to or after phase n. The probability that the service consists of exactly the first s phases (or stages) is $p_1 \cdots p_{s-1}(1 - p_s)$, where $p_0 = p_n = 1$. The total time to complete the service is a *Cox random variable*. Its distribution is

$$F(t) = \sum_{s=1}^{n} p_1 \cdots p_{s-1}(1 - p_s) H(t|\mu_1, \ldots, \mu_s), \quad t \geq 0, \tag{3.8}$$

where $H(t|\mu_1, \ldots, \mu_s)$ is the hypo-exponential distribution of completing s phases.

Cox distributions are a subclass of phase-type distributions (the distributions of absorption times for Markov processes). Note that a Cox distribution is a mixture of hypo-exponential distributions. This implies that a mixture of Cox distributions is also a Cox distribution. Another useful property is that a sum of independent Cox random variables is again a Cox random variable. Because of these properties, the time to complete a complex job consisting of independent series-parallel subtasks with Cox distributions can be modeled by a Cox distribution (a mixture models parallel subtasks, and a sum models subtasks in series).

An important feature of Cox distributions is that they form a dense subset within the set of all distributions of nonnegative random variables. This means that any general service time distribution can be approximated by a Cox distribution. We will now describe how to use this property for modeling networks with nonexponential service times.

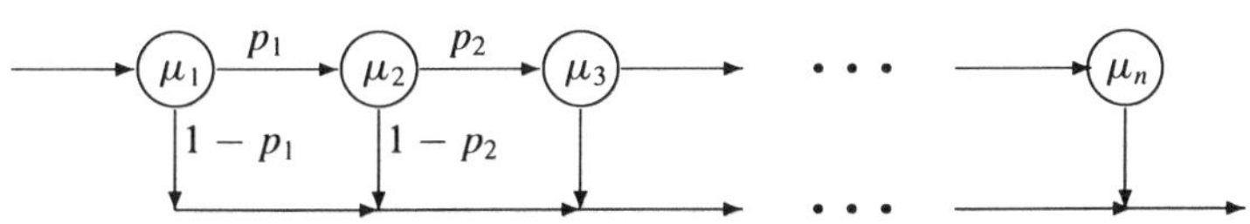

FIGURE 3.1. Phases of a Cox Service Time

Consider the multiclass Whittle network defined in Section 3.1 with the following additional assumptions. Suppose that the service time requirement for each α-unit at node j has a Cox distribution with parameters $\mu_{\alpha js}$, $p_{\alpha js}$, for $s = 1, \ldots, n$. The number of phases n may depend on α and j. To incorporate these Cox services into the network state, we assign the class label αjs to an α-unit that is in phase s of its service at node j. We let $x_{\alpha js}$ denote the number of such units at node j, and represent the state of the network by the vector $x = (x_{\alpha js} : \alpha js \in M, \ j \neq 0)$, where M denotes the set of all class labels αjs.

Assume that each node j is one of the following types.

• A processor sharing node with service rates

$$\phi_{\alpha js}(x) = x_{\alpha js} x_j^{-1} \mu_{\alpha js}.$$

The rate $\mu_{\alpha js}$ is apportioned equally among the $x_{\alpha js}$ αj-units at the node in phase s of their service.

• Infinite-server node with service rates $\phi_{\alpha js}(x) = x_{\alpha js}\mu_{\alpha js}$.

Also, in case the network is open, assume the arrival rates from outside are $\lambda_{\alpha 0}\mu_0(|x|)$.

The routing rates $\lambda_{\alpha j, \beta k}$ must also be augmented to contain the phase parameter. From the definition of a Cox distribution, it is clear that the new routing rates for the network should be

$$\lambda_{\alpha js, \alpha j(s+1)} = 1 - p_{\alpha js}, \quad 1 \leq s < n,$$

$$\lambda_{\alpha jn, \beta k1} = \lambda_{\alpha j, \beta k},$$

with the rest of the rates being 0. An easy check shows that traffic equations for these rates have a solution

$$w_{\alpha js} = w_{\alpha j} \prod_{\ell=1}^{s-1} p_{\alpha j\ell}, \quad \alpha js \in M,$$

where $w_{\alpha j}$ is a solution to the traffic equations for $\lambda_{\alpha j, \beta k}$.

Then from Corollary 3.11, it follows that an invariant measure for the network is

$$\pi(x) = \prod_{n=0}^{|x|-1} \mu_0(n) \prod_{j=1}^{m} \gamma_j \prod_{\alpha, s} \frac{1}{x_{\alpha js}!} (w_{\alpha js}/\mu_{\alpha js})^{x_{\alpha js}} \tag{3.9}$$

where $\gamma_j = x_j!$ or 1 according to whether node j is a PS node or an IS node, and $\mu_0(\cdot) \equiv 1$ if the network is closed.

One can use this result for modeling networks with general service times. Specifically, consider the network with the modification that the service times have general distributions. Since these distributions can be approximated by Cox distributions, the invariant measure above for the approximating Cox distributions can be used as an approximation for the network with general service times. Unfortunately, such approximations tend to be difficult for large numbers of phases and customer types.

3.5 Networks with Constraints

Sections 2.4 and 2.5 discussed reversible processes with constraints. The central idea was that a reversible Markov process restricted to a subset of its state space is also a reversible Markov process, and its stationary distribution is a truncation of the original distribution to the subset. In this section, we expand on this theme for Whittle networks that are locally reversible.

We begin with a general result concerning locally reversible Markov processes with constraints. Consider a Markov process X on a countable space $\mathbb{E}$ with transition rates $q(x, y)$. Without loss in generality, assume that X is ergodic, and let π denote its stationary distribution. Fix a subset $\tilde{\mathbb{E}} \subset \mathbb{E}$, and let $\tilde{X}$ be a Markov process on $\tilde{\mathbb{E}}$ whose transition rates, denoted by $\tilde{q}(x, y)$, are the rates $q(x, y)$ restricted to $\tilde{\mathbb{E}}$ (transitions from states inside $\tilde{\mathbb{E}}$ to states outside of $\tilde{\mathbb{E}}$ are "blocked" or suppressed). Assume that $\tilde{X}$ is irreducible.

Definition 3.13. Suppose the restricted process $\tilde{X}$ is ergodic and its stationary distribution is

$$\tilde{\pi}(x) = \pi(x)/\sum_{y\in\tilde{\mathbb{E}}} \pi(y), \qquad x \in \tilde{\mathbb{E}}.$$

This is the conditional probability stationary distribution of X conditioned that it is in $\tilde{\mathbb{E}}$. We say that $\tilde{X}$ is a *truncation of the process X to $\tilde{\mathbb{E}}$.*

This truncation property is typically not true. It obviously holds if and only if $\tilde{\pi}$ satisfies the balance equations for $\tilde{X}$, which is equivalent to π satisfying

$$\pi q(x, \tilde{\mathbb{E}}) = \pi q(\tilde{\mathbb{E}}, x), \qquad x \in \tilde{\mathbb{E}}, \tag{3.10}$$

where $\pi q(A, B) \equiv \sum_{x\in A} \pi(x) \sum_{y\in B} q(x, y)$. Another characterization is as follows.

Proposition 3.14. *The process $\tilde{X}$ is a truncation of X to $\tilde{\mathbb{E}}$ if and only if*

$$\pi q(x, \tilde{\mathbb{E}}^c) = \pi q(\tilde{\mathbb{E}}^c, x), \qquad x \in \tilde{\mathbb{E}}. \tag{3.11}$$

In particular, if X is reversible on $\tilde{\mathbb{E}}$ or on $\tilde{\mathbb{E}}^c$, then $\tilde{X}$ is a truncation of X to $\tilde{\mathbb{E}}$.

PROOF. The balance equations that π satisfies, which are $\pi q(x, \mathbb{E}) = \pi q(\mathbb{E}, x)$, can be written as

$$\pi q(x, \tilde{\mathbb{E}}) + \pi q(x, \tilde{\mathbb{E}}^c) = \pi q(\tilde{\mathbb{E}}, x) + \pi q(\tilde{\mathbb{E}}^c, x), \qquad x \in \mathbb{E}.$$

From this it follows that (3.11) is equivalent to (3.10). This equivalence establishes the first assertion. The second assertion of the theorem follows since X being reversible on $\tilde{\mathbb{E}}$ or on $\tilde{\mathbb{E}}^c$ would imply (3.10) or (3.11), respectively. □

We will now apply this result to networks. For the rest of this section, assume the Markov process X represents an open or closed Jackson or Whittle network process with service and routing intensities $\phi_j(\cdot)$ and λ_{jk}, $j, k \in M$ ($0 \in M$ if the network is open). We will consider a modification of this process in which the numbers of units in a sector $J \subset M$ are restricted. Specifically, assume that the state space of the network is

$$\tilde{\mathbb{E}} = \{x \in \mathbb{E} : (x_j : j \in J) \in A\}, \tag{3.12}$$

where A is the set of "allowable" values of the vector $(x_j : j \in J)$.

The following are typical examples:

• *Nodes with Finite Capacities.* $\tilde{\mathbb{E}} = \{x \in \mathbb{E} : x_j \le \ell_j,\ j \in J\}$, where ℓ_j is the capacity of node j. In this case, whenever some node $k \in J$ is such that $x_k = \ell_k$, then no additional units can enter that node until a unit departs from it. This is called *communication blocking*.

• *Sectors with Capacity or Load Constraints.*

$$\tilde{\mathbb{E}} = \{x \in \mathbb{E} : x_1 + x_2 \le \ell_{12}, x_1 + x_4 \le x_2 + x_5\},$$

where $J = \{1, 2, 4, 5\}$. In this case, sector $\{1, 2\}$ cannot contain more than ℓ_{12} units and sector $\{1, 4\}$ cannot contain more units than sector $\{2, 5\}$.

• *Resource Constraints.* $\tilde{\mathbb{E}} = \{x \in \mathbb{E} : \sum_{j \in J} r_{ij} x_j \le r_i, i \in I\}$. Here each unit at node j requires r_{ij} units of a resource i, and there are only r_i units of the resource available. Typical resources are space, computer memory, manufacturing tools, and money. Other common constraints can be formulated by functions of the state x.

For the next result, assume that $\tilde{X}$ is the resulting network process on the restricted state space $\tilde{\mathbb{E}}$ as in (3.12). Define

$$\bar{J} = J \cup \{k \notin J : \lambda_{jk} \text{ or } \lambda_{kj} > 0 \text{ for some } j \in J\}.$$

Theorem 3.15. *If the rates $\{\lambda_{jk}\}$ are reversible on $\bar{J}$, then the network process $\tilde{X}$ is a truncation of X to $\tilde{\mathbb{E}}$.*

Proof. It suffices to verify the balance equations (3.10). But these will follow upon showing that

$$\sum_k \pi \tilde{q}(x, T_{jk}x) = \sum_k \pi \tilde{q}(T_{jk}x, x), \quad j \in M,\ x \in \tilde{\mathbb{E}}. \tag{3.13}$$

To this end, first consider the case $j \in \bar{J}$. Then (3.13) is equivalent to

$$\sum_{k \in \bar{J}} \pi \tilde{q}(x, T_{jk}x) + \sum_{k \in \bar{J}^c} \pi q(x, T_{jk}x) = \sum_{k \in \bar{J}} \pi \tilde{q}(T_{jk}x, x) + \sum_{k \in \bar{J}^c} \pi q(T_{jk}x, x). \tag{3.14}$$

The first and third sums are equal since an easy check shows that the kth term in these sums are equal by the assumption that λ_{jk} is reversible on $\bar{J}$. Also, this assumption and Proposition 3.14 applied to λ_{jk} ensure that

$$w_j \sum_{k \in \bar{J}^c} \lambda_{jk} = \sum_{k \in \bar{J}^c} w_k \lambda_{kj}, \quad j \in \bar{J}^c.$$

Using this, one can show that the second and fourth sums in (3.14) are equal. Thus (3.14) holds for $j \in \bar{J}$.

Next, consider the case $j \in \bar{J}^c$. Then $\lambda_{jk} = \lambda_{kj} = 0$ for $k \in J$, and so (3.13) is equivalent to

$$\sum_{k \in J^c} \pi q(x, T_{jk}x) = \sum_{k \in J^c} \pi q(T_{jk}x, x).$$

But this equation holds because Proposition 3.14 applied to the reversible rates $\{\lambda_{jk}\}$ on J ensures that

$$w_j \sum_{k \in J^c} \lambda_{jk} = \sum_{k \in J^c} w_k \lambda_{kj}, \quad j \in J^c.$$

This completes the proof of (3.13). □

3.6 Networks with Blocking and Rerouting

The last section discussed networks with blocking whose invariant measures agree with those of the original network. We now discuss a variation of this theme in which a Markov network process with blocking plus "rerouting" has invariant measures that agree with the original network process.

We begin with a basic result for Markov chains, which underlies our results for networks. Let $\{p_{jk} : j, k \in J\}$ denote irreducible Markov transition probabilities on a countable set J. Consider a Markov chain on a subset I of J that moves as follows. Whenever it is in state $j \in I$, a sequence of states is selected by the probabilities p_{jk} until a state $k \in I$ is selected. That is, a sequence of states $k_1, \ldots, k_\ell, k$ is selected with probability $p_{jk_1} p_{k_1 k_2} \cdots p_{k_\ell k}$, where $k_i \notin I$, $i = 1, \ldots, \ell$, and $k \in I$. Then the chain moves from j to k. Let r_{jk} denote the transition probability of the chain moving from j to k. This Markov chain on I with transition probabilities $\{r_{jk}\}$ can be interpreted as the Markov chain with transition probabilities p_{jk} *restricted to the subset I by rerouting*.

It follows, by conditioning on the first state selected, that

$$r_{jk} = p_{jk} + \sum_{i \notin I} p_{ji} \eta_{ik}, \quad j, k \in I. \tag{3.15}$$

Here η_{ik} is the probability that, for a Markov chain on J with transition probabilities $\{p_{jk}\}$, the first entry into I starting from $i \notin I$ occurs in state $k \in I$. These absorption probabilities are the solution to the equations

$$\eta_{jk} = p_{jk} + \sum_{i \notin I} p_{ji} \eta_{ik}, \quad j \notin I,\ k \in I. \tag{3.16}$$

Proposition 3.16. *If $\{\pi_j : j \in J\}$ is an invariant measure for $\{p_{jk}\}$, then $\{\pi_j : j \in I\}$ is an invariant measure for $\{r_{jk}\}$.*

PROOF. Suppose that $\{\pi_j : j \in J\}$ is a positive measure that satisfies

$$\pi_j = \sum_{k \in J} \pi_k p_{kj}, \quad j \in J. \tag{3.17}$$

We will show that $\{\pi_j : j \in I\}$ satisfies

$$\pi_j = \sum_{k \in I} \pi_k r_{kj}, \quad j \in I. \tag{3.18}$$

By (3.17), we know that

$$\pi_j = \sum_{k \in I} \pi_k p_{kj} + \sum_{k \notin I} \pi_k p_{kj}, \quad j \in I. \tag{3.19}$$

Now, from (3.16), (3.17), and (3.15), we have

$$\begin{aligned}
\sum_{k \notin I} \pi_k p_{kj} &= \sum_{k \notin I} \pi_k (\eta_{kj} - \sum_{i \notin I} p_{ki} \eta_{ij}) \\
&= \sum_{i \notin I} \sum_{k \in J} \pi_k p_{ki} \eta_{ij} - \sum_{k \notin I} \pi_k \sum_{i \notin I} p_{ki} \eta_{ij} \\
&= \sum_{k \in I} \pi_k \sum_{i \notin I} p_{ki} \eta_{ij} \\
&= \sum_{k \in I} \pi_k (r_{kj} - p_{kj}).
\end{aligned}$$

For the second equality, k is changed to i and (3.17) is applied. Substituting the preceding expression in (3.19) yields (3.18). □

We will now consider the notion of blocking and rerouting in networks. Suppose that $\{X_t : t \geq 0\}$ is the generalization of Jackson and Whittle processes discussed in Proposition 1.23. Namely, X is a Markov process with transition rates

$$q(x, y) = \begin{cases} \phi_j(x)\lambda_{jk}(x) & \text{if } y = T_{jk}x \text{ for some } j \neq k \text{ in } M \\ 0 & \text{otherwise,} \end{cases}$$

where $\lambda_{jk}(x)$ is a routing rate as a function of the state x. For simplicity, assume that $\lambda_{jk}(x)$ is the probability of a unit moving from node j to node k, and so $\sum_k \lambda_{jk}(x) = 1$. Assume that the ϕ_j are Φ-balanced, and that there is a positive function Λ on $\mathbb{E}$ such that

$$\Lambda(x) = \sum_k \Lambda(T_{jk}x)\lambda_{kj}(T_{jk}x), \quad j \in M, \ x \in \mathbb{E} \text{ with } x_j \geq 1. \tag{3.20}$$

These assumptions imply, by Proposition 1.23, that an invariant measure of the process is

$$\pi(x) = \Lambda(x)\Phi(x), \quad x \in \mathbb{E}. \tag{3.21}$$

We will consider this network with the following blocking and rerouting protocol. Suppose the network is in state $x \in \mathbb{E}$, and a unit departs from node j. The disposition of the rest of the units in the network is given by the vector $x - e_j$. For each such vector, there is a partition of the node set M such that only movements of units under the probabilities $\{\lambda_{jk}(x)\}$ between nodes in the same subset of the partition are admissible. We let $I(x - e_j)$ denote a typical partition subset. The partition may consist of the singleton M, but we disregard the degenerate case where the partition is the singleton M for each $x - e_j$ (then there is no blocking).

Now, we assume that the unit departing from node j in some partition subset $I \equiv I(x - e_j)$ selects a sequence of nodes according to the probabilities $\{\lambda_{jk}(x)\}$ until a state $k \in I$ is selected. That is, the unit selects a sequence of states $k_1, \ldots, k_\ell, k$ with the probability $\lambda_{jk_1}(x)\lambda_{k_1k_2}(x)\cdots\lambda_{k_\ell k}(x)$, where $k_i \notin I$, $i = 1, \ldots, \ell$, and $k \in I$. Then the unit moves from node j to node k. This selection process is done instantaneously. Let $r_{jk}(x)$ denote the probability that the unit moves from j to k according to this blocking-rerouting procedure.

As in the case of the Markov chain model above, it follows that

$$r_{jk}(x) = \lambda_{jk}(x) + \sum_{i \notin I} \lambda_{ji}(x)\eta_{ik}(x), \qquad j, k \in I. \tag{3.22}$$

Here $\eta_{ik}(x)$ is the probability that, for a Markov chain on M with transition probabilities $\{\lambda_{jk}(x)\}$, the first entry into I starting from $i \notin I$ occurs in state $k \in I$. These absorption probabilities are the solution to the equations

$$\eta_{jk}(x) = \lambda_{jk}(x) + \sum_{i \notin I} \lambda_{ji}(x)\eta_{ik}(x), \qquad j \notin I,\ k \in I. \tag{3.23}$$

The resulting network process $\{\tilde{X}_t : t \geq 0\}$ is a Markov process. Its transition rates are

$$\tilde{q}(x, y) = \phi_j(x) r_{jk}(x)$$

if $y = T_{jk}x$ for some $j \neq k$ in M and j, k are in the same subset of nodes with admissible transitions; and $\tilde{q}(x, y) = 0$ otherwise. Suppose that $\tilde{X}$ is irreducible on a space $\tilde{\mathbb{E}} \subset \mathbb{E}$. We interpret the network process $\tilde{X}$ as a *restriction of X under blocking and rerouting*.

Theorem 3.17. *An invariant measure for $\tilde{X}$ is $\tilde{\pi}(x) = \Phi(x)\Lambda(x)$, $x \in \tilde{\mathbb{E}}$.*

PROOF. To prove the assertion, it suffices by Proposition 1.23 to show that equation (3.20) holds for $x \in \tilde{\mathbb{E}}$. This equation can be written as

$$\Lambda(x + e_j) = \sum_{I(x)} 1(j \in I(x)) \sum_k \Lambda(x + e_k)\lambda_{kj}(x + e_k),$$

where $x + e_j \in \tilde{\mathbb{E}}$ with $x_j \geq 1$ and $j \in M$. The first sum is over all subsets $I(x)$ that partition M. To prove the preceding equation, it suffices to show that, for each subset $I(x)$,

$$\pi_j = \sum_{k \in I(x)} \pi_k r_{kj}, \qquad j \in I(x), \tag{3.24}$$

where $\pi_j \equiv \Lambda(x + e_j)$ and $r_{jk} \equiv r_{jk}(x + e_j)$.

Now, setting $p_{jk} \equiv \lambda_{jk}(x + e_j)$, it follows that the relations (3.22) and (3.23) are the same as (3.15) and (3.16), respectively. Consequently, the Markov chain with transition probabilities $\{r_{jk} : j, k \in I(x)\}$ is a restriction with rerouting of the Markov chain with transition probabilities $\{p_{jk} : j, k \in M\}$. Also, (3.20) with the preceding notation implies that $\{\pi_j : j \in M\}$ is an invariant measure for $\{p_{jk} : j, k \in M\}$. Then the desired expression (3.24) follows by Proposition 3.16. □

3.7 Bottlenecks in Closed Jackson Networks

We now switch from blocking to bottlenecks. In this section, we address the question: How does the stationary distribution of a closed Jackson network change as the number of units in the network tends to infinity? We show that (a) the number of units in the nodes with the largest traffic intensity tends to infinity, and (b) the distribution of the numbers of units in the remaining nodes converges to the distribution of an open Jackson network.

Consider a closed Jackson network with ν units and load-independent service rates $\phi_j(x) = \mu_j$, for $j \in M = \{1, \dots, m\}$. From Theorem 1.12, its stationary distribution is

$$\pi(x) = c_\nu \prod_{j \in M} r_j^{x_j}, \quad |x| = \nu,$$

where $r_j = w_j/\mu_j$ is the traffic intensity, and the w_j's satisfy the traffic equations

$$w_j \sum_{k \in M} \lambda_{jk} = \sum_{k \in M} w_k \lambda_{kj}, \quad j \in M.$$

We will consider the convergence of this stationary distribution as $\nu \to \infty$.

Let $(X_1^\nu, \dots, X_m^\nu)$ denote a random vector with the distribution π that represents the numbers of units at the nodes in steady state. The superscript ν highlights the number of units ν in the network, which we now treat as a variable. The nodes with the largest traffic intensity would be the bottlenecks when ν is large. In other words, the heaviest traffic would be in the sector

$$J \equiv \{j \in M \ : \ r_j = r \equiv \max\{r_1, \dots, r_m\}\}.$$

The traffic in the complement $K \equiv M \backslash J$ would be lighter. Recall that $x_J = (x_j : j \in J)$ denotes the state of the nodes in J and $x(J) \equiv \sum_{j \in J} x_j$ is the total number of units in J. For each $k \in K$, the ratio $\rho_k \equiv r_k/r$ is the traffic intensity at node k relative to the traffic intensity r at the nodes in J. Assume K is not empty; an empty K is not of interest.

The following result says that the distribution of X_K^ν converges to the distribution of an open Jackson network on K as $\nu \to \infty$. Also, the number of units $X(J)^\nu$ in the bottleneck sector J converges to infinity. This implies that, for a closed Jackson network with a large number of units and load-independent service rates, the distribution of its nonbottleneck nodes can be approximated by a product-form distribution as in an open network.

Theorem 3.18. *Under the preceding assumptions, $X(J)^\nu$ converges in distribution to ∞ as $\nu \to \infty$, and*

$$\lim_{\nu \to \infty} P\{X_K^\nu = x_K\} = \prod_{k \in K} (1 - \rho_k)\rho_k^{x_k}, \quad x_K \geq 0.$$

PROOF. It suffices to show that, for any vector $x_K \geq 0$ and integer $\ell \geq x(K)$,

$$\lim_{\nu \to \infty} P\{X(J)^\nu \geq \ell - x(K), X_K^\nu = x_K\} = \prod_{k \in K} (1 - \rho_k)\rho_k^{x_k}. \tag{3.25}$$

From the distribution π above, it follows that

$$\begin{aligned}
P\{X(J)^{\nu} &\geq \ell - x(K), X_K^{\nu} = x_K\} \\
&= \sum_y \pi(y) 1(y(J) \geq \ell - x(K), y_K = x_K) \\
&= c_\nu \sum_{n=\ell-x(K)}^{\nu} \sum_{x_J} 1(x(J) = n) \prod_{j \in J} r_j^{x_j} \prod_{k \in K} r_k^{x_k} \\
&= c_\nu r^{-\nu} \sum_{n=\ell-x(K)}^{\nu} |J|^n \prod_{k \in K} \rho_k^{x_k}.
\end{aligned}$$

Here $|J|$ is the number of nodes in J. Using similar reasoning,

$$c_\nu^{-1} = r^{-\nu} \sum_{n=0}^{\nu} |J|^n a_{\nu-n},$$

where

$$a_i = \sum_{x_K} 1(x(K) = i) \prod_{k \in K} \rho_k^{x_k}.$$

Combining the preceding displays, we have

$$P\{X(J)^{\nu} \geq \ell - x(K), X_K^{\nu} = x_K\} = \frac{\sum_{n=\ell-x(K)}^{\nu} |J|^n}{\sum_{n=0}^{\nu} |J|^n a_{\nu-n}} \prod_{k \in K} \rho_k^{x_k}. \tag{3.26}$$

Now, supposing that the nodes are labeled such that $K = \{1, 2, \ldots, |K|\}$, then

$$\begin{aligned}
a_i &= \sum_{x_1=0}^{i} \sum_{x_2=0}^{i-x_1} \cdots \sum_{x_{|K|}=0}^{i-x_1-\cdots-x_{|K|-1}} \prod_{k \in K} \rho_k^{x_k} \\
&\to \prod_{k \in K} (1 - \rho_k)^{-1}, \quad \text{as } i \to \infty.
\end{aligned}$$

In light of this, letting $\nu \to \infty$ in equation (3.26) yields (3.25). □

3.8 Modeling Whittle Networks by Locations of the Units

We have been representing networks by the numbers of units at their nodes. Another approach is to depict the evolution of a network by the locations of its units. In this section, we describe this approach for closed and finite-capacity Whittle networks with processor-sharing nodes, and comment on its applicability to other types of networks.

Consider an m-node Whittle network that is closed with ν units or open with capacity ν. As in Chapter 1, let λ_{jk} denote the routing rates of the individual units and let $\phi_j(x)$ denote the service rate when there are $x = (x_1, \ldots, x_m)$ units at the respective nodes. Assume the rates λ_{jk} are irreducible on the node set M, where

$M \equiv \{1, \ldots, m\}$ or $\{0, 1, \ldots, m\}$ according as the network is closed or open. Let w_j denote the stationary distribution of the routing process with rates λ_{jk}. Then by Theorem 1.15, we know that the process $\{X_t : t \geq 0\}$ that represents the numbers of units at the nodes has the stationary distribution

$$\pi(x) = c\Phi(x) \prod_{j \in M} w_j^{x_j}.$$

We will now analyze the network in terms of the locations of the units. In case the network is closed, we label the units as $1, \ldots, \nu$. In case the network is open, we assume that the indices $1, \ldots, \nu$ are labels or tokens that the units in the network carry as follows. Whenever there are $n < \nu$ units in the network, a unit entering the network selects one of the $\nu - n$ unused labels with equal probability. The unit retains the label until it exits the network, and then the label becomes available for another unit. The unit carrying the label i is called unit i.

We assume that the services at each node are under a processor sharing discipline in which each unit at a node receives the same service treatment. Then the time to a potential departure of a unit i from a node j has an exponential distribution with rate $\phi_j(x)x_j^{-1}$. Here $1/x_j$ is the probability that unit i is one selected to depart from the x_j units at node j.

We will represent the network by the stochastic process $Y(t) \equiv (Y_1(t), \ldots, Y_\nu(t))$, where $Y_i(t)$ denotes the node location of unit i at time t. A typical state of the process Y is a vector $y = (y_1, \ldots, y_\nu)$ in M^ν. Whenever Y is in state y, a transition is triggered by some unit i moving from its current node y_i to some node k. Let $T_k^i y$ denote the resulting state, which is y with y_i replaced by k. Also, let $\mathbf{n}(y) \equiv (n_1(y), \ldots, n_\nu(y))$, where

$$n_j(y) \equiv \sum_{i=1}^{\nu} 1(y_i = j), \quad j \in M, \ y \in M^\nu,$$

which is the number of units in node j.

Under the preceding assumptions, Y is a Markov process and its transition rates are

$$q_Y(y, y') = \begin{cases} \lambda_{y_i k}(i)\phi_{y_i}(\mathbf{n}(y))n_{y_i}(y)^{-1} & \text{if } y' = T_k^i y \text{ for some } i \text{ and } k \\ 0 & \text{otherwise.} \end{cases}$$

The proof (Exercise 6) of the following result is similar to the proof of Theorem 1.15 for Whittle processes.

Theorem 3.19. *The location process Y defined above is ergodic, and its stationary distribution is*

$$\pi_Y(y) = \frac{1}{\nu!}\Phi(\mathbf{n}(y)) \prod_{j \in M} w_j^{n_j(y)} n_j(y)!, \quad y \in M^\nu.$$

This distribution satisfies the partial balance equations

$$\pi_Y(y) \sum_{k \in M} q(y, T_k^i y) = \sum_{k \in M} \pi_Y(T_k^i y) q(T_k^i y, y), \quad 1 = 1, \ldots, \nu.$$

The marginal distributions of π_Y are also related to those of π. Indeed, the stationary distribution of the location process of the ith unit is (Exercise 6)

$$\pi_{Y_i}(j) = \nu^{-1} L_j, \quad j \in M, \tag{3.27}$$

where L_j denotes the expected number of units at node j (L_j can be computed from the jth marginal distribution of π).

The location process Y does not have an exact analogue for an unlimited capacity open Whittle network, since the convention of labeling a fixed number of units does not apply to a varying and unlimited number of units. For such a network, however, we can say the following. If the associated network process X is ergodic and stationary, then conditional stationary probabilities for the unit locations are

$$P\{\text{the unit locations are } y_1, \ldots, y_\nu \mid |X_0| = \nu\} = \pi_Y(y).$$

3.9 Partially Balanced Networks

We saw in Theorems 1.12–1.15 that Jackson and Whittle networks satisfy certain partial balance equations. These balance equations are a coarser version of the detailed balance equations that reversible Markov processes satisfy. Since reversible Markov processes have canonical representations of their transition rates and invariant measures, a natural question is: Do analogous canonical representations hold for partially balanced Markov processes and networks? The answer is no, because this would be tantamount to asking for a canonical representation of invariant measures for any Markov process. However, there are some partial results along these lines that we now present.

Throughout this section, we assume that $\{X_t : t \geq 0\}$ is an irreducible Markov jump process on a countable state space $\mathbb{E}$ with transition rates $q(x, y)$ and an invariant measure π. For subsets A and B of $\mathbb{E}$, we write

$$\pi q(A, B) \equiv \sum_{x \in A} \sum_{y \in B} \pi(x) q(x, y).$$

When the process is ergodic, $\pi q(A, B)$ is the probability flux between A and B, or the average number of jumps that X makes from A to B per unit time.

Recall that q is reversible with respect to π if it satisfies the detailed balance equations $\pi q(x, y) = \pi q(y, x)$, $x, y \in \mathbb{E}$. We will now consider the following general partial balance condition.

Definition 3.20. For each $x \in \mathbb{E}$, let $\mathbb{E}_\gamma(x)$ and $\mathbb{E}'_\gamma(x)$, $\gamma \in \Gamma$, be two partitions of $\mathbb{E}$. The q is *partially balanced over* $\{\mathbb{E}_\gamma, \mathbb{E}'_\gamma\}$ *with respect to* π if π is a positive measure on $\mathbb{E}$ such that

$$\pi q(x, \mathbb{E}_\gamma(x)) = \pi q(\mathbb{E}'_\gamma(x), x), \quad x \in \mathbb{E}, \ \gamma \in \Gamma. \tag{3.28}$$

A measure π satisfying this definition is an invariant measure since it satisfies the total balance equations $\pi q(x, \mathbb{E}) = \pi q(\mathbb{E}, x)$, which are the sum of (3.28) over γ. Note that partial balance partitions always exist: The degenerate case with the coarsest partitions $\mathbb{E}_\gamma(x) = \mathbb{E}'_\gamma(x) \equiv \mathbb{E}$ is always possible. The opposite extreme is when the two partitions consist of single-point sets (the finest partitions), which corresponds to detailed balance.

Partial balance equations (3.20) are potentially easier to solve for π than the total balance equations, provided that one can find them. Partial balance also yields insights into what subsets of transitions are balanced. There are generally many partial balance partitions for a process and it is of interest to identify the finest ones possible. Although the index set Γ in the definition above is the same for both partitions, some of the sets in a partition may be empty and the number of nonempty sets in the two partitions may be different. In some cases, the two partitions are the same. For instance, the partial balance equations in Theorem 1.15 for the Jackson and Whittle network processes are based on the balance partitions

$$\mathbb{E}_j(x) = \mathbb{E}'_j(x) = \{T_{jk}x \in \mathbb{E} : k \in M\}, \quad j \in M.$$

A canonical representation of partial balance transition rates is as follows. It is useful for checking whether certain partitions of a process are partial balance partitions. Its proof (Exercise 3) is similar to that of Theorem 1.5.

Proposition 3.21. *The transition rate function q is partially balanced over $\{\mathbb{E}_\gamma, \mathbb{E}'_\gamma\}$ if and only if it is of the form*

$$q(x, y) = r(x, y)/\pi(x), \quad x, y \in \mathbb{E},$$

where π is a positive measure on $\mathbb{E}$ and r is a nonnegative function on $\mathbb{E}^2$ that satisfies

$$\sum_{y\in\mathbb{E}_\gamma(x)} r(x, y) = \sum_{y\in\mathbb{E}'_\gamma(x)} r(y, x), \quad x \in \mathbb{E}.$$

In this case, π is an invariant measure for q.

The next result is a key tool for identifying or constructing partially balanced processes. It is the basis of quasi-reversible network processes, which we discuss in Chapter 8. Suppose the transition rates of the process X have the form

$$q(x, y) = q_I(h(x), h(y))\tilde{q}(x, y), \quad x, y \in \mathbb{E}, \tag{3.29}$$

where h is a function from $\mathbb{E}$ to a countable set I, and q_I and $\tilde{q}$ are transition rates for irreducible Markov jump processes on I and $\mathbb{E}$ respectively. Let $I_\gamma(i)$ and $I'_\gamma(i)$, $i \in I$ and $\gamma \in \Gamma$, denote partial balance partitions for q_I with respect to a measure π_I on I. Associated with $I_\gamma(i)$, define partitions on $\mathbb{E}$ by

$$\begin{aligned} \mathbb{E}_{\gamma i}(x) &= \{y \in \mathbb{E} : h(y) = i \in I_\gamma(h(x))\}, \\ \mathbb{E}_\gamma(x) &= \cup_{i\in I}\mathbb{E}_{\gamma i}, \qquad\qquad x \in \mathbb{E}, i \in I, \gamma \in \Gamma. \end{aligned}$$

Let $\mathbb{E}'_{\gamma i}(x)$ and $\mathbb{E}'_\gamma(x)$ be similar partitions associated with $I'_\gamma(i)$.

Theorem 3.22. *For the Markov process X with transition rates (3.29), suppose $\tilde{q}$ is partially balanced over $\{\mathbb{E}_{\gamma i}, \mathbb{E}'_{\gamma i} : \gamma \in \Gamma, i \in I\}$ with respect to $\tilde{\pi}$, and*

$\tilde{\pi}\tilde{q}(x, \mathbb{E}_{\gamma i}(x))$ is independent of i for each x and γ. Then q is partially balanced over $\{\mathbb{E}_\gamma, \mathbb{E}'_\gamma\}$ with respect to

$$\pi(x) \equiv \pi_I(h(x))\tilde{\pi}(x), \quad x \in \mathbb{E}.$$

PROOF. The assertion follows since, for each x and γ,

$$\begin{aligned}
\pi q(x, \mathbb{E}_\gamma(x)) &= \sum_{i \in I_\gamma(h(x))} \sum_{y \in \mathbb{E}_{\gamma i}(x)} \pi_I q_I(h(x), h(y))\tilde{\pi}\tilde{q}(x, y) \\
&= \sum_{i \in I_\gamma(h(x))} \pi_I q_I(h(x), i)\tilde{\pi}\tilde{q}(x, \mathbb{E}_{\gamma i}(x)) \\
&= \sum_{i \in I_\gamma(h(x))} \pi_I q_I(i, h(x))\tilde{\pi}\tilde{q}(\mathbb{E}'_{\gamma i}(x), x) \\
&= \sum_{i \in I_\gamma(h(x))} \sum_{y \in \mathbb{E}'_{\gamma i}(x)} \pi_I q_I(h(y), h(x))\tilde{\pi}\tilde{q}(y, x) \\
&= \pi q(\mathbb{E}'_\gamma(x), x).
\end{aligned}$$

The second and fourth equalities follow since $h(y) = i$ when $y \in \mathbb{E}_{\gamma i}(x)$. And the third equality follows by the partial balance of q_I and $\tilde{q}$ and the assumption that $\tilde{\pi}\tilde{q}(x, \mathbb{E}_{\gamma i}(x))$ is independent of i. □

The preceding result, loosely speaking, says the following. Suppose q is the product of q_I and $\tilde{q}$, and one knows or can obtain invariant measures π_I and $\tilde{\pi}$ for them separately. If in addition, $\tilde{\pi}\tilde{q}(x, \mathbb{E}_{\gamma i}(x))$ is independent of i, then an invariant measure of q is the product of π_I and $\tilde{\pi}$. This "divide and conquer" strategy is predicated on obtaining π_I and $\tilde{\pi}$.

The following is an immediate consequence of Theorem 3.22, where h is the identity function.

Corollary 3.23. *Suppose the transition rates of the Markov process X are*

$$q(x, y) = q_1(x, y)q_2(x, y), \quad x, y \in \mathbb{E},$$

where q_1 and q_2 are irreducible transition rates on $\mathbb{E}$. Let $\{\mathbb{E}_\gamma, \mathbb{E}'_\gamma\}$ be partial balance partitions for q_1 with respect to π_1. Suppose q_2 is reversible with respect to π_2, and $\pi_2 q_2(x, y)$ is the same for each x, γ and $y \in \mathbb{E}_\gamma(x) \cup \mathbb{E}'_\gamma(x)$. Then q is partially balanced over $\{\mathbb{E}_\gamma, \mathbb{E}'_\gamma\}$ with respect to

$$\pi(x) \equiv \pi_1(x)\pi_2(x), \quad x \in \mathbb{E}.$$

We end this section with a network example of the preceding result.

Example 3.24. *Partially Balanced Network.* Suppose that X is a slight extension of a Whittle network process with transition rates

$$q(x, T_{jk}x) = \lambda_{jk}\phi_{jk}(x), \quad x \in \mathbb{E}.$$

The departure intensities $\phi_{jk}(x)$ are now allowed to depend on k. The network may be closed or open with finite or unlimited capacity.

Suppose that λ_{jk} are irreducible transition rates on the node set M. Let $\{I_\gamma, I'_\gamma\}$ be partial balance partitions for λ_{jk} with respect to a positive measure w_j on M, where $w_0 = 1$ if the network is open. Associated with $I_\gamma(j)$, define partitions on $\mathbb{E}$ by

$$\mathbb{E}_{\gamma j}(x) = \{T_{jk}x \in \mathbb{E} : k \in I_\gamma(j)\},$$
$$\mathbb{E}_\gamma(x) = \cup_{j\in M}\mathbb{E}_{\gamma j}(x), \quad x \in \mathbb{E},\ j \in M, \gamma \in \Gamma.$$

Let $\mathbb{E}'_{\gamma j}(x)$ and $\mathbb{E}'_\gamma(x)$ be similar partitions associated with $I'_\gamma(j)$.

Assume that ϕ_{jk} are *Φ-balanced departure–arrival intensities* in the sense that Φ is a positive function on $\mathbb{E}$ such that for $x \in \mathbb{E}$ and $j, k \in M$ with $T_{jk}x \in \mathbb{E}$,

$$\Phi(x)\phi_{jk}(x) = \Phi(T_{jk}x)\phi_{kj}(T_{jk}x).$$

In addition, assume that these quantities are the same for each $k \in I_\gamma(j) \cup I'_\gamma(j)$ and any j, x and γ. Then q is partially balanced over $\{\mathbb{E}_\gamma, \mathbb{E}'_\gamma\}$ with respect to

$$\pi(x) \equiv \Phi(x)\prod_{j=1}^{m} w_j^{x_j}, \quad x \in \mathbb{E}. \tag{3.30}$$

To see this, note that $q_1(x, T_{jk}x) \equiv \lambda_{jk}$ are transition rates for a Whittle network process with departure rates $\phi_j \equiv 1$. Then it follows similarly to Theorem 1.15 that q_1 is partially balanced over $\{\mathbb{E}_{\gamma j}, \mathbb{E}'_{\gamma j}\}$ with respect to the measure $\prod_{j=1}^m w_j^{x_j}$. Also, the transition rate function $q_2 \equiv \phi_{jk}(x)$ is reversible with respect to Φ. Thus it follows by Corollary 3.23 that q is partially balanced over $\{\mathbb{E}_\gamma, \mathbb{E}'_\gamma\}$ with respect to the measure (3.30). □

3.10 Exercises

1. Suppose the multiclass network process X in Section 3.1 represents an open network, and ϕ_0 has the form $\phi_0(|x|)$, where $|x| = \sum_{\alpha j} x_{\alpha j}$. Show that the service rates are balanced by the function

$$\Phi(x) = \prod_{i=1}^{n} \phi_0(i-1)/\phi_{(\beta k)_i}(x^i), \quad x \in \mathbb{E},$$

where $x^0, \ldots, x^n$ is a direct path from $x^0 = 0$ to $x^n = x$ with $n = |x|$ and $x^i = x^{i-1} + e_{(\beta k)_i}$.

2. Show that the service rates in Exercise 1 are a special case of the sector-dependent service rates in Example 3.3. Show that the latter rates are balanced by Φ given by (3.3).

3. Prove Proposition 3.21.

4. *Networks with infinite number of nodes or classes.* Consider a Whittle network with the modification that it has an infinite number of nodes labeled $1, 2, \ldots$, where the number of units is still finite for the unlimited capacity open network case. Assume that its routing rates λ_{jk} satisfy $\sum_{k=1}^{\infty}\lambda_{jk} < \infty$ and they do not have transient states. What additional assumption on the service rates is needed (for the closed or open cases) in order for the network process to be a well-defined Markov process? Under these assumptions, an invariant measure for the network (as in Theorem 1.15) is $\pi(x) = \Phi(x)\prod_{j=1}^{\infty} w_j^{x_j}$. In case the network has unlimited capacity and the $w_j < 1$ for each j, is another assumption needed in order for the infinite product in this distribution to be positive? If so, what is it? Specify the comparable assumptions needed to model a multiclass Whittle network with a "finite" number of nodes and an infinite number of classes.

5. *Networks with infinite number of units.* It is possible to define a Whittle process with an infinite number of nodes and units, where each node contains a finite number of units. Specify additional assumptions on the routing and service rates that yield a well-defined Markov network process with invariant measure $\pi(x) = \Phi(x)\prod_{j=1}^{\infty} w_j^{x_j}$.
6. Prove Theorem 3.19 and expression (3.27). Hint: Use the properties $\mathbf{n}(y) = \mathbf{n}(y_2, \ldots, y_\nu) + e_{y_1}$ and

$$\sum_y (\ldots) = \sum_x \sum_y (\ldots) 1(\mathbf{n}(y) = x).$$

7. In the setting of Theorem 3.19, consider the modification in which each unit i has irreducible routing rates $\lambda_i(j,k)$ on M that are a function of i. Let $w_i(j)$ denote the stationary distribution of $\lambda_i(j,k)$. Show that the stationary distribution of the location process Y, under this modification, is

$$\pi_Y(y) = c\Phi(\mathbf{n}(y)) \prod_{i=1}^{\nu} w_i(y_i) \prod_{j \in M} n_j(y)!, \quad y \in M^\nu.$$

3.11 Bibliographical Notes

The first models of multiclass networks were the BCMP model developed by Baskett et al. (1975), and the model with deterministic routing developed by Kelly (1976,1979). These references also discussed networks with Cox and general service times. The Cox distribution was introduced by Cox (1955). Further insights on networks with general service times and "insensitivity" is covered in Schassberger (1978) and Whittle (1986b). For more details on Cox and general phase-type distributions, see Asmussen (1987) and Neuts (1994). Another useful approximation for networks with general service times is the queueing network analyzer (QNA) model discussed in Kühn (1979) and Whitt (1983).

This book does not cover multiclass networks with FCFS single servers and general service time distributions that depend on the customer class. Such networks in heavy traffic can be approximated by fluid models or by diffusion processes. Sample references on this are Harrison and Reiman (1981), Rieman (1984), Dai and Harrison (1992), Dai (1995), Mandelbaum and Pats (1998), Bramson (1998), and Williams, (1998). Another approach to critically loaded networks is to approximate their probability distributions via the Kolmogorov backward and forward difference-differential equations using singular perturbation techniques as in Knessl and Tier (1995).

The results on blocking and rerouting are from Economou and Fakinos (1998) and Proposition 3.16 is from Kemeny and Snell (1976). Other types of blocking are considered for instance in Akyildiz and von Brand (1989); and Perros (1994) surveys several approximations for blocking. The elementary model of bottlenecks that we discuss has more interesting variations as in Malyshev and Yakovlev (1996). The section on partial balance is based on ideas in Whittle (1985,1986a,1986b).

A sample of network models with special themes or structures that we did not cover are in Kumar and Kumar (1994), Gross and Harris (1985), Stecke and Solberg (1985), Buzacott and Yao (1986), Boucherie and van Dijk (1990), Kelly and Williams (1995), and Glasserman et al. (1996).

4

Network Flows and Travel Times

This chapter addresses the following questions about movements of units in stationary Jackson and Whittle processes. What flows of units between nodes are Poisson processes? When a unit moves from one node to another, what is the probability distribution of the locations of the other units in the network? What is the distribution of the time it takes for a typical unit to traverse a series of nodes?

The answers to these questions require an understanding of Palm probabilities for Markov processes at their transition times. The theory for these probabilities is a self-contained, elementary part of the theory of Palm probabilities. We cover this subtheory in Section 4.6, and give a more comprehensive study of Palm probabilities in Chapter 5. Another key tool for our analysis is a generalization of Lévy's formula for expectations of functionals of a Markov process. This formula is the topic of Section 4.2.

4.1 Point Process Notation

In this section, we introduce the notation and a few properties of point processes that we use throughout the rest of the book. Additional material on point processes is in Chapters 5 and 9. Point processes in time (i.e., on the real line $\mathbb{R}$ or half line $\mathbb{R}_+$) describe such events as customer arrival times, or times at which customers move from a node j to a node k. Point processes in a space are natural for modeling such things as mobile customers in regions of the plane $\mathbb{R}^2$. We will also use "space–time" point processes, where each point is a pair of numbers in time and a space (e.g., the arrival time of a customer to a system and the location where it enters).

To describe point processes on general spaces, we will use the following notation, which is now standard in applied probability. Let $\mathbb{E}$ denote a complete, separable metric space (a Polish space), and let $\mathcal{E}$ denote its family of Borel sets. We will describe sets of points in $\mathbb{E}$ by counting measures. For most of our applications, $\mathbb{E}$ will be a Euclidean space. We refer to $\mathbb{E}$ simply as a *space*, and denote other spaces of this type by $\mathbb{E}'$. (With a slight abuse of notation, we will also continue using $\mathbb{E}$ as the state space for Markov processes; the nature of $\mathbb{E}$ should be clear from the context.)

Suppose that $x_1, \ldots, x_k$ are locations of points (or unit masses) in $\mathbb{E}$. There may be more than one point at a location, and the order of the subscripts on the locations is invariant under permutations. These points are described by the *counting measure* ν on $\mathbb{E}$ defined by

$$\nu(A) = \sum_{n=1}^{k} 1(x_n \in A), \qquad A \in \mathcal{E},$$

where $\nu(A)$ denotes the number of point in A. Let $\mathbb{M}$ denote the set of all such counting measures on $\mathbb{E}$ that are finite on compact sets. Endow $\mathbb{M}$ with the σ-field $\mathcal{M}$ on $\mathbb{M}$ generated by the sets $\{\nu \in \mathbb{M} : \nu(A) = n\}$, for $A \in \mathcal{E}$ and $n \in \mathbb{Z}_+$.

Definition 4.1. A *point process* N on $\mathbb{E}$ is a measurable map from a probability space $(\Omega, \mathcal{F}, \mathcal{P})$ to the space $(\mathbb{M}, \mathcal{M})$. The quantity $N(A)$ is the number of points in the set $A \in \mathcal{E}$. We express N as

$$N(A) = \sum_{n} 1(X_n \in A), \qquad A \in \mathcal{E}, \tag{4.1}$$

where the X_n's denote the *locations* of the points of N. The summation is $\sum_{n=1}^{N(\mathbb{E})}$, where $N(\mathbb{E})$ may be finite or infinite.

The space of measures $\mathbb{M}$ and related measure theory technicalities are not used explicitly in the following discussion. The results are understandable simply by thinking of $\mathbb{E}$ as a Euclidean space and N as a counting process on it. When $\mathbb{E} = \mathbb{R}$ (or $\mathbb{R}_+$), we denote the point locations by T_n instead of X_n. Also, unless specified otherwise, we assume the points are ordered such that

$$\ldots < T_{-2} < T_{-1} < T_0 \leq 0 < T_1 < T_2 < \ldots.$$

Note that these times are subscripted such that $T_0 \leq 0 < T_1$. We also write $N(a, b] = N((a, b])$ for $a < b$, and $N(t) = N(0, t]$, for $t > 0$. We will frequently refer to integrals with respect to such point processes in time. The integral of a real-valued process $\{Y_t : t \in \mathbb{R}\}$ with respect to N is simply the summation

$$\int_B Y_t N(dt) = \sum_n Y_{T_n} 1(T_n \in B).$$

The *probability distribution* of a point process N (i.e., $P\{N \in \cdot\}$) is determined by its finite-dimensional distributions

$$P\{N(A_1) = n_1, \ldots, N(A_k) = n_k\}, \qquad A_1, \ldots, A_k \in \mathcal{E}.$$

It suffices to define these probabilities on sets A_i that generate $\mathcal{E}$. For instance, when $\mathbb{E} = \mathbb{R}$, intervals of the form $(a, b]$ generate $\mathcal{E}$. The point process N is said to be *simple* if $P\{N(\{x\}) \leq 1 \text{ for all } x \in \mathbb{E}\} = 1$ (i.e., the point locations are distinct). The *mean measure* of N is $\mu(A) = EN(A)$, $A \in \mathcal{E}$, which may be infinite.

The most prominent point processes are Poisson processes defined as follows.

Definition 4.2. A point process N on $\mathbb{E}$ is a *Poisson process with mean measure* μ if it satisfies the following conditions.
(a) N has *independent increments*: The quantities $N(A_1), \ldots, N(A_k)$ are independent for disjoint sets $A_1, \ldots, A_k$ in $\mathcal{E}$.
(b) For each $A \in \mathcal{E}$, the quantity $N(A)$ is a Poisson random variable with mean $\mu(A)$; and μ is finite on compact sets.

In this definition, if $\mu(\{x\}) > 0$, then the number of points $N(\{x\})$ exactly at x has a Poisson distribution with mean $\mu(\{x\})$. On the other hand, if $\mu(\{x\}) = 0$, then N cannot have more than one point at the location x. This occurs, for instance, when $\mu(A) = \int_A r(x)dx$, where $r(x)$ is the *intensity* or rate of N at the location x, and dx denotes the Lebesgue measure.

Some properties of networks involve point processes on product spaces. To define a point process N on a product space $\mathbb{E} \times \mathbb{E}'$, it suffices to specify its values $N(A \times B)$ for product sets $A \times B \in \mathcal{E} \times \mathcal{E}'$. An important case is as follows.

Definition 4.3. A point process N on $\mathbb{R} \times \mathbb{E}$ is a *space–time point process*. We denote its points by the pairs (T_n, X_n).

In this definition, it is possible for $N(I \times \mathbb{E})$ to be infinite for finite intervals I, and so it is not appropriate to assume the T_n's are ordered. We also sometimes use the nonnegative time axis $\mathbb{R}_+$ instead of $\mathbb{R}$. A space–time process is a natural arrival process for a service system or particle system, where T_n is the arrival time of the nth customer or particle, and X_n is the location in a space $\mathbb{E}$ where it enters. The preceding definition allows for an infinite number of particles to enter a system at a specified time. For instance, an infinite number of points may be present at time 0. Although the T_n's are not ordered, one can say that the nth particle with space–time entry (T_n, X_n) has $\sum_{k \neq n} 1(T_k \leq T_n)$ predecessors.

Suppose N is a space–time point process that is Poisson. Consider the special case in which its mean measure has the form

$$EN((s, t] \times B) = a(t - s)F(B), \quad s < t,$$

where $a > 0$ and F is a probability measure on $\mathbb{E}'$. In this case, the points $\{T_n\}$ form a Poisson process on $\mathbb{R}$ with rate a, and the $\{X_n\}$ are independent and independent of $\{T_n\}$, and each X_n has the distribution F. We say that the space–time Poisson process N has a *rate* a and *space distribution* F.

4.2 Extended Lévy Formula for Markov Processes

Our study of movements in networks will involve functionals of Markov processes. This section covers expressions for their expected values.

We will use the following notation throughout this chapter. We assume that $\{X_t : t \in \mathbb{R}\}$ is a Markov jump process on a countable state space $\mathbb{E}$ with transition rates $q(x, y)$. The time axis is the entire real line $\mathbb{R}$, because this is natural for analyzing stationary systems. The results here automatically apply to processes defined only on the nonnegative time axis. The process X has piecewise constant sample paths, and we assume its sample paths $\{x_t : t \in \mathbb{R}\}$ are in the set D of all functions from $\mathbb{R}$ to $\mathbb{E}$ that are right continuous and have limits from the left. The left-hand limit of X at time t is $X_{t-} \equiv \lim_{s \uparrow t} X_s$. Frequent reference will be made to time shifts of the process X defined as follows.

Definition 4.4. For each $t \in R$, the *time-shift operator* S_t on D is a mapping $S_t : D \to D$ defined by $S_t z \equiv \{z(s + t) : s \in \mathbb{R}\}$, for $z \in D$. The process X with its time parameter *shifted by the amount* t is $S_t X \equiv \{X_{s+t} : s \in \mathbb{R}\}$.

The time-shifted process $S_t X$ is what an observer of X would see at time t: With t as the time origin, the future would evolve as X_{t+u} for $u \geq 0$ and the past would be seen as X_{t-u} for $u > 0$. The stochastic process X is *stationary* if the distribution of the process $S_t X$ is independent of t.

We denote the transition times of X by

$$\ldots < \tau_{-2} < \tau_{-1} < \tau_0 \leq 0 < \tau_1 < \tau_2 < \ldots .$$

For this section, we let N denote the point process of these times; that is

$$N(A) \equiv \sum_n 1(\tau_n \in A) = \sum_{t \in A} 1(X_t \neq X_{t-}), \quad A \subset \mathbb{R}.$$

Reference to a time set $A \subset \mathbb{R}$ means that A is a Borel subset of $\mathbb{R}$ (this avoids introducing a symbol for the Borel sets). In later sections, we use N to denote other point processes. Finally, we say that an expectation of the form $E \int_A Y_s \, ds$ exists if $\int_A E|Y_s| \, ds$ is finite.

This section focuses on functionals of the Markov process X associated with its transitions times τ_n. As a preliminary example, suppose that $h(x, y)$ is a *value* (cost or utility) of a transition of the process from state x to state y. Then the value of the transitions in the time set A is

$$\int_A h(X_{t-}, X_t) N(dt) = \sum_n h(X_{\tau_{n-1}}, X_{\tau_n}) 1(\tau_n \in A).$$

The expectation of this functional can be expressed as follows.

Example 4.5. *Lévy's Formula*. For $h : \mathbb{E}^2 \to \mathbb{R}$ and $A \subset \mathbb{R}$,

$$E \int_A h(X_{t-}, X_t) N(dt) = E \int_A \sum_{y \neq X_t} q(X_t, y) h(X_t, y) \, dt, \tag{4.2}$$

provided the right side exists. This formula follows from the extended Lévy formula in Theorem 4.6 below with $g(t, S_t X) = h(X_{t-}, X_t)1(t \in A)$. Note that if the process X is stationary with distribution π, then (4.2) becomes

$$E \int_{(a,b]} h(X_{t-}, X_t) N(dt) = (b-a) \sum_x \pi(x) \sum_{y \neq x} q(x, y) h(x, y).$$

In particular, the expected number of transitions of X from x to y per unit time is

$$E \int_{(0,1]} 1(X_{t-} = x,\ X_t = y) N(dt) = \pi(x) q(x, y).$$

An example of (4.2) is that, for $f : \mathbb{E} \to \mathbb{R}$,

$$E \int_A f(X_t) N(dt) = E \int_A \sum_{y \neq X_t} q(X_t, y) f(y)\, dt,$$

provided the right side exists. Another example is Dynkin's formula; see Exercise 1. □

We now consider the situation for the process X in which a transition at time t is associated with a real-valued quantity $g(t, S_t X)$, which is a function of t and $S_t X$. This quantity is a function of the future, as well as the past, of the process, since the time-shifted process $S_t X$ is the entire sample path of X "centered" at t. Such functionals are the basis of Palm probabilities for Markov processes, which we discuss later in this chapter. For $g : \mathbb{R} \times D \to \mathbb{R}$, $t \in \mathbb{R}$, and $x, y \in \mathbb{E}$, we define

$$G(t, x, y) \equiv E[g(\tau_{n+1}, S_{\tau_{n+1}} X) \mid \tau_n = t, X_{\tau_n} = x, X_{\tau_{n+1}} = y].$$

We assume this expectation exists; it does not depend on n because of the Markovian structure of X. The following is an extended Lévy formula.

Theorem 4.6. *For $g : \mathbb{R} \times D \to \mathbb{R}$ and $u \in \mathbb{R}$,*

$$E \int_{\mathbb{R}} g(t, S_t X) N(dt) = E \int_{\mathbb{R}} \sum_{y \neq X_t} q(X_t, y) G(T_t, X_t, y)\, dt, \tag{4.3}$$

provided the last expectation exists. Here $T_t \equiv \sup\{s \le t : X_s \neq X_t\}$ is the time of the last transition of X before or at time t.

Proof. It suffices to show that, for each n,

$$E \int_{(\tau_n, \tau_{n+1}]} g(t, S_t X) N(dt) = E \int_{(\tau_n, \tau_{n+1}]} \sum_{y \neq X_t} q(X_t, y) G(T_t, X_t, y)\, dt.$$

Clearly, N has exactly one point at τ_{n+1} in the interval $(\tau_n, \tau_{n+1}]$, and $(T_t, X_t) = (\tau_n, X_{\tau_n})$, for $\tau_n < t < \tau_{n+1}$. Therefore, the preceding display is equal to

$$E[g(\tau_{n+1}, S_{\tau_{n+1}} X)] = E[\sum_{y \neq X_{\tau_n}} q(X_{\tau_n}, y) G(\tau_n, X_{\tau_n}, y)(\tau_{n+1} - \tau_n)]. \tag{4.4}$$

We will use the properties that the sojourn times of the Markov process X in the states it visits are exponentially distributed, and the sequence of states it visits forms a Markov chain. In particular,

$$P\{\tau_{n+1} - \tau_n > t \mid X_{\tau_n}, \tau_n\} = e^{-q(X_{\tau_n})t}$$
$$P\{X_{\tau_{n+1}} = y \mid X_{\tau_n}, \tau_n\} = q(X_{\tau_n}, y)q(X_{\tau_n})^{-1},$$

where $q(x) = \sum_y q(x, y)$.

Conditioning the right side of (4.4) on X_{τ_n}, τ_n and using the expression $E[\tau_{n+1} - \tau_n \mid X_{\tau_n}, \tau_n] = q(X_{\tau_n})^{-1}$, it follows that the right side of (4.4) equals

$$\begin{aligned} &E[\sum_{y \neq X_{\tau_n}} q(X_{\tau_n}, y)G(\tau_n, X_{\tau_n}, y)q(X_{\tau_n})^{-1}] \\ &= E[\sum_{y \neq X_{\tau_n}} G(\tau_n, X_{\tau_n}, y)P\{X_{\tau_{n+1}} = y \mid X_{\tau_n}, \tau_n\}] \\ &= E[G(\tau_n, X_{\tau_n}, X_{\tau_{n+1}})] \\ &= E[g(\tau_{n+1}, S_{\tau_{n+1}}X)]. \end{aligned}$$

This proves (4.4), which in turn completes the proof of (4.3). □

The extended Lévy formula obviously reduces as follows for stationary processes and time-homogeneous functions.

Corollary 4.7. *Suppose the Markov process X is stationary, and denote its distribution by $\pi(x) \equiv P\{X_t = x\}$, $x \in \mathbb{E}$. Then, for $g : D \to \mathbb{R}$,*

$$\begin{aligned} &E\int_{(a,b]} g(S_tX)N(dt) \qquad (4.5) \\ &\quad = (b-a)\sum_x \pi(x)\sum_{y\neq x} q(x, y)E[g(S_{\tau_1}X) \mid X_{\tau_0} = x, X_{\tau_1} = y] \end{aligned}$$

provided the right side of this equality exists.

The preceding results also apply when the state space of the Markov process X is an uncountable Euclidean or Polish space. The only difference is that the transition rate $q(x, y)$ for countable states is replaced by a transition kernel $q(x, B)$ for a transition from x into a set B. Then sums involving q are replaced by integrals. For instance, the right side of (4.2) becomes $E[\int_A \int_{\mathbb{E}} q(X_t, dy)h(X_t, y)\, dt]$. Otherwise, the proofs are the same.

4.3 Poisson Functionals of Markov Processes

There are a variety of point processes associated with the transition times of a Markov process. A typical example is the point process of times at which units move from node j to node k in a Jackson process. In this section, we develop general criteria for such a point process to be a Poisson process. We apply the results in Section 4.5 to characterize Poisson flows in networks.

As in the preceding section, assume that $\{X_t : t \in \mathbb{R}\}$ is a Markov jump process on a countable state space $\mathbb{E}$ with transition rates $q(x, y)$. With no loss in generality, assume that X is ergodic and let π denote its stationary distribution. We will consider certain types of transitions of X as follows.

Definition 4.8. Suppose $\mathcal{T}_0$ is a subset of $\mathbb{E}^2$ that does not contain pairs with equal entries. A *$\mathcal{T}_0$-transition of X* is a transition from a state x to another state y, for some $(x, y) \in \mathcal{T}_0$. The point process N of times at which these transitions occur is defined by

$$N(A) = \sum_n 1((X_{\tau_n -}, X_{\tau_n}) \in \mathcal{T}_0, \tau_n \in A), \quad A \subset \mathbb{R}. \tag{4.6}$$

The $N(A)$ is the number of $\mathcal{T}_0$-transitions of X in the time set A.

The subscript 0 on the set $\mathcal{T}_0$ refers to a transition defined only by the values of the process at the transition time viewed as a time origin. The second half of this chapter deals with more general transitions called $\mathcal{T}$-transitions that may involve more information about the process. There are many examples of point processes of $\mathcal{T}_0$-transitions since any subset $\mathcal{T}_0$ determines one. For instance, if $\mathbb{E}$ is the nonnegative integers and one is interested in the number of jumps whose size exceeds b, then this point process is determined by $\mathcal{T}_0 = \{(x, y) \in \mathbb{E}^2 : |x - y| > b\}$. Note that N is the point process of all transition times τ_n of X if $\mathcal{T}_0 = \mathbb{E}^2 \backslash \{(x, x) : x \in \mathbb{E}\}$.

Throughout the rest of this section, we assume that N is the point process of $\mathcal{T}_0$-transitions of X for a fixed transition set $\mathcal{T}_0$. Although N is a function of the transition set $\mathcal{T}_0$, we suppress the $\mathcal{T}_0$ in its definition. Our interest is in criteria for N to be a Poisson process.

First, we relate the independent increments property needed for N to be a Poisson process to the following notion.

Definition 4.9. *The future of N is independent of the past of X*, denoted by $N_+ \perp X_-$, if $\{N(A) : A \subset (t, \infty)\}$ is independent of $\{X_s : s \leq t\}$, for each $t \in \mathbb{R}$.

In this definition, $\{X_s : s \leq t\}$ can be replaced simply by X_t since X is Markovian. Similarly, $N_- \perp X_+$ denotes that the past of N is independent of the future of X.

Theorem 4.10. *If $N_+ \perp X_-$ or $N_- \perp X_+$, then N is a Poisson process (not necessarily time homogeneous).*

PROOF. By a characterization of Poisson processes, the N will be a Poisson process if it is simple, has no fixed atoms, and has independent increments. It is well known that the probability is 0 that the Markov process X has a jump at any specified time. Thus, N is simple with no fixed atoms.

It remains to show that N has independent increments. First, suppose $N_+ \perp X_-$. Then for any $s < t$ in $\mathbb{R}$,

$$\begin{aligned} P\{N(s, t] = n | N(A) : A \subset (\infty, s]\} &= E[P\{N(s, t] = n | X_r : r \leq s\}] \\ &= P\{N(s, t] = n\}. \end{aligned}$$

Using this, one can show by induction that N has independent increments on any number of disjoint time sets. Similarly, the independent increments property of N also follows when $N_- \perp X_+$, since

$$\begin{aligned} P\{N(s,t] = n | N(A) : A \subset (t,\infty)\} &= E[P\{N(s,t] = n | X_u : u > t\}] \\ &= P\{N(s,t] = n\}. \end{aligned}$$ □

Our next step in analyzing the point process N of $\mathcal{T}_0$-transitions is to obtain an expression for its mean measure. Applying Lévy's formula (4.2) to the definition of $N(A)$, we have

$$EN(A) = \int_A E\alpha(X_t)\,dt, \tag{4.7}$$

where

$$\alpha(x) = \sum_y q(x,y) 1((x,y) \in \mathcal{T}_0).$$

The $\alpha(X_t)$ is the "conditional intensity" of N given X_t in the sense that

$$E[N(t, t+dt] | X_t] = \alpha(X_t)\,dt.$$

The function $\alpha(x)$ (which is also a function of the transition set $\mathcal{T}_0$) plays a key role in our analysis.

We are now ready to present our first criterion for N to be a Poisson process.

Theorem 4.11. *The N is a Poisson process with rate a and $N_+ \perp X_-$ if and only if*

$$\alpha(x) = a, \quad x \in \mathbb{E}. \tag{4.8}$$

PROOF. Suppose (4.8) holds. Fix an $s \in \mathbb{R}$. Then by (4.7), it follows that the process

$$M_t = N(s, s+t] - \int_s^{s+t} \alpha(X_u)\,du, \quad t \geq 0,$$

is an $\mathcal{F}_t^X$-martingale, where $\mathcal{F}_t^X$ is the σ-field generated by $\{X_u : s \leq u \leq s+t\}$. By Watanabe's characterization of Poisson processes, the N is an $\mathcal{F}_t^X$-Poisson process, or a *Markov-modulated Poisson process*. This means that, conditioned on X being in state x in a time interval $(a, b]$, the N is a Poisson process on that interval with rate $\alpha(x)$. The process $A_t = \int_s^{s+t} \alpha(X_u)\,du$ is the "compensator" of $N(s, s+\cdot]$. Then under the assumption (4.8), it follows that $N(s, s+\cdot]$ is an $\mathcal{F}_t^X$-Poisson process with rate a. In particular, $N(s, s+t]$ is independent of $\mathcal{F}_s^X$ for each t. Since these observations hold for each s, it follows that N is a Poisson process with rate a and $N_+ \perp X_-$.

Conversely, assume the preceding conclusion is true. Then, for any x and $t \geq 0$, it follows by (4.7) that

$$at = EN(0,t] = E[N(0,t] | X_0 = x] = \int_0^t E[\alpha(X_s) | X_0 = x]\,ds.$$

The integrand is continuous in s since X is a Markov process. Taking the derivative of the preceding equation with respect to t yields

$$a = E[\alpha(X_t)|X_0 = x], \quad t \geq 0.$$

Then, using the first jump time $\tau_1 = \inf\{t > 0 : X_t \neq X_0\}$, we can write

$$a = \alpha(x)P\{\tau_1 > t|X_0 = x\} + E[\alpha(X_t)1(\tau_1 \leq t)|X_0 = x].$$

Letting $t \downarrow 0$ yields (4.8). □

The preceding criterion for the point process N of $\mathcal{T}_0$-transitions to be a Poisson process is what one might anticipate, and it is often tacitly assumed when one defines a particular Markov process. Another less intuitive and more useful criterion for N to be Poisson is as follows.

Theorem 4.12. *Suppose the Markov process X is ergodic and stationary. Let π denote its stationary distribution, and define*

$$\bar{\alpha}(x) = \pi(x)^{-1}\sum_y \pi(y)q(y,x)1((y,x) \in \mathcal{T}_0), \quad x \in \mathbb{E}.$$

Then N is a Poisson process with rate a and $N_- \perp X_+$ if and only if

$$\bar{\alpha}(x) = a, \quad x \in \mathbb{E}. \tag{4.9}$$

Proof. Consider the time reversal of X, which is $\bar{X}_t \equiv \lim_{s\uparrow -t} X_{-t}$ (this is the process $\{X_{-t} : t \in \mathbb{R}\}$ modified to have right-continuous paths). Since X is stationary, we know by Theorem 2.5 on time reversals that $\bar{X}$ is also a stationary Markov process with transition rates

$$\bar{q}(x,y) = \pi(x)^{-1}\pi(y)q(y,x), \quad x, y \in \mathbb{E},$$

and its stationary distribution is the same as the stationary distribution π for X. Now, consider the point process

$$\bar{N}(A) = \sum_{t\in A} 1((\bar{X}_t, \bar{X}_{t-}) \in \mathcal{T}_0), \quad A \subset \mathbb{R}.$$

By Lévy's formula and the definitions of $\bar{q}$ and $\bar{\alpha}$, we have

$$E\bar{N}(A) = E\int_A \sum_y \bar{q}(\bar{X}_t, y)1((y, \bar{X}_t) \in \mathcal{T}_0)\,dt = E\int_A \bar{\alpha}(\bar{X}_t)\,dt.$$

Note that $\bar{N}(A) = N(-A)$, for each A, and so $\bar{N}$ is the time reversal of N. Consequently, $\bar{N}$ is a Poisson process with rate a if and only if N is. Furthermore, since $\bar{N}$, $\bar{X}$ are time reversals of N, X, it follows that N is a Poisson process with rate a and $N_- \perp X_+$ if and only if $\bar{N}$ is a Poisson process with rate a and $\bar{N}_+ \perp \bar{X}_-$. But the latter is equivalent to (4.9) by Theorem 4.11. This proves the assertion. □

Here is a classic application of the preceding theorem.

Example 4.13. *Birth–Death Process with Poisson Departures.* Suppose the Markov process X represents a birth–death process with birth rate λ and death rate μ_x, when there are x units in the system. This might represent a queueing process with arrival rate λ and departure rates μ_x. An example is an $M/M/s$ queueing system with s independent servers $(1 \leq s \leq \infty)$ that have exponential service times with rate μ, and $\mu_x = \mu \min\{x, s\}$.

The transition rates of the process X are

$$q(x, y) = \lambda 1(y = x + 1) + \mu_x 1(y = x - 1).$$

Implicit in the description of the process, the point process of arrivals is Poisson with rate λ regardless of whether or not X is ergodic. This also follows formally by the structure of q and Theorem 4.11 since

$$\alpha(x) = \sum_y q(x, y)1(y = x + 1) = q(x, x + 1) = \lambda.$$

Next, consider the point process N of times of departures from the system, which are $\mathcal{T}_0$-transition times, where $\mathcal{T}_0 = \{(x + 1, x) : x \geq 0\}$. Its associated α function is

$$\alpha(x) = \sum_y q(x, y)1(y = x - 1) = \mu_{x-1}.$$

This depends on x, and so Theorem 4.11 does not ensure that N is Poisson. However, let us now assume that X is ergodic and stationary. Its stationary distribution is

$$\pi(x) = c\lambda^x/(\mu_1 \cdots \mu_x), \quad x \geq 1,$$

provided $c^{-1} = 1 + \sum_{x \geq 1} \lambda^x/(\mu_1 \cdots \mu_x)$ is finite, which we assume is true. In this case,

$$\begin{aligned}\bar{\alpha}(x) &= \pi(x)^{-1} \sum_y \pi(y) q(y, x) 1(y = x + 1) \\ &= \pi(x)^{-1} \pi(x + 1) q(x + 1, x) = \lambda.\end{aligned}$$

Thus, Theorem 4.12 ensures that N is a Poisson process with rate λ. In particular, this result yields the Burke-Reich property that the departure processes for $M/M/s$ queues are Poisson. □

We now consider a queueing system with non-Poisson departures, but the times at which certain batches depart form a Poisson process.

Example 4.14. *A Batch Service System.* Consider a Markovian queueing process whose state is the number of customers in the system and whose transition rates are

$$q(x, y) = \lambda 1(y = x + 1) + \mu 1(y = \max\{0, x - K\}).$$

Here λ, μ, and K are positive, and $\lambda < K\mu$. This represents a system in which customers arrive by a Poisson process with rate λ and are served in batches as follows. Whenever there are $x \geq K$ customers in the system, batches of K customers depart at the rate μ; and whenever $x < K$ customers are present, all of the customers depart at the rate μ. The system is ergodic and its stationary distribution is $\pi(x) = r^x(1-r)$, $x \geq 0$, where r is the unique root in (0, 1) of the equation

$$\mu r^{K+1} - (\lambda + \mu)r + \lambda = 0.$$

This assertion follows by showing that this distribution satisfies the balance equations.

Now, assume the process is stationary. Let N denote the point process of times at which batches of size K depart from the system. Then N is a Poisson process with rate $\mu + \lambda(1 - r^{-1})$. This follows by Theorem 4.12 since

$$\begin{aligned}\bar{\alpha}(x) &= \pi(x)^{-1}\pi(x+K)q(x+K, x)\\ &= \mu r^K = \mu + \lambda(1 - r^{-1}).\end{aligned}$$

□

4.4 Multivariate Compound Poisson Processes

We have been studying the point process N that records times of $\mathcal{T}_0$-transitions of the Markov process X. This section addresses similar issues for multivariate Poisson and compound Poisson processes associated with $\mathcal{T}_0$-transitions of X.

We begin with a multivariate analogue of Theorem 4.12. This result is useful for determining when several flows in a network are independent Poisson processes. Suppose that N_i is a point process of $\mathcal{T}_0{}^i$-transitions of X, for $i = 1, \dots n$. For $x \in \mathbb{E}$ and $\mathbf{u} \in \{0, 1\}^n$, define

$$\bar{\alpha}(x, \mathbf{u}) \equiv \pi^{-1}(x) \sum_y \pi(y)q(y, x)1\Big(1\big((x, y) \in \mathcal{T}_0{}^i\big) = u_i,\ 1 \leq i \leq n\Big).$$

Also, let e_i denote the n-dimensional unit vector with a 1 in position i.

Theorem 4.15. *Suppose the Markov process X is ergodic and stationary, and its stationary distribution is π. Then $N_1, \dots, N_n$ are independent Poisson processes with respective rates $a_1, \dots, a_n$ such that $(N_1, \dots, N_n)_- \perp X_+$ if and only if, for each $x \in \mathbb{E}$ and $\mathbf{u} \in \{0, 1\}^n$,*

$$\bar{\alpha}(x, \mathbf{u}) = \begin{cases} a_i & \text{if } \mathbf{u} = e_i \text{ for some } 1 \leq i \leq n \\ 0 & \text{otherwise.} \end{cases} \tag{4.10}$$

This result is a special case of Theorem 4.19 below for compound point processes. The criterion (4.10) is the multivariate analogue of the criterion (4.9) in Theorem 4.12 for single point processes. A reader interested in seeing how the preceding result characterizes Poisson flows in networks can skip to the next section and read the rest of this section later.

Our aim now is to study multivariate compound point processes associated with $\mathcal{T}_0$-transitions. Let N denote a point process of $\mathcal{T}_0$-transitions of the Markov process X. Assume that whenever X makes an $\mathcal{T}_0$-transition from x to y, a mark $h(x, y)$

in a complete, separable metric space $\mathbb{E}'$ is assigned to the transition. Then the times at which the marks are recorded and the associated marks are modeled by the *space–time point process* M on $\mathbb{R} \times \mathbb{E}'$ defined, for $A \subset \mathbb{R}$, $B \in \mathcal{E}'$, by

$$M(A \times B) = \sum_{t \in A} 1(h(X_{t-}, X_t) \in B)1((X_{t-}, X_t) \in \mathcal{T}_0). \tag{4.11}$$

The $M(A \times B)$ is the number of $\mathcal{T}_0$-transitions in the time set A at which a mark in the set B is recorded. Clearly $N(A) = M(A \times \mathbb{E}')$, which means that M contains the point process N of $\mathcal{T}_0$-transitions at which marks are recorded. Since the probability is 0 that X takes a transition at any fixed time, M is a simple point process without fixed atoms and $M(\{t\} \times \mathbb{E}') = 0$ or 1, for each $t \in \mathbb{R}$.

Our interest is in criteria under which M is a space–time Poisson process with rate a and space distribution F. That is, it is a Poisson process with mean of the form

$$EM((s, t] \times B) = a(t - s)F(B),$$

where $a \geq 0$ and F is a probability measure on $\mathbb{E}'$. In this case, N is a Poisson process with rate a, and the marks are independent of N and each one has the distribution F.

Let $M_+ \perp X_-$ denote that *the future of M is independent of the past of X*; that is, $\{M(A \times B) : A \subset (t, \infty), B \in \mathcal{E}'\}$ is independent of $\{X_s : s \leq t\}$, for each $t \in \mathbb{R}$. In addition, for $x \in \mathbb{E}$ and $B \in \mathcal{E}'$, define

$$\alpha(x, B) = \sum_y q(x, y)1(h(x, y) \in B)1((x, y) \in \mathcal{T}_0),$$

$$\bar{\alpha}(x, B) = \frac{1}{\pi(x)} \sum_y \pi(y)q(y, x)1\big(h(y, x) \in B\big)1\big((y, x) \in \mathcal{T}_0\big).$$

The next three results are analogues of Theorems 4.10, 4.11, and 4.12. Their proofs are left as exercises for the reader.

Theorem 4.16. *If $M_+ \perp X_-$ or $M_- \perp X_+$, then M is a Poisson process.*

Theorem 4.17. *The M is a space–time Poisson process with rate a and space distribution F such that $M_+ \perp X_-$ if and only if*

$$\alpha(x, B) = aF(B), \quad x \in \mathbb{E},\ B \in \mathcal{E}'.$$

Theorem 4.18. *Suppose the Markov process X is ergodic and stationary with stationary distribution π. Then M is a space–time Poisson process with rate a and space distribution F such that $M_- \perp X_+$ if and only if*

$$\bar{\alpha}(x, B) = aF(B), \quad x \in \mathbb{E},\ B \in \mathcal{E}'.$$

We now characterize multivariate compound Poisson processes associated with $\mathcal{T}_0$-transitions. Consider the n-dimensional random measure $(M_1, \ldots, M_n)$ defined by

$$M_i(A) = \sum_{t \in A} h_i(X_{t-}, X_t)1((X_{t-}, X_t) \in \mathcal{T}_0), \quad A \subset \mathbb{R},\ 1 \leq i \leq n, \tag{4.12}$$

where h_i is a real-valued function on $\mathbb{E}^2$. This is an n-dimensional *compound Poisson process with rate a and atom distribution F* on $\mathbb{R}^n$ if it has independent increments in time and, for any $B_1, \dots, B_n$ and $s < t$ in $\mathbb{R}$,

$$P\{M_1(s,t] \in B_1, \dots, M_n(s,t] \in B_n\}$$
$$= \sum_{k=0}^{\infty} F^{k*}(B_1 \times \cdots \times B_n) a^k (t-s)^k \exp(-a(t-s))/k!.$$

The independent increments in time means that, for any disjoint time sets $A_1, \dots, A_n$, the vectors $(M_1(A_j), \dots, M_n(A_j))$, $1 \le j \le n$, are independent.

For such a process, it follows that each M_i is a compound Poisson process with rate $a_i = a[1 - F_i(0)]$ and atom distribution F_i, where F_i is the ith marginal distribution of F. Also, the M_i's are independent if and only if F is a product of its marginal distributions.

The following is an analogue of Theorem 4.18. A corresponding analogue of Theorem 4.17 is in Exercise 6.

Theorem 4.19. *Suppose the Markov process X is ergodic and stationary. Let π denote and its stationary distribution and, for $x, y \in \mathbb{E}$ and $B \subset \mathbb{R}^n$, define*

$$H(x,y) = (h_1(x,y), \dots, h_n(x,y)),$$
$$\bar{\alpha}(x,B) = \pi^{-1}(x) \sum_y \pi(y) q(y,x) 1\big((y,x) \in \mathcal{T}_0\big) 1\big(H(y,x) \in B\big).$$

Then $(M_1, \dots, M_n)$ is a compound Poisson process with rate a and atom distribution F such that $(M_1, \dots, M_n)_- \perp X_+$ if and only if

$$\bar{\alpha}(x,B) = aF(B), \quad x \in \mathbb{E},\ B \subset \mathbb{R}^n. \tag{4.13}$$

PROOF. Let M denote the space–time point process defined by (4.11) with $\mathbb{E}' = \mathbb{R}^n$ and $h = H$. We can also express M as

$$M(A \times B) = \sum_k 1\big(T_k \in A, (Y_k^1, \dots, Y_k^n) \in B\big),$$

where the T_k's denote the times of $\mathcal{T}_0$-transitions and $Y_k^i = h_i(X_{T_k-}, X_{T_k})$. Using this notation, we can write

$$M_i(A) = \sum_k Y_k^i 1(T_k \in A).$$

Now, by the preceding expressions and Exercise 3, it follows that $(M_1, \dots, M_n)$ is a compound Poisson process with rate a and atom distribution F such that $(M_1, \dots, M_n)_- \perp X_+$ if and only if M is a Poisson process on $\mathbb{R} \times \mathbb{R}^n$ such that $M_- \perp X_+$ and

$$E[M((s, s+t] \times B)] = atF(B), \quad \text{for each } s, t, \text{ and } B.$$

But the latter is equivalent to (4.13) by Theorem 4.18. □

The preceding result also applies to one-dimensional compound Poisson processes. Here is an application.

Example 4.20. *Busing System With Compound Poisson Departures.* Consider a Markovian queueing process whose state is the number of customers in the system, and its nonzero transition rates are

$$\begin{aligned} q(0,1) &= \lambda(1-p), \qquad\qquad q(x,x+1)=\lambda, \quad x\ge 1,\\ q(x,x-n) &= \mu p^{n-1}(1-p), \quad n=1,\dots,x-1,\\ q(x,0) &= \mu p^{x-1}, \quad x\ge 1. \end{aligned}$$

Here λ and μ are positive and $0 < p < 1$. This represents a system in which customers arrive by a Poisson process with rate λ and are served in batches as follows. When there are customers in the system, "buses" arrive at a rate μ to take them immediately from the system. Busing is common in computer systems and material handling systems. The number of customers each bus can take is a random variable with the geometric distribution $p^{n-1}(1-p)$, $n \ge 1$. When x units are present, the actual number that departs in a batch has the truncated geometric distribution $p^{n-1}(1-p)$, for $n < x$, and p^{n-1}, for $n = x$. Also, when there are no customers in the queue and a customer arrives, then with probability p there is a bus available to take the customer without delay. The process is ergodic and its stationary distribution is

$$\pi(x) = \pi(0)(1-p)\lambda^x/(\mu+p\lambda)^x, \quad x \ge 1, \tag{4.14}$$

provided $\lambda < \mu + p\lambda$, which we assume is true. One can prove this π is the stationary distribution by verifying that it satisfies the balance equations.

Assuming the process is stationary, consider the compound departure process

$$M(A) = \sum_{t\in A} \max\{0, X_t - X_{t-}\}, \quad A \subset \mathbb{R}.$$

This describes the total number of departures in the time set A; it records both the times at which batches of customers depart and the batch sizes. Then M is a compound Poisson process with rate $\lambda(1-p)$ and geometric atom distribution $r^{n-1}(1-r)$, $n \ge 1$, where $r = p\lambda/(\mu+p\lambda)$. This follows by Theorem 4.19 since, for each $x \in \mathbb{E}$ and $n \ge 1$,

$$\begin{aligned} \bar{\alpha}(x,n) &= \pi(x)^{-1}\sum_y \pi(y)q(y,x)[1(x=0, y=n) + 1(x\ge 1, y=x+n)]\\ &= \lambda(1-p)r^{n-1}(1-r). \end{aligned}$$

□

4.5 Poisson Flows in Jackson and Whittle Networks

For this section, we assume that X is an ergodic Jackson or Whittle process as in Chapter 1 that represents an open m-node network with unlimited capacity. We now apply the results in the preceding section to identify Poisson flows for this process. Recall that its transition rates are $q(x, T_{jk}x) = \lambda_{jk}\phi_j(x)$, for $j \ne k$ in

$M = \{0, 1, \ldots, m\}$, and its stationary distribution is

$$\pi(x) = c\Phi(x)\prod_{j=1}^{m} w_j^{x_j}, \quad x \in \mathbb{E}.$$

For each $j \neq k$ in M, we define the point process N_{jk} by

$$N_{jk}(A) = \sum_{t \in A} 1(X_t = T_{jk}X_{t-}), \quad A \subset \mathbb{R}.$$

The $N_{jk}(A)$ records the number of times units move from node j to node k in the time set A. We first consider the arrival and exit processes for the network.

Theorem 4.21. *Suppose X is a stationary Whittle process. Then the following statements are equivalent.*
(1) *The network's arrival processes $N_{01}, \ldots, N_{0m}$ are independent Poisson processes with respective rates $\lambda_{01}, \ldots, \lambda_{0m}$, and $(N_{01}, \ldots, N_{0m})_+ \perp X_-$.*
(2) *The network's exit processes $N_{10}, \ldots, N_{m0}$ are independent Poisson processes with respective rates $w_1\lambda_{10}, \ldots, w_m\lambda_{m0}$, and $(N_{10}, \ldots, N_{m0})_- \perp X_+$.*
(3) $\phi_0(\cdot) \equiv 1$.
These statements hold if X is a stationary Jackson process.

PROOF. Consider (2) in the setting of Theorem 4.15 applied to $N_{10}, \ldots, N_{m0}$. In this case,

$$\bar{\alpha}(x, \mathbf{u}) = \frac{1}{\pi(x)} \sum_y \pi(y)q(y, x)1\Big(1(y = x + e_j) = u_j, 1 \leq j \leq m\Big).$$

Then by the structure of π and the Φ-balance property, for $x \in \mathbb{E}$ and $1 \leq j \leq m$,

$$\bar{\alpha}(x, e_j) = \pi^{-1}(x)\pi(x + e_j)\lambda_{j0}\phi_j(x + e_j) = w_j\lambda_{j0}\phi_0(x).$$

Also, since only one unit may move at a time in the network, it follows that $\bar{\alpha}(x, \mathbf{u}) = 0$ when $\mathbf{u}$ is not a unit vector. Thus, (2) is equivalent to (3) by Theorem 4.15. Similarly, one can prove (1) is equivalent to (3), as we suggest in Exercise 8, by verifying (4.40) in Exercise 7. Finally, if X is a Jackson process, then (3) is true by assumption, and so (1) and (2) are true. □

The preceding result shows that Poisson arrival processes to the network beget Poisson exit processes. We now see that some internal flows in the network may also be Poisson processes. Suppose $\hat{M}$ is a sector of the network such that each unit exiting $\hat{M}$ never returns to $\hat{M}$. This holds if the routing process on M with rates λ_{jk} has the property that whenever it exits $\hat{M}$ it must enter the outside node 0 before it can return to $\hat{M}$ again. We will consider the flows N_{jk}, for $j \in \hat{M}$ and $k \in \hat{M}^c \equiv M\backslash\hat{M}$.

Let $\hat{X}$ denote the process X restricted to the nodes in $\hat{M}$; that is, $\hat{X}_t = \hat{x}$ if $X_t = x$, where $\hat{x} = (x_j : j \in \hat{M})$ denotes the restriction of x to $\hat{M}$. Then $\hat{X}$ is a network process with node set $\hat{M} \cup \{0\}$ and state space $\hat{\mathbb{E}} = \{\hat{x} : x \in \mathbb{E}\}$.

In case X is a Whittle process, we will make the following additional assumptions on the service rates ϕ_j, which are automatically satisfied when X is a Jackson process.

- ϕ_j is a function $\hat{\phi}_j(\hat{x})$ of only $\hat{x}$ if $j \in \hat{M}$.
- ϕ_k is a function $\hat{\phi}^c_k(\hat{x}^c)$ of only $\hat{x}^c$ if $0 \neq k \in \hat{M}^c$.
- $\phi_0(x) = \hat{\phi}^c_0(\hat{x}^c)$ for some function $\hat{\phi}^c_0$.
- $\{\hat{\phi}_j : j \in \hat{M} \cup \{0\}\}$ are $\hat{\Phi}$-balanced (all these functions are defined on $\hat{\mathbb{E}}$), where $\hat{\phi}_0(\cdot) \equiv 1$. Similarly, $\{\hat{\phi}^c_j : j \in \hat{M}^c \cup \{0\}\}$ are $\hat{\Phi}^c$-balanced.

These assumptions ensure that ϕ_j are Φ-balanced, where $\Phi(x) = \hat{\Phi}(\hat{x})\hat{\Phi}^c(\hat{x}^c)$.

Theorem 4.22. *Suppose X is a stationary Jackson process or a stationary Whittle process that satisfies the assumptions above. Then $\{N_{jk} : j \in \hat{M}, k \in \hat{M}^c\}$ are independent Poisson processes with respective rates $\{w_j\lambda_{jk} : j \in \hat{M}, k \in \hat{M}^c\}$. Furthermore, $(N_{jk} : j \in \hat{M}, k \in \hat{M}^c)_- \perp \hat{X}_+$.*

PROOF. Under the assumptions, $\hat{X}$ is an open stationary Whittle process on the node set $\hat{M}$. Indeed, its service rates $\hat{\phi}_j$ are $\hat{\Phi}$-balanced, and its routing rates are clearly

$$\hat{\lambda}_{j0} = \sum_{k\in\hat{M}^c} \lambda_{jk}, \quad \text{and} \quad \hat{\lambda}_{jk} = \lambda_{jk} \quad j, k \in \hat{M}.$$

Recall that, associated with the process X, there are w_j's that satisfy the traffic equations

$$w_j \sum_{k\in M} \lambda_{jk} = \sum_{k\in M} w_k\lambda_{kj}, \qquad j \in M.$$

Under the assumption that a unit exiting $\hat{M}$ cannot return to $\hat{M}$, the preceding equations are

$$w_j \sum_{k\in\hat{M}\cup\{0\}} \hat{\lambda}_{jk} = \sum_{k\in\hat{M}\cup\{0\}} w_k\hat{\lambda}_{kj}, \qquad j \in \hat{M} \cup \{0\}.$$

Consequently, the parameters w_j associated with $\hat{X}$ are the same as those for the larger process X. Thus, $\hat{X}$ is an ergodic Whittle process whose routing rates and w_j's are as above.

Now, by Theorem 4.21 and the assumption $\hat{\phi}_0(\cdot) \equiv 1$, we know that the exit processes $\hat{N}_{j0} = \sum_{k\in\hat{M}^c} N_{jk}$ for X are independent Poisson processes with respective rates $w_j\hat{\lambda}_{j0}$, $j \in J$. Next, observe that, for each $j \in J$, the N_{jk}'s form a partition of $\hat{N}_{j0}$ in which each point of $\hat{N}_{j0}$ is assigned to the process N_{jk} with probability $\hat{\lambda}_{jk}/\hat{\lambda}_{j0}$, for $k \in \hat{M}^c$. Thus, by the basic theorem on such partitions of Poisson processes (see Theorem 9.17), the N_{jk}'s are independent Poisson processes with respective rates

$$(\hat{\lambda}_{jk}/\hat{\lambda}_{j0})w_j\hat{\lambda}_{j0} = w_j\lambda_{jk}, \quad j \in J, k \in \hat{M}^c.$$

We also know by Theorem 4.21 that $(\hat{N}_{j0} : j \in \hat{M})_- \perp \hat{X}_+$, and therefore $(N_{jk} : j \in \hat{M}, k \in \hat{M}^c)_- \perp \hat{X}_+$. □

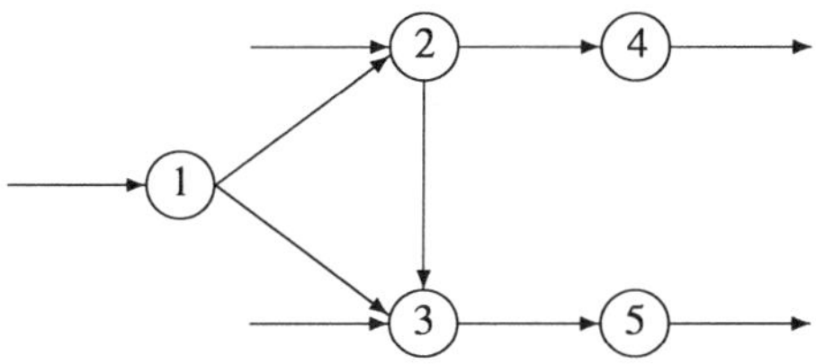

FIGURE 4.1. Open Acyclic Jackson Network

Example 4.23. *Open Acyclic Jackson Network*. Suppose X represents an open Jackson network as shown in Figure 4.1. The flows in the network are acyclic as shown by the arrows. Consequently, each unit can visit a node at most once.

Then for each node j, it follows from Theorem 4.22 with $\hat{M} = j$ that the flows N_{jk} for each k are independent Poisson processes with rates $w_j \lambda_{jk}$. In other words, the flow between each pair of nodes in the network is a Poisson process. This property is true for any network in which each unit can visit a node at most once. While some of the flows are independent, some of them are dependent. For instance, by Theorem 4.22, N_{23} and N_{24} are independent; and N_{40} and N_{50} are independent. However, N_{23} and N_{50} are not independent. In general, flows $N_{j_1 k_1}, \ldots, N_{j_n k_n}$ are independent if each unit appearing in one of these flows cannot appear in any of the others. To compute the rates of these Poisson processes one must obtain the w_j's from the traffic equations $w_j = \sum_k w_k p_{kj}$, where $w_0 = 1$ and $p_{jk} = \lambda_{jk} / \sum_\ell \lambda_{j\ell}$. An easy check shows that the solution for this example is

$$w_1 = p_{01}, \quad w_2 = p_{02} + w_1 p_{12}, \quad w_3 = p_{03} + w_1 p_{13}, \quad w_4 = w_2 p_{24}, \quad w_5 = w_3.$$

In particular, N_{35} has the rate

$$\lambda_{35}[\lambda_{03} + \lambda_{01}\lambda_{13}/(\lambda_{12} + \lambda_{13})]/[\lambda_{01} + \lambda_{02} + \lambda_{03})]. \qquad \square$$

4.6 Palm Probabilities for Markov Processes

In this section, we describe Palm probabilities of a stationary Markov process associated with certain point processes of its transition times. The following example illustrates the need for Palm probabilities and the types of issues we will address for networks.

Example 4.24. *Palm Probabilities for an $M/M/1$ System*. Suppose X_t represents the number of units in a stationary $M/M/1$ queueing system at time t. Consider the probability that the system contains $x + 1$ units at some time t "conditioned" that there is an arrival at time t. This probability is not a conventional conditional probability, since the probability of an arrival at any instant is 0. Therefore, it is natural to express this probability as the limiting conditional probability

$$P_N\{X_0 = x + 1\} = \lim_{s \uparrow t} P\{X_t = x + 1 | X_t = X_s + 1\}.$$

This is the Palm probability of the process given that there is an arrival at time 0. We define Palm probabilities by (4.17) below. The subscript N on P_N stands for the point process of arrival times. The stationarity of X ensures that the probability on the right side is independent of t.

From the discussion below, it follows that the preceding limiting probability can be expressed in terms of the stationary distribution of X, which is $\pi(x) = \rho^x(1-\rho)$, where ρ is the arrival rate divided by the service rate. Specifically, by expressions (4.16) and (4.19) below, it follows that

$$P_N\{X_0 = x+1\} = \pi(x)q(x, x+1) / \sum_y \pi(y)q(y, y+1) = \pi(x).$$

This says that the distribution of the number of units an arrival sees in the system in equilibrium is the same as the stationary distribution of X. In other words, *arrivals see time averages*.

Another quantity of interest is the time W that a typical arrival in equilibrium spends in the system (waiting and in service). This sojourn time is naturally described with respect to the Palm probability given that an arrival occurs at time 0. That is,

$$P_N\{W \leq t\} = \lim_{s \uparrow t} P\{W \leq t | X_t = X_s + 1\}.$$

From Exercise 9, this distribution is an exponential distribution with rate $\mu - \lambda$. Our aim is to develop similar results for networks. □

Throughout the rest of this section, we assume that the Markov process $\{X_t : t \in \mathbb{R}\}$ is stationary and ergodic, and we let π denote its stationary distribution. We will consider Palm probabilities of this process associated with certain types of transitions occurring at time 0. Such a transition may involve only the values of the process at a transition time, which we have been calling a $\mathcal{T}_0$-transition (the arrival event in the preceding example is a special case). More common are transitions that involve information about the past or future of the process. An example is a transition in a network process in which a unit arrives at a node j and no more units enter node j until that arrival exits node j.

We will describe a transition of the process X in terms of a set of sample paths or "trajectories" $\mathcal{T}$ as follows. Recall that a sample path of X is an element of the set D, and τ_n's denote the transition times of X. Also, $S_t X = \{X_{s+t} : s \in \mathbb{R}\}$ is the process X with its time parameter shifted by the amount t.

Definition 4.25. Suppose $\mathcal{T}$ is a subset of D such that $z(0) \neq z(0-)$, $z \in \mathcal{T}$. A *$\mathcal{T}$-transition of X occurs at time t* if $S_t X \in \mathcal{T}$. The point process N of times at which $\mathcal{T}$-transitions occur is defined by

$$N(A) = \sum_n 1(\tau_n \in A,\ S_{\tau_n} X \in \mathcal{T}), \quad A \subset \mathbb{R}.$$

We also write

$$N(A) = \sum_n 1(T_n \in A), \quad A \subset \mathbb{R},$$

where

$$\dots < T_{-2} < T_{-1} < T_0 \le 0 < T_1 < T_2 \dots$$

denote the *times of the $\mathcal{T}$-transitions*. Although N is a function of $\mathcal{T}$, we suppress the $\mathcal{T}$ for simplicity.

Keep in mind that the $\mathcal{T}$-transition times $\{T_n : n \in \mathbb{Z}\}$ are contained in the set of all transition times $\{\tau_n : n \in \mathbb{Z}\}$ of X. Also, a transition time τ_n of X is a $\mathcal{T}$-transition if the sample path of X centered at that time, which is $S_{\tau_n}X$, is in $\mathcal{T}$. Note that a $\mathcal{T}_0$-transition is a special case of a $\mathcal{T}$-transition. For instance, the arrival transition in the $M/M/1$ example above is a $\mathcal{T}_0$-transition, where $\mathcal{T}_0 = \{(x, x+1) : x \ge 0\}$, and this arrival transition is also a $\mathcal{T}$-transition, where $\mathcal{T} = \{z \in D : z(0) = z(0-)+1\}$. Another $\mathcal{T}$-transition for the $M/M/1$ queue is a transition at which a unit exits the system and there are no more arrivals during the next b time units. In this case, $\mathcal{T} = \{z \in D : z(0) = z(0-) - 1, z(t) \le z(t-), t \in (0, b]\}$, and this $\mathcal{T}$-transition is not a $\mathcal{T}_0$-transition.

Hereafter, we assume that $\mathcal{T}$ is a fixed subset of D and that N is the point process of $\mathcal{T}$-transitions of X. Since X is stationary, the distribution of the time-shifted process $S_t X$ is independent of t. From this and the preceding definition of N, it follows that N is a *stationary point process*; that is, the distribution of the time-shifted process $\{N(A+t) : A \subset \mathbb{R}\}$ is independent of t. This stationarity implies that

$$EN(A) = |A| EN(0, 1], \tag{4.15}$$

where $|A|$ is the Lebesgue measure of A. The expectation $\lambda_{\mathcal{T}} \equiv EN(0, 1]$ is called the *intensity of the $\mathcal{T}$-transitions*. By the extended Lévy formula (4.3), this intensity is

$$\begin{aligned} \lambda_{\mathcal{T}} &= E\sum_n 1(\tau_n \in (0, 1],\ S_{\tau_n}X \in \mathcal{T}) \qquad (4.16)\\ &= \sum_x \pi(x) \sum_{y \ne x} q(x, y) P\{S_{\tau_1}X \in \mathcal{T} | X_{\tau_0} = x, X_{\tau_1} = y\}. \end{aligned}$$

We will only consider $\mathcal{T}$-transitions whose intensity $\lambda_{\mathcal{T}}$ is finite and positive. Since $\mathcal{T}$-transitions are contained in all transitions of X, it follows by Lévy's formula that

$$\lambda_{\mathcal{T}} \le E\sum_n 1(\tau_n \in (0, 1]) = \sum_x \pi(x) \sum_{y \ne x} q(x, y).$$

The last quantity is the intensity of all transitions of X. Thus, the $\mathcal{T}$-transitions will have a finite intensity when the intensity of all transitions is finite, which is true for most applications.

The type of Palm probability that we use in this chapter is as follows.

Definition 4.26. The *Palm probability* P_N of the stationary Markov process X given that a $\mathcal{T}$-transition occurs at time 0 is defined by

$$P_N\{X \in \mathcal{T}'\} = \lambda_{\mathcal{T}'}/\lambda_{\mathcal{T}}, \quad \mathcal{T}' \subset \mathcal{T}. \tag{4.17}$$

We also call P_N the *Palm probability of X for the point process N.*

Note that $P_N\{X \in \mathcal{T}\} = 1$, which is consistent with saying that a $\mathcal{T}$-transition occurs at time 0. This justifies confining attention to events of the form $X \in \mathcal{T}'$, where $\mathcal{T}'$ is a subset of $\mathcal{T}$ rather than D. Equivalently, one could define P_N by

$$P_N\{X \in \mathcal{T}'\} = \lambda_{\mathcal{T}'\cap\mathcal{T}}/\lambda_{\mathcal{T}}, \quad \mathcal{T}' \subset D.$$

Keep in mind that N is a special type of point process of X whose time points are contained in the transition times of X. The intensities (4.16) of such point processes are tractable, but the intensities of point processes whose time points are not transition times of X are less tractable. Palm probabilities (4.17) for $\mathcal{T}$-transitions are adequate for the network analysis in this chapter. More involved applications, however, require the use of general Palm probabilities for point processes associated with stationary processes; see Chapter 5. In this general setting, the point process times need not be $\mathcal{T}$-transition times of a process as above.

Loosely speaking, the Palm probability (4.17) is the portion of $\mathcal{T}$-transitions that are also $\mathcal{T}'$-transitions. Another representation of the probability P_N in terms of the transition times τ_n is

$$P_N\{X \in \mathcal{T}'\} = \frac{E\sum_n 1(S_{\tau_n}X \in \mathcal{T}',\, \tau_n \in (0,1])}{E\sum_n 1(S_{\tau_n}X \in \mathcal{T},\, \tau_n \in (0,1])}, \quad \mathcal{T}' \subset \mathcal{T}.$$

This is the expected number of times an observer sees X in $\mathcal{T}'$ at transitions during $(0, 1]$ divided by the expected number of times the observer sees X in $\mathcal{T}$ at the transitions. The time interval $(0, 1]$ can be replaced by any time set A because the stationarity of N ensures that $EN(A) = |A|\lambda_{\mathcal{T}}$.

Expression (4.16) for intensities of transitions yields the following formulas. The probability that X has a transition from x to y at a $\mathcal{T}$-transition is

$$P_N\{X_{0-} = x, X_0 = y\} = \lambda_{\mathcal{T}}^{-1}\pi(x)q(x, y)P\{S_{\tau_1}X \in \mathcal{T}|X_{\tau_0} = x, X_{\tau_1} = y\}.$$

Also, when it exists,

$$E_N[f(X)] = \lambda_{\mathcal{T}}^{-1} \sum_{(x,y)\in\mathcal{T}} \pi(x)q(x, y)E[f(S_{\tau_1}X)|X_{\tau_0} = x, X_{\tau_1} = y]. \quad (4.18)$$

The following result shows that Palm probabilities are limits of conditional probabilities as we stated in Example 4.24.

Proposition 4.27. *For any $t \in \mathbb{R}$ and $\mathcal{T}' \subset \mathcal{T}$,*

$$P_N\{X \in \mathcal{T}'\} = \lim_{s\uparrow t} P\{S_tX \in \mathcal{T}'|S_tX \in \mathcal{T}, X_t \neq X_s\}. \quad (4.19)$$

PROOF. Since $\mathcal{T}' \subset \mathcal{T}$, it follows that expression (4.19) can be written as $P_N\{X \in \mathcal{T}'\} = \alpha_{\mathcal{T}'}/\alpha_{\mathcal{T}}$, where

$$\alpha_{\mathcal{T}} \equiv \lim_{s\uparrow t}(t-s)^{-1}P\{S_tX \in \mathcal{T}, X_t \neq X_s\}.$$

Therefore, to prove (4.19), it suffices to show $\alpha_{\mathcal{T}} = \lambda_{\mathcal{T}}$, for any $\mathcal{T} \subset D$. But this is true, since conditioning on X_s, X_t and using the stationarity of X, the definition of $q(x, y)$, and expression (4.16),

$$\alpha_{\mathcal{T}} = \lim_{s \uparrow t}(t - s)^{-1} \sum_{x} P\{X_s = x\} \sum_{y \neq x} [P\{X_t = y | X_s = x\}$$
$$\times P\{S_t X \in \mathcal{T} | X_s = x, X_t = y\}]$$
$$= \sum_{x} \pi(x) \sum_{y \neq x} q(x, y) P\{S_{\tau_1} X \in \mathcal{T} | X_{\tau_0} = x, X_{\tau_1} = y\}$$
$$= \lambda_{\mathcal{T}}.$$

□

The representation (4.19) of a Palm probability as a limit is consistent with the usual way of interpreting a probability conditioned on a continuous random variable. Expression (4.19) gives insight into the meaning of a Palm probability and it may be useful in theoretical studies. To evaluate Palm probabilities, however, one typically uses (4.18) or the Campbell–Mecke formulas in Chapter 5. The rest of the results in this section are also useful in this regard.

In some instances, it is convenient to obtain Palm probabilities by appealing to time reversals as follows.

Proposition 4.28. *Suppose $\bar{X}$ is the time reversal of X; that is, $\bar{X}_t = \lim_{s \uparrow -t} X_s$. Let $\bar{P}_{\bar{N}}$ denote the Palm probability of $\bar{X}$, where $\bar{N}$ denotes the point process of $\bar{\mathcal{T}}$-transitions of $\bar{X}$ and $\bar{\mathcal{T}} = \{z \in D : \{\bar{z}(t)\} \in \mathcal{T}\}$. Then*

$$P_N\{X \in \mathcal{T}'\} = \bar{P}_{\bar{N}}\{\bar{X} \in \bar{\mathcal{T}}'\}, \quad \mathcal{T}' \subset \mathcal{T}.$$

Proof. Let $N_{\mathcal{T}'}$ denote the point process of $\mathcal{T}'$-transitions of X. Then $N_{\mathcal{T}'}$ also equals the point process of $\bar{\mathcal{T}}'$-transitions of $\bar{X}$. Also, by their definitions, $N = \bar{N}$. Therefore,

$$P_N\{X \in \mathcal{T}'\} = E[N_{\mathcal{T}'}(0, 1]]/E[\bar{N}(0, 1]] = \bar{P}_{\bar{N}}\{\bar{X} \in \bar{\mathcal{T}}'\}, \quad \mathcal{T}' \subset \mathcal{T}.$$

□

We next consider Palm probabilities of events at the $\mathcal{T}$-transition times T_n. Since the Markov process X is stationary in the time parameter t, intuition might suggest that the sequence $\{X_{T_n} : n \in \mathbb{Z}\}$ of X-values at $\mathcal{T}$-transitions is stationary. This sequence, however, is not stationary under P—but it is stationary under the Palm probability P_N. A generalization of this property is as follows. This result is a special case of Theorem 6.9 for general Palm probabilities, and the strong law of large numbers, Theorem 6.1, for stationary sequences. Recall that $S_{T_n} X = \{X_{T_n + t} : t \in \mathbb{R}\}$.

Theorem 4.29. *For $h : D \to \mathbb{E}$, define*

$$Y_n = h(S_{T_n} X), \quad n \in \mathbb{Z}.$$

The sequence $\{Y_n : n \in \mathbb{Z}\}$ is stationary under P_N, and it satisfies the strong law of large numbers

$$\lim_{n \to \infty} n^{-1} \sum_{k=1}^{n} Y_k = E_N(Y_1), \quad \textit{w.p.1 under } P_N. \tag{4.20}$$

As an example of this result, the sequence $\{T_{n+1} - T_n : n \in \mathbb{Z}\}$ of times between $\mathcal{T}$-transitions of X is stationary under P_N. This sequence, however, is not stationary under P.

We now describe a large family of functionals of the process X at the times T_n that are stationary processes under P. The following result is a direct consequence of the Campbell–Mecke formula; see Theorem 6.13 and Example 6.15.

Theorem 4.30. *For $f : \mathbb{R} \times D \to \mathbb{R}$, define*

$$Y_t = \sum_n f(t - T_n, S_{T_n} X), \quad t \in \mathbb{R}.$$

The process $\{Y_t \in \mathbb{R}\}$ is stationary under P and it satisfies the strong law of large numbers

$$\lim_{t\to\infty} t^{-1} \int_0^t Y_s \, ds = EY_0, \quad \text{w.p.1 under } P. \tag{4.21}$$

Furthermore,

$$EY_0 = \lambda_{\mathcal{T}} E_N \int_0^{T_1} f(t, X)\, dt. \tag{4.22}$$

4.7 Sojourn and Travel Times of Markov Processes

In this section, we use Palm probabilities to describe expected sojourn and travel times of Markov processes.

Consider the stationary Markov process X that we have been studying in the preceding sections. Sojourn times of X have the following properties.

Theorem 4.31. *Let N denote the point process of entrance times $\{T_n : n \in \mathbb{Z}\}$ of X into a proper subset B of $\mathbb{E}$. Let W_n denote the sojourn time of X in B starting at time T_n. The sequence $\{W_n : n \in \mathbb{Z}\}$ is stationary under P_N, and it satisfies the strong law of large numbers (4.20). Furthermore,*

$$E_N(W_0) = \lambda^{-1} P\{X_0 \in B\}, \tag{4.23}$$

where $\lambda = \sum_{x \notin B} \pi(x) \sum_{y \in B} q(x, y)$.

Proof. The times T_n are $\mathcal{T}$-transition times, where $\mathcal{T} = \{z \in D : z(0-) \notin B, z(0) \in B\}$. The intensity of these transitions equals the λ as specified. Clearly, $W_n = h(S_{T_n} X)$, where $h(z) = \inf\{t > 0 : z(t) \notin B\}$ (the first entrance time to B^c). Then the first assertion follows from Theorem 4.29.

To prove (4.23), consider the process

$$Y_t \equiv 1(X_t \in B) = \sum_n 1(T_n \le t < T_n + W_n), \quad t \in \mathbb{R}.$$

Using the function $f(t - T_n, S_{T_n} X) \equiv 1(0 \le t - T_n < W_n)$, it is clear that Y satisfies the hypotheses of Theorem 4.30. Consequently, Y is a stationary process and

$$EY_0 = \lambda E_N \int_0^{T_1} 1(0 \le t < W_0)\, dt = \lambda E_N(W_0).$$

Also, since Y is stationarity, $EY_0 = P\{X_0 \in B\}$. These observations yield (4.23). □

We now consider an elementary travel time problem for the Markov process X. Let B and B' be nonempty, disjoint subsets of $\mathbb{E}$ whose union is not $\mathbb{E}$. Our interest is in determining the expected time it takes the process X to travel from B to B'. This problem is an example of general travel time problems for networks that we consider later in Section 6.6.

We first consider some related passage-time probabilities. Let $\alpha(x)$ denote the probability that conditioned on $X_0 = x$, the process enters B' before it enters B. Analogously, looking backward in time, let $\bar{\alpha}(x)$ denote the probability that conditioned on $X_0 = x$, the process X exited B prior to time 0, more recently than it exited B'. We will also use the probability $p(x, y) \equiv q(x, y)/\sum_{y'} q(x, y')$ of a transition of X from x to y.

Proposition 4.32. *The probabilities $\alpha(x)$ are the solution to the following equations: $\alpha(x) = 1$ or 0 according to whether x is in B' or in B, and*

$$\alpha(x) = \sum_{y \in B'} p(x, y) + \sum_{y \notin B'} p(x, y)\alpha(y), \qquad x \notin B \cup B'.$$

The probabilities $\bar{\alpha}(x)$ are the solution to the preceding equations, where the roles of B and B' reversed and $p(x, y)$ is replaced by

$$\bar{p}(x, y) = \pi(y)\pi(x)^{-1} p(y, x).$$

PROOF. The first assertion follows by conditioning on the probability $p(x, y)$ that the first jump of X is from x to y (a standard argument for Markov processes). To prove the second assertion, consider the time reversal of X, which is defined by $\bar{X}_t = \lim_{s \uparrow -t} X_s$, $t \in \mathbb{R}$. Clearly, $\bar{\alpha}(x)$ is the probability that conditioned on $\bar{X}_0 = x$, the process $\bar{X}$ enters B before it enters B'. Also, from Theorem 2.5 on time reversals, we know that $\bar{p}(x, y)$ is the probability of a transition of $\bar{X}$ from x to y. Then the second assertion follows from the first assertion applied to $\bar{X}$. □

We are now ready to determine the expected travel time from B to B'. Let N denote the point process of times $\{T_n : n \in \mathbb{Z}\}$ at which X exits B and subsequently enters B' before returning to B. Let W_n denote the sojourn time of X in the set $\mathbb{E}\backslash B \cup B'$ in the time interval $[T_n, T_{n+1})$. This W_n is the nth *travel time from B to B'* beginning at time T_n.

Corollary 4.33. *The sequence $\{W_n : n \in \mathbb{Z}\}$ of travel times from B to B' is stationary under P_N, and it satisfies the strong law (4.20). Furthermore,*

$$E_N(W_0) = \lambda^{-1} \sum_{x \notin B \cup B'} \pi(x)\alpha(x)\bar{\alpha}(x), \tag{4.24}$$

where $\lambda = \sum_{x\in B}\pi(x)\sum_{y\notin B} q(x,y)\alpha(y)$ *and* $\alpha(x)$, $\bar{\alpha}(x)$ *are the probabilities described in Proposition 4.32.*

PROOF. The point process N of times at which X begins traveling from B to B' is given by

$$N(A) = \sum_n 1(\tau_n \in A, X_{\tau_{n-1}} \in B, X_{\tau_n} \notin B, \eta_{B'}(\tau_n) < \eta_B(\tau_n)),$$

where $\eta_B(t) \equiv \inf\{u > t : X_u \in B\}$ is the first time after t that X enters B. By the extended Lévy formula (4.3), the intensity of N is

$$\lambda = EN(0,1] = \sum_{x\in B}\pi(x)\sum_{y\in B^c} q(x,y)P\{\eta_{B'}(\tau_1) < \eta_B(\tau_1) \mid X_{\tau_0} = x, X_{\tau_1} = y\}.$$

Clearly, the last conditional probability is the probability $\alpha(y)$ of entering B' before B conditioned on starting in state y.

Now, the travel time can be expressed as $W_n = \eta_{B'}(T_n)$, which is the time it takes for X to enter B' starting at time T_n. This travel time is of the form $W_n = h(S_{T_n}X)$, and so the first assertion of the corollary follows by Theorem 4.29.

Next, proceeding as in the proof of Theorem 4.31, we consider the process $Y_t = \sum_n 1(T_n \le t < T_n + W_n)$ and deduce by Theorem 4.30 that Y is a stationary process and $EY_0 = \lambda E_N(W_0)$. Now, another representation for Y_t is

$$Y_t = 1\left(X_t \notin B \cup B', \eta_{B'}(t) < \eta_B(t), \bar{\eta}_{B'}(t) < \bar{\eta}_B(t)\right),$$

where $\bar{\eta}_B(t) \equiv \sup\{s < t : X_s \in B\}$ is the last exit time of X from B prior to time t. Then

$$\begin{aligned} EY_0 &= P\{X_0 \notin B \cup B', \eta_{B'}(0) < \eta_B(0), \bar{\eta}_{B'}(0) < \bar{\eta}_B(0)\} \\ &= \sum_{x\notin B\cup B'} \pi(x)P\{\eta_{B'}(0) < \eta_B(0) | X_0 = x\} \\ &\qquad\qquad \times P\{\bar{\eta}_{B'}(0) < \bar{\eta}_B(0) | \bar{X}_0 = x\}. \end{aligned}$$

The last equality used the Markovian property that conditioned on the present state, the past and future of the process are independent. Clearly, the product of the last two conditional probabilites equals $\alpha(x)\bar{\alpha}(x)$, where these probabilities are described in Proposition 4.32. Thus, the preceding expression for EY_0 combined with $EY_0 = \lambda E_N(W_0)$ from above yield (4.24). □

4.8 Palm Probabilities of Jackson and Whittle Networks

The rest of this chapter discusses properties of Jackson and Whittle networks that involve Palm probabilities. This section characterizes network transitions under which a moving unit sees a time average. This property is applied to obtain distributions of sojourn times at nodes.

For this discussion, we assume that X is an ergodic, stationary Jackson or Whittle process that represents an open or closed network. As usual, we denote its transition rates by $q(x, T_{jk}x) = \lambda_{jk}\phi_j(x)$, and then its stationary distribution is

$$\pi(x) = c\Phi(x)\prod_{j=1}^{m} w_j^{x_j}, \quad x \in \mathbb{E}.$$

In addition, in case X is a Whittle process, we assume its service rates are of the form

$$\phi_j(x) = \Phi(x - e_j)/\Phi(x), \quad x \in \mathbb{E}. \tag{4.25}$$

This condition, which is automatically satisfied for Jackson processes, yields simpler formulas for the Palm probabilities we now derive.

We first consider the disposition of the units at a $\mathcal{T}$-transition of the network. In any transition, exactly one unit moves from some node j to a node k and the other units do not move. Such a transition can be expressed as a transition from $x + e_j$ to $x + e_k$, where x is the vector of unmoved units. Let $\mathbb{E}'$ denote the set of all such vectors of unmoved units. The probability distribution $\pi_{\mathcal{T}}$ of the unmoved units at a $\mathcal{T}$-transition is defined by

$$\pi_{\mathcal{T}}(x) = \sum_{j,k\in M} P_N\{X_{0-} = x + e_j, X_0 = x + e_k\}, \quad x \in \mathbb{E}'. \tag{4.26}$$

Here P_N is the Palm probability of the point process N of $\mathcal{T}$-transitions. We are interested in finding conditions under which this probability simplifies as follows.

Definition 4.34. We say that a *moving unit sees a time average* (MUSTA) at a $\mathcal{T}$-transition if $\pi_{\mathcal{T}} = \pi'$, where π' is the distribution defined on $\mathbb{E}'$ as follows.
• $\pi' = \pi$ and $\mathbb{E}' = \mathbb{E}$ if the network is open with unlimited capacity.
• $\pi' = \pi_{\nu-1}$ and $\mathbb{E}' = \{x : |x| = \nu - 1\}$ if the network is closed with ν units.
• $\pi' = \pi_{\nu-1}$ and $\mathbb{E}' = \{x : |x| \le \nu - 1\}$ if the network is open with capacity ν.
In the last two cases, $\pi_{\nu-1}$ is the stationary distribution of a closed network with $\nu - 1$ units, or an open network with capacity $\nu - 1$, respectively.

The MUSTA property $\pi_{\mathcal{T}} = \pi'$ implies that $\pi_{\mathcal{T}}$ is independent of $\mathcal{T}$ since π' is. In the first case where $\pi' = \pi$, MUSTA says that a moving unit sees the disposition of the unmoved units as if they came from the same type of open network with unlimited capacity. An example of this is that arrivals to an $M/M/1$ queue see time averages; recall Example 4.24. Similarly, in the second and third cases where $\pi' = \pi_{\nu-1}$, MUSTA says that a moving unit in a $\mathcal{T}$-transition sees the unmoved units distributed according to the equilibrium distribution of a network with one less unit. In each of these three cases, we have

$$\pi'(x) = c'\Phi(x)\prod_{j=1}^{m} w_j^{x_j}, \quad x \in \mathbb{E}',$$

where c' is the normalization constant. Thus, the moving unit sees an average (the stationary distribution) of the same type of network as its parent process. Related terms in the literature are ASTA (arrivals see time averages), ESTA (events see time averages), and PASTA (Poisson arrivals see time averages).

To characterize the MUSTA property, we will use the following representation of the rate of $\mathcal{T}$-transitions for the network process X.

Proposition 4.35. *The rate of $\mathcal{T}$-transitions is*

$$\lambda_{\mathcal{T}} = \sum_{x \in \mathbb{E}'} \pi'(x) \gamma_{\mathcal{T}}(x), \tag{4.27}$$

where

$$\gamma_{\mathcal{T}}(x) = \frac{c}{c'} \sum_{j,k} w_j \lambda_{jk} P\{S_{\tau_1} X \in \mathcal{T} | A_{jk}(x)\},$$

and $A_{jk}(x) = \{X_{\tau_0} = x + e_j, X_{\tau_1} = x + e_k\}$, for $x \in \mathbb{E}'$, $j, k \in M$.

PROOF. Using the notation above on unmoved units at a transition,

$$\begin{aligned} \lambda_{\mathcal{T}} &= \sum_{x \in \mathbb{E}} \sum_{j,k} \pi(x) q(x, T_{jk}x) P\{S_{\tau_1} X \in \mathcal{T} \mid X_{\tau_0} = x, X_{\tau_1} = T_{jk}x\} \\ &= \sum_{x \in \mathbb{E}'} \sum_{j,k} [\pi(x + e_j) q(x + e_j, x + e_k)] \\ &\qquad \times P\{S_{\tau_1} X \in \mathcal{T} | A_{jk}(x)\}. \end{aligned} \tag{4.28}$$

Using $\Phi(x + e_j)\phi_j(x + e_j) = \Phi(x)$ from (4.25), and the expressions above for q, π, and π', it follows that the term in brackets in (4.28) equals

$$c\Phi(x + e_j) w_j \prod_{\ell=1}^{m} w_\ell^{x_\ell} \lambda_{jk} \phi_j(x + e_j) = \frac{c}{c'} w_j \lambda_{jk} \pi'(x).$$

Then substituting the last expression back into (4.28) yields (4.27). □

The distribution of the process X under the Palm probability of the point process N of $\mathcal{T}$-transitions is as follows.

Corollary 4.36. *The Palm probability P_N of X is given by*

$$P_N\{X \in \mathcal{T}'\} = \sum_{x \in \mathbb{E}'} \pi'(x) \gamma_{\mathcal{T}'}(x) / \sum_{y \in \mathbb{E}'} \pi'(y) \gamma_{\mathcal{T}}(y), \quad \mathcal{T}' \subset \mathcal{T}. \tag{4.29}$$

PROOF. This follows from Proposition 4.35 since $P_N\{X \in \mathcal{T}'\} = \lambda_{\mathcal{T}'}/\lambda_{\mathcal{T}}$. □

The following is a characterization of the MUSTA property.

Theorem 4.37. *The $\mathcal{T}$-transition has the MUSTA property if and only if $\gamma_{\mathcal{T}}(x)$ is independent of x. In this case,*

$$P_N\{X \in \mathcal{T}'\} = \sum_{x \in \mathbb{E}'} \pi'(x) \sum_{j,k \in M} p(j,k) P\{S_{\tau_1} X \in \mathcal{T}' | A_{jk}(x)\}, \tag{4.30}$$

where

$$p(j,k) = w_j \lambda_{jk} / \sum_{j',k' \in M} w_{j'} \lambda_{j'k'}, \quad j, k \in M.$$

PROOF. In light of (4.29), the distribution (4.26) of the unmoved units at a $\mathcal{T}$-transition is

$$\pi_{\mathcal{T}}(x) = \sum_{j,k \in M} P_N\{X_{0-} = x + e_j,\, X_0 = x + e_k\}$$
$$= \pi'(x)\gamma_{\mathcal{T}}(x) / \sum_{y \in \mathbb{E}'} \pi'(y)\gamma_{\mathcal{T}}(y).$$

Then $\pi_{\mathcal{T}} = \pi'$ (the $\mathcal{T}$-transition has the MUSTA property) if and only if $\gamma_{\mathcal{T}}(x)$ is independent of x. In this case, a little thought shows that (4.29) reduces to (4.30). □

The probability $p(j, k)$ above has an interpretation in terms of the Markov routing process on M whose transition rates are λ_{jk}. Namely, $p(j, k)$ is the Palm probability of a route transition from j to k given there is a transition. Loosely speaking, $p(j, k)$ is the stationary probability that the routing process goes from j to k at a transition.

Example 4.38. *Simple Network Transitions.* Consider a transition of the network in which a unit moves from node j to node k for some (j, k) in a set $\chi \subset M^2$. We call this a *simple network transition in* χ. This is a $\mathcal{T}$-transition, where

$$\mathcal{T} = \{\{x_t\} \in D : x_0 = T_{jk}x_{0-}, \text{ for some } (j, k) \in \chi\}.$$

No other sample path information aside from χ is needed to describe this transition. In this case,

$$\gamma_{\mathcal{T}}(x) = \frac{c}{c'} \sum_{(j,k) \in \chi} w_j \lambda_{jk},$$

which is independent of x. Thus, there is MUSTA at any simple network transition. □

Some elementary applications of Theorem 4.37 are as follows.

Example 4.39. *Palm Probabilities for Action at a Node.* Consider a fixed node $j \neq 0$ in the network. Let P_j denote the Palm probability of the network transition at which a unit enters node j. This is a simple network transition in $M \times \{j\}$, and so it has the MUSTA property as we saw in the preceding example. Let X_t^j denote the number of units at node j at time t. Then it follows by (4.30) that the probability that an arrival to j sees n customers there is

$$P_j\{X_0^j = n + 1\} = \sum_{x \in \mathbb{E}'} \pi'(x) 1(x_j = n). \tag{4.31}$$

Next, consider the sojourn time W_j at node j of a unit that arrives in equilibrium. Then by (4.30), its distribution is

$$P_j\{W_j \le t\} = \sum_{x \in \mathbb{E}'} \pi'(x) \sum_{i \in M} p(i, j) P\{W_j \le t | A_{ij}(x)\}, \quad t \in \mathbb{R}.$$

A more specific evaluation of this probability requires knowledge about how node j processes units. Assume node j serves units on a first-come, first-served basis and that a unit's sojourn time has a distribution $F_j(t|n)$ that depends only on the number of units n at the node at the beginning of the sojourn (independent of later arrivals and the rest of the network). Then the preceding expression is

$$P_j\{W_j \le t\} = \sum_{x \in \mathbb{E}'} \pi'(x) F_j(t|x_j + 1). \tag{4.32}$$

Now, let us consider these probabilities as seen by a unit departing from node j. Let $\bar{P}_j$ denote the Palm probability of a network transition at which a unit exits node j. In addition to the assumptions above on the services, suppose units exit node j in the same order in which they arrive. For a unit exiting node j, consider the time $\bar{W}_j$ it just spent at node j. We could use the formula (4.30) as we did above to obtain a general expression for $\bar{P}_j\{\bar{W}_j \le t\}$, but it would be difficult to evaluate (try it for an $\cdot/M/s$ node). Instead, we will take another approach and prove

$$\bar{P}_j\{\bar{W}_j \le t\} = P_j\{W_j \le t\}. \tag{4.33}$$

This says that the *backward-looking* sojourn time is equal in distribution to the *forward-looking* sojourn time distribution.

Consider the process $\bar{X}_t = \lim_{s \uparrow -t} X_s$, which is the time reversal of X. This is an ergodic, stationary Markov process with the same stationary distribution π as X. Furthermore, $\bar{X}$ is the same type of network process as X, since its transition rates are

$$\bar{q}(x, T_{jk}x) = \pi(x)^{-1}\pi(T_{jk}x)q(T_{jk}x, x) = \bar{\lambda}_{jk}\phi_j(x),$$

where $\bar{\lambda}_{jk} = w_j^{-1} w_k \lambda_{kj}$. The w_j's that satisfy the traffic equations for λ_{jk} also satisfy the traffic equations for $\bar{\lambda}_{jk}$, since the latter is the time reversal of the former. In addition, the process $\bar{X}$ has the property that units exit node j in the same order as they arrive.

Because of this structure of $\bar{X}$, it is clear that $\bar{W}_j$ is the sojourn time of an arrival into j for the process $\bar{X}$. Also, $\bar{P}_j$ is a Palm probability of $\bar{X}$. By Proposition 4.28, we know that $\bar{P}_j\{\bar{X} \in \mathcal{T}\} = P_j\{X \in \bar{\mathcal{T}}\}$, where $\bar{\mathcal{T}} = \{\{x_t\} \in D : \{\bar{x}_t\} \in \mathcal{T}\}$. These observations prove the assertion (4.33) that the backward-looking sojourn time in j is the same as the forward-looking sojourn time. A similar argument shows that the probability that a departing unit sees n units at j is $\bar{P}_j\{\bar{X}_0^j = n\} = P_j\{X_0^j = n\}$. □

Example 4.40. *Sojourn Times in a $\cdot/M/s$ Node.* Suppose the process X represents an open Jackson network with unlimited capacity. Assume node j is a $\cdot/M/s$ node, where each of the s servers works at rate μ. Let W_j^* denote the length of time an arrival to j must wait in the queue before its service. Since $\pi' = \pi$ is a product form and j is an s-server node, it follows by (4.31) that

$$P_j\{X_0^j = n\} = \pi_j(n),$$

where

$$\pi_j(n) = c(w_j/\mu)^n/n!, \quad n \le s,$$

$$\pi_j(n) = \pi(s)(w_j/s\mu)^{n-s}, \quad n > s.$$

Then

$$P_j\{W_j^* = 0\} = P_j\{X_0^j < s\} = \sum_{n=0}^{s-1} \pi_j(n).$$

Also, conditioning on X_0, we have

$$P_j\{W_j^* > t\} = \sum_{n=s}^{\infty} \pi_j(n) P\{W_j^* > t | X_0 = n+1\}.$$

Clearly, $P\{W_j^* > t | X_0 = n+1\} = P\{N(t) < n-s\}$, where N is a Poisson process with rate $s\mu$, and $N(t) < n-s$ is the event that there are fewer than $n-s$ service completions in time t (the arrival at time 0 is still in the queue at time t). Substituting this Poisson probability in the preceding equation and using a little algebra, it follows that

$$P_j\{W_j^* > t\} = P_j\{X_0^j \geq s\} P\{\xi^* > t\}, \quad t \geq 0,$$

where ξ^* is an exponential random variable with rate $s\mu - w_j$ that represents the waiting time given that a unit has to wait.

Next, consider the sojourn time W_j at node j of an arrival. From (4.32), it follows that

$$P_j\{W_j \leq t\} = \sum_{n=0}^{\infty} \pi_j(n) F_j(t|n+1).$$

Then one can show, as described in Exercise 9, that

$$P_j\{W_j \leq t\} = P\{\xi \leq t\} P_j\{X_0^j < s\} + P_j\{X_0^j \geq s\} P\{\xi + \xi^* \leq t\},$$

where ξ is an exponential service time with rate μ that is independent of the exponential waiting time ξ^*. Note that in case $s = 1$, it follows (Exercise 9) that the distribution $P_j\{W_j \leq t\}$ is exponential with rate $\mu - w_j$. This exponential sojourn time is a Jackson network version of the exponential sojourn time in a single $M/M/1$ queue as in Exercise 9. Also, from the preceding example, we know that the backward-looking sojourn time distribution is the same as its forward-looking sojourn time. In other words, a unit departing from node j looking backward in time, and not knowing the past, can only surmise that its sojourn time was exponentially distributed with rate $\mu - w_j$. □

4.9 Travel Times on Overtake–Free Routes

In this section we continue our study of the stationary network process X, which represents an open or closed Jackson or Whittle network. The previous section described sojourn times at isolated nodes in the network. The focus now is on the joint distribution of sojourn times at nodes on a certain type of route in the network.

We begin with a preliminary example of the main result below.

Example 4.41. *Sojourns in Tandem Jackson Networks with Single-Server Nodes.* Suppose the process X represents an open tandem network with unlimited capacity in which all units enter at node 1 and proceed to nodes $2, \ldots, m$ in that order. Assume each node j is a single-server node with service rate μ_j and it operates under a first-in, first-out discipline. Let P_1 denote the Palm probability of the transition in which a unit enters node 1 at time 0. For such an arrival into node 1, let $W_1, \ldots, W_m$ denote its sojourn times at the respective nodes. From Theorem 4.22, it follows that the flow of units between each pair of nodes j and $j+1$ is a Poisson process, and these flows are clearly dependent. Then each node j in equilibrium operates like an $M/M/1$ process. Consequently, the sojourn time of a unit at each node, as we saw in Example 4.39, is exponentially distributed. Furthermore, by Theorem 4.43 below, the sojourn times $W_1, \ldots, W_m$ under P_1 are independent exponential random variables with respective rates $\mu_1 - w_1, \ldots, \mu_m - w_m$. It is surprising that these times are independent since the node populations are dependent and the flows between the nodes are dependent. □

Our terminology for routes will be as follows. A *simple route* of the network is a vector $r = (r_1, \ldots, r_\ell)$ of nodes in $\{1, \ldots, m\}$ such that $\lambda_{r_1 r_2} \cdots \lambda_{r_{\ell-1} r_\ell} > 0$. A unit *traverses the route* r if it enters node r_1 and then proceeds to nodes $r_2, \ldots, r_\ell$ in that order in its next $\ell - 1$ moves. We will consider a unit's sojourn times at nodes on the following type of overtake–free route in which units that traverse the route finish it in the same order in which they start it. Furthermore, a unit's sojourn time at any node on the route is not affected, even indirectly, by the presence of units that start the route later than it did.

Definition 4.42. A simple route $r = (r_1, \ldots, r_\ell)$ is *overtake–free* if it satisfies the following conditions:
(a) The nodes $r_1, \ldots, r_\ell$ are distinct, and each one serves units on a first-in, first-out basis. The service times at node r_i are independent exponentially distributed with rate $\phi_{r_i}(x_{r_i}) = \mu_{r_i}$, independent of x_{r_i}.
(b) For $s < \ell$, each feasible path from r_s to any $i \in \{r_1, \ldots, r_\ell\}$ must pass through r_{s+1}. That is, if $\lambda_{r_s j_1} \lambda_{j_1 j_2} \cdots \lambda_{j_n i} > 0$, then $r_{s+1} \in \{j_1, \ldots, j_n, i\}$.
(c) The service intensities of the nodes satisfy the following conditions, which automatically hold for Jackson networks. For each $s = 1, \ldots, \ell - 1$, let B_s denote the set of all nodes on a feasible path from r_s to r_{s+1} that contains r_s, r_{s+1} only at the beginning and end nodes, respectively, and B_s contains r_s but not r_{s+1}. Think of B_s as the set of nodes "between r_s and r_{s+1}". For each $j \in B_s$, the service rate $\phi_j(x)$ is independent of x_k, for $k \notin B_s$. And for each $k \notin B_1 \cup \cdots \cup B_{\ell-1}$, the rate $\phi_k(x)$ is independent of x_j, for $j \in B_1 \cup \cdots \cup B_{\ell-1}$.

We are now ready for our main result on travel times. Consider an overtake–free route $r = (r_1, \ldots, r_\ell)$ of the stationary network process X. Let P_{r_1} denote the Palm probability of a network transition in which a unit enters node r_1 at time 0 and traverses the route r. Let $W_{r_1}, \ldots, W_{r_\ell}$ denote the sojourn times at the respective nodes on the route for that unit. Finally, let $F(t|\mu, n)$ denote the Erlang distribution with parameters μ and n. Its density is

$$\frac{dF(t|\mu, n)}{dt} = \mu(\mu t)^{n-1}e^{-\mu t}/(n-1)!, \quad t \geq 0.$$

Theorem 4.43. *If the process X represents an open network with unlimited capacity, then the sojourn times $W_{r_1}, \ldots, W_{r_\ell}$ under the Palm probability P_{r_1} are independent exponential random variables with respective rates $\mu_{r_1} - w_{r_1}, \ldots, \mu_{r_\ell} - w_{r_\ell}$. If X represents a closed network with ν units (or a ν-capacity open network), then, for $t_1, \ldots, t_\ell$ in $\mathbb{R}$,*

$$P_{r_1}\{W_{r_1} \leq t_1, \ldots, W_{r_\ell} \leq t_\ell\} = \sum_{x \in \mathbb{E}'} \pi_{\nu-1}(x) \prod_{i=1}^{\ell} F(t_i|\mu_{r_i}, x_{r_i} + 1). \tag{4.34}$$

PROOF. For simplicity, renumber the nodes such that $(r_1, \ldots, r_\ell) = (1, \ldots, \ell)$. Let $F(t_1, \ldots, t_\ell) = P_1\{W_1 \leq t_1, \ldots, W_\ell \leq t_\ell\}$. Now, the two assertions of the theorem can be stated as the single assertion that

$$F(t_1, \ldots, t_\ell) = \sum_{x \in \mathbb{E}'} \pi'(x) \prod_{i=1}^{\ell} F(t_i|\mu_i, x_i + 1). \tag{4.35}$$

This is the second assertion with $\pi' = \pi_{\nu-1}$, and it is the first assertion with $\pi' = \pi$ since that assertion is

$$F(t_1, \ldots, t_\ell) = \prod_{i=1}^{\ell}(1 - e^{-(\mu_i - w_i)t_i}) = \sum_{x \in \mathbb{E}} \pi(x) \prod_{i=1}^{\ell} F(t_i|\mu_i, x_i + 1).$$

The last equality follows as in Example 4.40 since each node i on the route is a single-server with rate μ_i.

We will prove (4.35) by induction on the route length ℓ for all networks of the type we are considering. Clearly (4.35) is true for $\ell = 1$ by (4.32).

Now assume (4.35) is true for routes of length $1, \ldots, \ell - 1$, for some ℓ. Consider a route of length ℓ for the network process X. Let P_2 denote the Palm probability of a network transition in which a unit traversing the route departs from node 1 and enters node 2 at time 0. Let $W_1^*, \ldots W_\ell^*$ denote the sojourn times $W_1, \ldots, W_\ell$ viewed by that unit. That is

$$F(t_1, \ldots, t_\ell) = P_2\{W_1^* \leq t_1, \ldots, W_\ell^* \leq t_\ell\}.$$

It follows by Theorem 4.37 that the transition of a unit moving from node 1 to node 2 has the MUSTA property, since it is a simple network transition as described in Example 4.38. Then expressing the last probability as in (4.30), we have

$$F(t_1, \ldots, t_\ell) = \sum_{x \in \mathbb{E}'} \pi'(x) P\{W_1^* \leq t_1, \ldots, W_\ell^* \leq t_\ell | A_{12}(x)\}.$$

Here $A_{12}(x) = \{X_{\tau_0} = x + e_1, X_{\tau_1} = x + e_2\}$ is the event that a unit moves from node 1 to node 2 at time τ_1 and the disposition of the unmoved units is x. Let $J = B_1$ be the set of nodes between 1 and 2 (recall the definition of an overtake–free route), and let $K = \{1, \ldots, m\} \backslash J$. By the assumed structure of the service intensities ϕ_j on an overtake–free route, we can factor π' as

$$\pi'(x) = c'\Phi(x)\prod_{j=1}^{m} w_j^{x_j} = \frac{c'}{c'_J c'_K}\pi'_J(x_J)\pi'_K(x_K), \tag{4.36}$$

where $x_J \equiv (x_j : j \in J)$ and $\pi'_J(x_J) = c'_J\Phi_J(x_J)\prod_{j\in J} w_j^{x_j}$ for x_J in $\mathbb{E}'_J = \{x_J : x \in \mathbb{E}'\}$. The π_K is defined similarly on the set $\mathbb{E}'_K(x_J)$ of all x_K such that $x \in \mathbb{E}'$ and, $|x_K| \le \nu - |x_J| - 1$ if the network is closed with ν units or open with capacity ν. Also, $\Phi(x) = \Phi_J(x_J)\Phi_K(x_K)$.

Then conditioned on the event $A_{12}(x)$, the W_1^* is independent of $W_2^*, \dots, W_\ell^*$; W_1^* depends only on x_J; and $W_2^*, \dots, W_\ell^*$ depends only on x_K. Using this and the factored form of π', we can write

$$F(t_1, \dots, t_\ell) = \frac{c'}{c'_J c'_K} G(t_1)H(t_2, \dots, t_\ell), \tag{4.37}$$

where

$$G(t_1) = \sum_{x_J \in \mathbb{E}'_J} \pi'_J(x_J)P\{W_1^* \le t_1 | A_{12}(x)\},$$

$$H(t_2, \dots, t_\ell) = \sum_{x_K \in \mathbb{E}'_K(x_J)} \pi'_K(x_K)P\{W_2^* \le t_2, \dots, W_\ell^* \le t_\ell | A_{12}(x)\}.$$

We now show that G and H have forms like (4.35). Let X_t^J be an open Whittle process on the nodes J with state space $\mathbb{E}_J = \{x_J : x \in \mathbb{E}\}$ and transition rates

$$q(x_J, T_{jk}x_J) = \lambda_{jk}^J \phi_j(x_J),$$

where

$$\lambda_{jk}^J = \lambda_{jk}, \quad \lambda_{0k}^J = \sum_{i \notin J} w_i \lambda_{ik}, \quad \lambda_{j0}^J = \sum_{\ell \notin J} \lambda_{j\ell}, \quad j, k \in J.$$

A solution to the traffic equations for λ_{jk}^J is $w_j^J = w_j$ $(j \in J \cup \{0\})$, where w_j $(j \in M)$ is a solution to the equations for λ_{jk}. Consequently, the stationary distribution of X^J is

$$\pi_J(x_J) = c_J \Phi_J(x_J)\prod_{j \in J} w_j^{x_j}.$$

Assume X^J is stationary and let $\bar{P}_1$ denote the Palm probability of a simple network transition of X^J at which a unit departs from node 1 at time 0. Let $\bar{W}_1^J$ denote the sojourn time of that unit at node 1. Then by the definition of G, expression (4.33) for the time-reversal process, and the induction hypothesis for one node, we have

$$\begin{aligned} G(t_1) &= \bar{P}_1\{\bar{W}_1^J \le t_1\} = P_1\{W_1^J \le t_1\} \\ &= \sum_{x_J \in \mathbb{E}'_J} \pi'_J(x_J)F(t_1 | \mu_1, x_1 + 1). \end{aligned} \tag{4.38}$$

Next, define an open Whittle process X_t^K on the node set K with state space $\mathbb{E}^K(x_J)$ and capacity $\nu - |x_J|$. Here x_J is fixed. Assume X^K is stationary and let P_2

be the Palm probability of the transition of X^K in which a unit enters node 2 at time 0 and traverses the route $2, \ldots, \ell$, which is overtake–free. Let $W_2^K, \ldots, W_\ell^K$ denote that unit's sojourn times at the respective nodes $2, \ldots, \ell$. Then by the definition of H and the induction hypothesis for overtake–free routes of length $\ell - 1$, we have

$$H(t_2, \ldots, t_\ell) = P_2\{W_2^K \leq t_2, \ldots, W_\ell^K \leq t_\ell\}$$
$$= \sum_{x_K \in \mathbb{E}'_K(x_J)} \pi'_K(x_K) \prod_{i=2}^{\ell} F(t_i|\mu_i, x_i + 1). \tag{4.39}$$

Substituting (4.38) and (4.39) in (4.37) yields

$$F(t_1, \ldots, t_\ell) = \frac{c'}{c'_J c'_K} \sum_{x_J \in \mathbb{E}'_J} \pi'_J(x_J) F(t_1|\mu_1, x_1 + 1)$$
$$\times \sum_{x_K \in \mathbb{E}'_K(x_J)} \pi'_K(x_K) \prod_{i=2}^{\ell} F(t_i|\mu_i, x_i + 1).$$

Finally, because of (4.36), we can bring these summations together as a sum on $x \in \mathbb{E}'$ to obtain (4.35). This completes the induction argument. □

4.10 Exercises

1. *Dynkin's Formula.* Use Lévy's formula (4.2) to prove that, for a function $f : \mathbb{E} \to \mathbb{R}$,

$$E[f(X_t) - f(X_0)] = E\{\int_0^t [\sum_{y \neq X_s} q(X_s, y)(f(y) - f(X_s))]\, ds\},$$

provided the last expectation exists. This formula also holds when t is replaced by a stopping time of X.

2. In the context of Theorem 4.10, suppose $N_+ \perp X_-$. Prove that N is a Poisson process with rate a if and only if

$$EN(s, t] = a(t - s), \quad s < t.$$

3. *Characterization of Compound Poisson Processes.* Suppose M is a point process on $\mathbb{R}^{n+1}$ with point locations $(T_k, Y_k^1, \ldots, Y_k^n)$, $k \geq 1$. Define random measures $M_1, \ldots, M_n$ on $\mathbb{R}$ by

$$M_j(A) = \sum_k Y_k^j 1(T_k \in A), \quad A \subset \mathbb{R},\ 1 \leq j \leq n.$$

Prove that $(M_1, \ldots, M_n)$ is a compound Poisson process with rate a and atom distribution F on $\mathbb{R}^{n+1}$ if and only if M is a Poisson process with

$$EM((0, t] \times B) = atF(B), \quad \text{for each } t \text{ and } B.$$

4. *Queue with Compound Poisson Arrivals and Poisson Departures.* Consider a Markovian queueing process whose state is the number of customers in the system and whose transition rates are

$$q(x, y) = \lambda_x p^{n-1}(1-p)1(y = x+n) + \mu_x 1(y = x-1 \geq 0).$$

Here λ_x and μ_x are positive and $0 < p < 1$. This process represents a system in which batches of customers arrive at the rate λ_x when there are x customers present, and the number of customers in a batch has a geometric distribution with parameter p. Also, customers depart at the rate μ_x when x are in the system. Show that the stationary distribution of the process is

$$\pi(x) = \pi(0)\lambda_0\mu_1^{-1}\cdots\mu_x^{-1}\prod_{k=1}^{x-1}(\lambda_k + p\mu_k), \quad x \geq 1,$$

provided the sum of these terms over x is finite, which we assume is true. Next, assume the process is stationary and $\lambda_0 = \lambda_x + p\mu_x = a$, for each $x \geq 1$. Show that the times of customer departures form a Poisson process with rate a.

5. For the process in Example 4.20, show that its stationary distribution is given by (4.14).

6. Consider the random measures $M_1, \ldots, M_n$ defined by (4.12). Show that $(M_1, \ldots, M_n)$ is an n-dimensional compound Poisson process with rate a and atom distribution F such that $(M_1, \ldots, M_n)_+ \perp X_-$ if and only if, for each $x \in E$ and $B_1 \times \cdots \times B_n \in \mathbb{R}^n$,

$$\sum_y q(x, y)1((x, y) \in \mathcal{T}_0)1(h_i(x, y) \in B_i, 1 \leq i \leq n) = aF(B_1 \times \cdots \times B_n).$$

7. Consider the point processes N_i of $\mathcal{T}_0{}^i$-transitions as in Theorem 4.15 and define

$$\alpha(x, \mathbf{u}) = \sum_y q(x, y)1\Big(1\big((x, y) \in \mathcal{T}_0{}^i\big) = u_i,\ 1 \leq i \leq n\Big).$$

Show that $N_1, \ldots, N_n$ are independent Poisson processes with respective rates $a_1, \ldots, a_n$ such that $(N_1, \ldots, N_n)_+ \perp X_-$ if and only if, for each $x \in E$ and $\mathbf{u} \in \{0, 1\}^n$,

$$\alpha(x, \mathbf{u}) = \begin{cases} a_i & \text{if } \mathbf{u} = e_i \text{ for some } 1 \leq i \leq n \\ 0 & \text{otherwise.} \end{cases} \tag{4.40}$$

8. In Corollary 4.21, show that (1) is equivalent to (3) by applying the result in the preceding exercise. Does this equivalence of (1) and (3) require the process to be stationary?

9. *Sojourn Times in $M/M/s$ Systems.* Consider a stationary $M/M/s$ queueing system with arrival rate λ and service rate μ. Its stationary distribution is

$$\pi(x) = c(\lambda/\mu)^x/x!, \quad x \leq s, \text{ and } \quad \pi(x) = \pi(s)(\lambda/s\mu)^{x-s}, \quad x > s.$$

Let P_N denote the Palm probability of the system given that a unit arrives to the system. Find the probability $P_N\{X_0 < s\}$ that an arrival does not have to wait in the queue for service. The Palm probability of the sojourn time in the system in equilibrium is

$$P_N\{W \le t\} = \sum_x \pi(x) P\{W \le t | X_{0-} = x, X_0 = x+1\}.$$

This is (4.32) for a network with one node. Explain why

$$P\{W \le t | X_{0-} = x, X_0 = x+1\} = \begin{cases} P\{\xi \le t\} & \text{if } x < s \\ P\{\xi + \sum_{i=1}^{x-s+1} \xi_i \le t\} & \text{if } x \ge s \end{cases}$$

where $\xi, \xi_1, \xi_2, \ldots$ are independent exponentially distributed random variables such that ξ has rate μ and the others have rate $s\mu$. Find the Laplace transform of the distribution $P_N\{W \le t\}$. Use it to justify

$$P_N\{W \le t\} = P\{\xi \le t\} P_N\{X_0 < s\} + P_N\{X_0 \ge s\} P\{\xi + W^* \le t\},$$

where W^* is an exponential random variable with rate $s\mu - \lambda$ independent of ξ. One can view ξ as the arrival's service time and W^* as its waiting time in the queue given that it has to wait. Show that in case $s = 1$, the distribution $P_N\{W \le t\}$ is exponential with rate $\mu - \lambda$.

4.11 Bibliographical Notes

Standard references for the theory of point processes are Kallenberg (1983), Daley and Vere-Jones (1988), Karr (1991), and Brandt and Last (1995). Lévy's formula appears in several texts, including the last reference and in Baccelli and Brémaud (1994). The extension of this formula in Theorem 4.6 has not appeared in the literature. The results on Poisson functionals of Markov processes and Poisson flows in networks are from Serfozo (1989). The key to these results is Watanabe's (1964) characterization of Poisson processes. Burke (1956) and Reich (1957) were the first to prove that output processes of certain stationary queueing systems are Poisson processes, and extensions of these results to networks using time-reversal reasoning are in Kelly (1979).

Palm probabilities for Markov processes based on the extended Lévy formula is a subtheory of Palm probabilities covered in Chapter 5 (see the references there for Palm probabilities). The distribution of Jackson networks at a transition and the MUSTA property were characterized by direct Markovian reasoning in Kelly (1979), Sevcik and Mitrani (1981), and Melamed (1982b). The approach in Section 4.8 using the formalism of Palm probabilities is from Serfozo (1993). Other variations on the theme of moving units seeing time averages are in Wolff (1982), König and Schmidt (1990), Melamed and Whitt (1990), and Brémaud et al. (1992). The distributions of sojourn times on overtake–free routes in Jackson networks were characterized using direct Markovian reasoning in Walrand and Varaiya (1980), Melamed (1982a), and Kelly and Pollett (1983). Related articles are Simon and Foley (1979) and Schassberger and Daduna (1983, 1987). The approach in Section 4.9 using Palm probabilities is from Kook and Serfozo (1993).

5

Little Laws

In a Markovian, regenerative, or stationary network, the average sojourn times of customers in a sector of the network can often be obtained from a Little law. Specifically, a Little law for a service system says that the average sojourn time W of a customer in the system and the average queue length L of the system are related by $L = \lambda W$, where λ is the average arrival rate of units to the system. This fundamental relation is a *law of averages* or law of large numbers when the quantities L, λ, W are "limits" of averages. It is also a *law of expectations* when the quantities are expected values. This chapter focuses on Little laws of averages, which are based on sample path analysis. The next chapter covers Little laws of expectations for stationary systems, which are based on Palm probability analysis.

In studying a system, one may want to use $L = \lambda W$ to obtain one of these values from the other two. Typically, λ and L are known or prescribed and one wants to determine that W exists and equals $\lambda^{-1}L$. On the other hand, in a simulation, one may want to estimate L in terms of λ and W. The law $L = \lambda W$ holds when each of the terms exists. In order to apply the law, it is therefore necessary to establish the existence of these quantities.

This chapter addresses the question: If any two of the limits L, λ, or W exist, then what additional conditions guarantee the existence of the other limit? Little laws for queues are special cases of more general laws for certain two-parameter utility processes. Since utility processes cover a rich area of applications and are not more complicated than queueing processes, we prove omnibus Little laws for utility processes and then apply them to queueing systems. Examples for networks are in several sections.

5.1 Little Laws for Markovian Systems

We begin this section by introducing the basic notation for this chapter. Then we present Little laws for Markovian systems. Their proofs are in Section 5.5.

We will consider a general service system or input–output system that processes discrete units (or customers). Suppose units arrive at the system at times $0 < T_1 \le T_2 \le \ldots$, where $T_n \to \infty$ w.p.1. We will often refer to the arrivals by the point process

$$N(t) = \sum_{n=1}^{\infty} 1\big(T_n \in (0, t]\big), \quad t \ge 0,$$

which denotes the number of arrivals in the time interval $(0, t]$. Note that $N(T_n) \ge n$; and $N(T_n) = n$ if $T_{n-1} < T_n < T_{n+1}$. The latter is true when customers arrive one at a time.

Let W_n denote the entire time the nth unit is in the system, including its service or multiple service times and any waiting times for service or hiatus times. Following tradition, we will often call the sojourn time W_n the *waiting time* of the nth unit. The nth unit departs from the system at time $T_n + W_n$ and never returns. The number of units that arrive in the time interval $(0, t]$ and are still in the system at time t is

$$X_t = \sum_n 1(T_n \le t < T_n + W_n), \quad t \ge 0. \tag{5.1}$$

This queue length process X_t has piecewise constant, right-continuous sample paths, and the number of its jumps up to time t is finite (it is bounded by $2N(t)$).

We make no specific assumptions on the processing of units or the dependencies among the variables T_n, W_n, X_t. We only assume these variables exist and are finite. In this system, units may arrive or depart in groups, a unit may be fed back for several services, the interarrival times may depend on the service times, etc. Also, this system may represent a special subpopulation of units in a larger system, such as the units in a sector of a network. The arrival process N is a function of the queue length process X (the arrivals are observable from X). On the other hand, the waiting times W_n may not be a function of X. They are, however, when units arrive and depart one at a time and units depart in the same order in which they arrive. When the queue length process alone does not contain enough information to determine W_n, one usually represents W_n or even N and X as functions of some general stochastic process that encompasses all the system dynamics.

We will present conditions that ensure the existence of the limits

$$\begin{aligned} L &= \lim_{t\to\infty} t^{-1} \int_0^t X_s \, ds && \text{average number in the system} \\ \lambda &= \lim_{t\to\infty} t^{-1} N(t) && \text{average arrival rate} \\ W &= \lim_{n\to\infty} n^{-1} \sum_{k=1}^{n} W_k && \text{average waiting time.} \end{aligned}$$

All the limit statements here are w.p.1. For simplicity, we will often omit the phrase w.p.1. We say that L *exists* if the limit L above exists and it is a positive, finite-valued random variable w.p.1. Similarly, we will refer to the *existence* of λ, W and analogous limits below. Some of the results herein also hold for L, W that may be zero or infinite, but we will not cover these degenerate cases.

To simplify notation, we will frequently write convergence statements like $\lim_{n\to\infty} a_n/c_n = 1$ w.p.1 for random variables a_n, c_n simply as $a_n \sim c_n$. This definition of asymptotic equivalence $\sim$ is consistent with the standard definition for nonrandom a_n, c_n.

The key idea behind the Little law $L = \lambda W$ is that the integral of X_t is simply another way of recording waiting times. Specifically, if the system is empty at times 0 and t ($X_0 = X_t = 0$), then the waiting time of the customers up to time t is

$$\int_0^t X_s\, ds = \sum_{n=1}^{N(t)} W_n.$$

Even at nonempty times t, many systems satisfy

$$t^{-1}\int_0^t X_s\, ds = t^{-1}\sum_{n=1}^{N(t)} W_n + o(1), \tag{5.2}$$

where $o(1) \to 0$ as $t \to \infty$. In this case, if the limits λ and W exist, then letting $t \to \infty$ in (5.2) yields $L = \lambda W$. Indeed, the right side of (5.2) converges to λW since $\sum_{n=1}^{N(t)} W_n \sim N(t)W$ and $N(t) \sim t\lambda$. We will focus on the relation (5.2) and justify the $o(1)$ term in (5.2) for various settings.

Since the proofs of Little laws are long, we present several of them in this and the next section and prove them later. The following are results for Markovian systems.

Theorem 5.1. *Suppose the queue length process is of the form $X_t = f(Y_t)$, where Y is an ergodic Markov jump process on a countable state space and f is a function on this space. Let $q(y, y')$ denote the transition rates of Y and let π denote its equilibrium distribution. Then the average queue length L and arrival rate λ, which may be infinite, are given by*

$$L = \sum_y \pi(y) f(y), \quad \lambda = \sum_y \pi(y) \sum_{y'} q(y, y') 1(f(y') = f(y) + 1).$$

If these quantities are finite and the process X may equal 0, then W exists and $L = \lambda W$.

Proof. The first assertion follows by the ergodic theorem for Markov processes. The second assertion follows by Theorem 5.24 below, which describes Little laws for regenerative processes. □

The preceding Little law is for limiting averages. The version of this law for expected values is as follows. This result is a special case of Theorem 6.22 in the next chapter, which is for stationary systems that need not be Markovian, and arrivals may occur in batches. The type of Palm probability here for Markovian systems was defined in Chapter 4; general Palm probabilities are covered in Section 6.1.

Theorem 5.2. *In the context of Theorem 5.1, suppose the Markov process Y is stationary and the queueing process X may equal* 0. *Assume that customers arrive one at a time for service and that each sojourn time W_n is a function of the queueing process after its arrival time T_n (i.e., $W_n = g(\{X_t : t \geq T_n\})$ for some g). If the expectations EX_0 and $\lambda \equiv EN(1)$ are finite, then*

$$EX_0 = \lambda E_N(W_1),$$

where $E_N(W_1)$ is the expected sojourn time of an arrival at time 0 *under the Palm probability P_N that there is an arrival at time* 0.

Here is our major application for networks.

Example 5.3. *Sojourn Times in Markovian Networks.* Suppose X is a Markov network process that records the numbers of units in an m-node network. For instance, X could be a Jackson or Whittle process. Assume X is ergodic and denote its equilibrium distribution by $\pi(x)$. Our interest is in sojourn times of units in a sector J of the network. The average sojourn time of units in J is

$$W_J = \lim_{n\to\infty} n^{-1} \sum_{i=1}^{n} W_i(J),$$

where $W_i(J)$ is the sojourn time in J of the ith unit to enter J. There is no restriction on the nodes at which the units enter and leave J, and a unit may have multiple visits to the nodes in J before it exits. Also, the units generally do not exit J in the same order in which they entered.

To evaluate W_J, we consider the process $\{X_t(J) : t \geq 0\}$ that denotes the number of units in J. Although $X(J)$ is not a Markov process, it is a function of the Markov process X. Namely, $X_t(J) = f(X_t)$, where $f(x) = \sum_{j\in J} x_j$. By Theorem 5.1, the average number of units in J is

$$L_J = \sum_x \sum_{j\in J} x_j \pi(x).$$

And the average arrival rate of units to J from outside of J is

$$\lambda_J = \sum_x \pi(x) \sum_{x'} q(x, x') 1(f(x) = f(x') + 1),$$

where $q(x, x')$ are the transition rates of X. Assume λ_J is finite and there is a state x of X such that $f(x) = 0$ (i.e., sector J is recurrently empty). Then by Theorem 5.1, the average sojourn time W_J exists and $L_J = \lambda_J W_J$.

Now, in addition to the assumptions above, suppose the Markov process X is stationary. Assume that units enter J one at a time, which means that the point process N of arrival times into J is simple. Also, assume that each W_n is a function of the queueing process after its arrival time T_n. Then Theorem 5.2 yields

$$E[X_0(J)] = \lambda_J E_N[W_1(J)],$$

where $E_N[W_1(J)]$ is the expected sojourn time of a unit in J that enters J at time 0.

It is sometimes of interest to differentiate between customer sojourn times we have been discussing and the customer "waiting times for service." To do this, one must know the service discipline at the nodes. For simplicity, assume that customers are served at each node j on a first-come, first-served basis and an arrival to node j does not wait for service when and only when there are less than s_j customers in the system (like an $\cdot/M/s_j$ node). The average *waiting time for services* that units endure in J is

$$W_J^* = \lim_{n\to\infty} n^{-1} \sum_{i=1}^{n} W_i(J)^*,$$

where $W_i(J)^*$ is the waiting time for services in J of the ith unit that enters J. Then under all the preceding assumptions in this example, the limit W_J^* exists and $L_J^* = \lambda_J W_J^*$, where λ_J is as above and L_J^* is the average number of units waiting for services in J, which is given by

$$L_J^* = \sum_x \sum_{j\in J} x_j \pi(x) 1(x_j \geq s_j).$$

In addition, $E[X_0(J)^*] = \lambda_J E_N[W_1(J)^*]$, where $E[X_0(J)^*]$ is the expected number of units waiting for services in J at time 0 and $E_N[W_1(J)^*]$ is the expected waiting time of a unit in J that enters J at time 0. □

5.2 Little Laws for General Queueing Systems

We now present basic Little laws for waiting times in general queueing systems, including non-Markovian networks. The proofs are in Section 5.5.

In addition to the notation above, we will use the time

$$\hat{T}_n = \max_{1\leq k\leq n} (T_k + W_k),$$

which is the time by which the first n units have departed. We call $\hat{T}_n$ the *thorough departure time of the first n units*. Let $\hat{N}(t)$ denote the number of $\hat{T}_n$'s in $(0, t]$ and let $\hat{\lambda} = \lim_{t\to\infty} t^{-1}\hat{N}(t)$. We call $\hat{N}$ the *thorough departure process*. Note that more than n units might depart from the system by time $\hat{T}_n$, and so $\hat{N}(t)$ is generally less than or equal to the actual number of departures up to time t. However, if the units depart in the same order in which they arrive, then $\hat{T}_n = T_n + W_n$ and $\hat{N}(t)$ is the actual number of departures up to time t.

The following result concerns the existence of the three limits L, λ, and W when two of them are known to exist.

Theorem 5.4. *The following statements are equivalent.*
(a) *The limits L, λ, and W exist, and $L = \lambda W$.*
(b) *L exists, $\hat{T}_n \sim T_n$ and either λ or $\hat{\lambda}$ exists.*
(c) *L and λ exist, and $n^{-1}W_n \to 0$.*

(d) *L, λ, and $\hat{\lambda}$ exist, and* $\lambda = \hat{\lambda}$.
(e) *λ and W exist.*
(f) *L and W exit, and* $\int_0^{\hat{T}_n} X_t\, dt \sim \sum_{k=1}^{n} W_k$.

Note that the thorough departure process $\hat{N}$ plays a prominent role. Keying the departures to $\hat{N}$ rather than the usual departure process simplifies the analysis and leads to more precise equivalent statements.

The equivalence of (a) and (e) implies that L exists automatically when λ and W exist. Unfortunately, the existence of W, which is the limit in a law of large numbers for the waiting time sequence, is usually difficult to verify because waiting times are usually intractable. There are a few exceptions, including systems with stationary waiting times as in Theorem 6.25. For cases in which L and λ exist, one can establish the existence of W and the relation $L = \lambda W$ by verifying any one of the conditions (b), (c), or (d). Condition (c) is often the easiest to verify.

A consequence of Theorem 5.4 is that $L = \lambda W$ is "universal" in the sense that it always holds when all three of the limits exist. In many systems, the three limits exist when any two of them exist. This three-for-the-price-of-two property is satisfied in systems with regular departures as we now describe.

Recall that the key relation (5.2) leading to $L = \lambda W$ is automatically true with $o(1) = 0$ at any time t when the system is empty and $X_0 = 0$. This suggests that a system that empties out occasionally is more likely to satisfy the relation. To formalize this idea, we will use the following notion.

Definition 5.5. The system is *recurrently empty* if there are strictly increasing random times $\tau_n \uparrow \infty$ such that $\tau_{n+1} \sim \tau_n$ and $X_t = 0$ for some t in each interval $[\tau_n, \tau_{n+1})$.

The τ_n will typically be times such as regeneration times that trigger special "cycles" in the system. The condition $\tau_{n+1} \sim \tau_n$, which is implied by $\tau_n/n \to c > 0$, is all that is needed here. Assuming a system is recurrently empty may be too strong in some cases. A weaker notion is as follows.

Definition 5.6. The departure times of the queueing system are *regular* with respect to strictly increasing random times $\tau_n \uparrow \infty$ if $\tau_{n+1} \sim \tau_n$ and $\zeta_n \sim \tau_n$, where

$$\zeta_n \equiv \max\{\hat{T}_k : \tau_{n-1} \le T_k < \tau_n\}.$$

The ζ_n is the thorough departure time for all arrivals during the time interval $[\tau_{n-1}, \tau_n)$.

Note that if the system is recurrently empty with respect to τ_n, then its departure times are regular with respect to τ_n. Indeed, all waiting times beginning in the interval $[\tau_{n-1}, \tau_n)$ terminate before time τ_{n+1}, since the system empties during $[\tau_n, \tau_{n+1})$, and so

$$\tau_n \sim \tau_{n-1} \le \zeta_n \le \tau_{n+1} \sim \tau_n.$$

On the other hand, there are many systems that are not recurrently empty, or never empty at all, but their departures are regular. Examples are easy to construct. Clearly, the departures are regular with respect to τ_n if $\tau_{n+1} \sim \tau_n$ and

$$\tau_n^{-1} \sum_k W_k 1(\tau_n \leq T_k < \tau_{n+1}) \to 0.$$

For systems with regular departures, Theorem 5.4 reduces to the following, which is a special case of Theorem 5.17.

Theorem 5.7. *Suppose the queueing system is recurrently empty or, more generally, its departure times are regular. If any two of the limits L, λ, and W exist, then the other limit exists and $L = \lambda W$.*

This result implies Theorem 5.1 since the Markovian system in Theorem 5.1 is recurrently empty and L and λ exist. This yields the existence of W and the relation $L = \lambda W$. Section 5.6 below shows that the assumption of regular departures is automatically satisfied for certain systems, such as Markovian and regenerative systems; also, see Exercise 4 in this chapter and Exercise 2 in Chapter 6.

The assumption that a system is recurrently empty or has regular departures is rather natural and not restrictive. It simply ensures that customers do not remain in the system for indefinitely long periods and that their waiting times are not extremely irregular. For instance, a system might have a protocol under which services are not performed when the queue length is below a specified level. Then some customers may get trapped in the system for irregularly long periods, especially under a "forgetful" or "layed-back" protocol. In these cases, the limit W may not exist and some W_n's may even be infinite.

5.3 Preliminary Laws of Large Numbers

This section contains several limit theorems that we will use to prove Little laws.

We begin by relating a law of large numbers for the point process $N(t)$ to a law of large numbers for its point locations T_n. As a preliminary, note that $N(T_n)$ is the "right-hand inverse" of T_n in the sense that $T_{N(T_n)} = T_n$. Consequently, the rate at which $N(t)$ tends to infinity should be the inverse of the rate at which T_n does. This property is formalized as follows.

Theorem 5.8. *For a positive λ, the following statements are equivalent.*

$$\lim_{n\to\infty} n^{-1} T_n = \lambda^{-1}. \tag{5.3}$$

$$\lim_{t\to\infty} t^{-1} N(t) = \lambda. \tag{5.4}$$

PROOF. Suppose that (5.3) holds. Using $T_{N(t)} \leq t < T_{N(t)+1}$, we have

$$T_{N(t)+1}^{-1} N(t) < t^{-1} N(t) \leq T_{N(t)}^{-1} N(t).$$

The supposition (5.3) along with $N(t) \uparrow \infty$ and $N(t)(N(t)+1)^{-1} \to 1$ ensure that the first and last terms in this display converge to λ. This proves (5.4).

Conversely, suppose (5.4) holds. Consider a fixed (random) $\tau \leq T_1$. Clearly $N(T_n - \tau) < n \leq N(T_n)$, and so

$$N(T_n)^{-1}T_n \leq n^{-1}T_n < N(T_n - \tau)^{-1}T_n.$$

The supposition (5.4) along with $T_n \uparrow \infty$ and $(T_n - \tau)^{-1}T_n \to 1$ ensure that the first and last terms in this display converge to λ^{-1}. This proves (5.3). □

Note that the preceding result applies to any point process on $\mathbb{R}_+$, since the increments of N may be any positive integer. We will frequently use this result without mention for $\hat{N}$ as well as for N.

We now consider laws of large numbers for more general processes. Assume that $\{Z(t) : t \geq 0\}$ is a nonnegative, nondecreasing real-valued stochastic process.

Lemma 5.9. *If the limit $Z \equiv \lim_{n\to\infty} T_n^{-1}Z(T_n)$ exists and $T_{n-1} \sim T_n$, then $\lim_{t\to\infty} t^{-1}Z(t) = Z$.*

Proof. Since $Z(t)$ is nondecreasing and $T_{N(t)} \leq t < T_{N(t)+1}$, we have

$$\frac{T_{N(t)}}{T_{N(t)+1}} \frac{Z(T_{N(t)})}{T_{N(t)}} < t^{-1}Z(t) \leq \frac{T_{N(t)+1}}{T_{N(t)}} \frac{Z(T_{N(t)+1})}{T_{N(t)+1}}.$$

Under the hypotheses, the right and left sides of this inequality converge to Z as $t \to \infty$, and hence so does $t^{-1}Z(t)$. □

We use the following result several times to establish limiting averages of processes. This result also establishes Little laws for one-parameter utility processes and sojourn and travel times for stochastic processes; see Sections 6.5–6.7. Here we refer to the limiting averages $\lambda \equiv \lim_{t\to\infty} t^{-1}N(t)$,

$$Z \equiv \lim_{t\to\infty} t^{-1}Z(t), \qquad \hat{Z} \equiv \lim_{n\to\infty} n^{-1}Z(T_n).$$

Theorem 5.10. *If any two of the limits Z, λ, and $\hat{Z}$ exist, then the other one exists and $Z = \lambda\hat{Z}$.*

Proof. If Z and λ exist, then by Theorem 5.8

$$n^{-1}Z(T_n) = (n^{-1}T_n)(T_n^{-1}Z(T_n)) \to \lambda^{-1}Z.$$

Thus, $\hat{Z}$ exists and $Z = \lambda\hat{Z}$. If Z and $\hat{Z}$ exist, then

$$n^{-1}T_n = (n^{-1}Z(T_n))(T_nZ(T_n)^{-1}) \to \hat{Z}Z^{-1},$$

and so Theorem 5.8 ensures that λ exists and $Z = \lambda\hat{Z}$. If Z and $\hat{Z}$ exist, then

$$T_n^{-1}Z(T_n) = nT_n^{-1}n^{-1}Z(T_n) \to \lambda\hat{Z}.$$

Thus, by Lemma 5.9, the limit Z exists and $Z = \lambda\hat{Z}$. □

The next result is useful for obtaining laws of large numbers for maxima of discrete- or continuous-time processes.

Lemma 5.11. *Suppose that c and c_n $(n \geq 1)$ are positive real numbers such that $n^{-1}c_n \to c$. Then $n^{-1}\max_{k\leq n} c_k \to c$.*

PROOF. Fix $\varepsilon > 0$, and choose n_1 such that $|n^{-1}c_n - c| < \varepsilon$, for each $n \geq n_1$. Since $c_n \sim nc \to \infty$, there is an $n_2 \geq n_1$ such that $c_{n_2} = \max_{k\leq n_2} c_k$. Next, choose $n_3 \geq n_2$ such that $n^{-1}c_{n_2} < \varepsilon$, for each $n \geq n_3$. Then, for each $n \geq n_3$, we have

$$c - \varepsilon \leq n^{-1}c_n \leq n^{-1}\max_{k\leq n} c_k$$
$$= \max\{n^{-1}c_{n_2}, n^{-1}\max_{n_2<k\leq n} c_k\} \leq c + \varepsilon.$$

This proves the assertion. □

5.4 Utility Processes

Little laws for queueing systems are special cases of Little laws for certain two-parameter utility processes. This section describes this relation, and then specifies the utility processes that we will study.

Example 5.12. *Waiting Times Modeled by a Utility Process.* For the queueing process X defined above, consider the two-parameter process

$$U(n,t) = \sum_{k=1}^{n}\int_0^t 1(T_k \leq s < T_k + W_k)\,ds, \qquad n \in \mathbb{Z}_+,\ t \in \mathbb{R}_+. \tag{5.5}$$

This represents the total sojourn time in the system during $(0, t]$ for the first n units. Since the queue length process is $X_t = \sum_n 1(T_n \leq t < T_n + W_n)$, the waiting time of units during $(0, t]$ of the $N(t)$ arrivals in that interval is

$$U(N(t), t) = \int_0^t X_s\,ds.$$

Also, the "total" waiting time for the first n arrivals is

$$U(n, \hat{T}_n) = \lim_{t\to\infty} U(n,t) = \sum_{k=1}^{n} W_k.$$

Then the average queue length and waiting times are given by the respective limits

$$L = \lim_{t\to\infty} t^{-1}U(N(t),t), \qquad W = \lim_{n\to\infty} n^{-1}U(n,\hat{T}_n).$$

These are *time averages* and *customer averages* of the waiting times. □

This example shows that it is natural to study waiting times and queue lengths in terms of two-parameter utility processes. Accordingly, we will consider Little laws for general utility processes defined as follows.

Consider a stochastic system like the service system above in which there is a nondecreasing sequence of times T_n at which some special event occurs, such as an order for a product. As above, assume $T_n \uparrow \infty$ and let $N(t)$ denote the number of

T_n's in the time interval $(0, t]$. Think of $N(t)$ as a general point process on $\mathbb{R}_+$; we make no assumptions about its distribution or structure (the increments of N are positive integer-valued random variables with arbitrary dependencies). Associated with this point process is a utility process $U(n, t)$ that measures some real-valued quantity (e.g. cost, value, time, stress), for $n \in \mathbb{Z}_+$, $t \in \mathbb{R}_+$. The quantity $U(n, t)$ is the cumulative utility up to time t associated with the first n times $T_1, \ldots, T_n$ (these times need not be in the interval $(0, t]$). The utility associated with T_n is not necessarily received at time T_n—it may be accumulated in bits or continuously, anywhere over the time horizon. For now, we place no monotonicity assumptions on the process $U(n, t)$. We will sometimes assume, however, that it is nonnegative and nondecreasing ($U(n, t) \le U(n', t')$ whenever $n \le n'$, $t \le t'$). Each parameter n or t may be continuous or discrete, but, for simplicity, we stick to the conventional setting in which n is discrete and t is continuous.

Now, the utility "associated" with the time interval $(0, t]$, or with the $N(t)$ time points, is defined by

$$U(t) \equiv U(N(t), t).$$

This utility may be accumulated after time t as well as before it. The infinite-horizon or complete utility for the first n times $T_1, \ldots, T_n$ is defined by

$$\hat{U}(n) \equiv \lim_{t \to \infty} U(n, t),$$

which is assumed to be finite w.p.1. The time at which the utility associated with $T_1, \ldots, T_n$ ceases to change is

$$\hat{T}_n = \inf\{t : U(n, t) = \hat{U}(n)\}.$$

We call $\hat{T}_n$ the nth *thorough termination time*. This terminology is consistent with the notion of a thorough departure time for the queueing system. We assume $\hat{T}_n$ is finite w.p.1. Then $\hat{U}(n) = U(n, \hat{T}_n)$. The case $\hat{T}_n = \infty$ involves technicalities that we will not cover. We shall consider the "time average" and "unit average" utilities

$$U = \lim_{t \to \infty} t^{-1} U(t), \qquad \hat{U} = \lim_{n \to \infty} n^{-1} \hat{U}(n), \quad \text{w.p.1.}$$

Our interest is in the relation $U = \lambda \hat{U}$, which is a generalization of $L = \lambda W$. With a slight abuse of notation, we use "U" in several ways ($U(n, t)$, $U(t)$, $\hat{U}(n)$, U, $\hat{U}$) to emphasize that these quantities are associated with a single utility process. Some cases of $U = \lambda \hat{U}$ have been studied using the abstract notation $H = \lambda G$, which is also related to expression (6.18) below for marked point processes.

A large class of utility processes are functionals of stochastic processes as follows.

Example 5.13. *Additive Utility Processes.* Suppose the system is described by a stochastic process $\{Y_t : t \ge 0\}$ and the times T_n. A natural utility process $U_n(t)$ associated with each time T_n is

$$U_n(t) = \int_0^t f(s, T_n, \{Y_u : u \ge s\})\, ds,$$

where f is a real-valued function of the current time s, the time T_n, and the future of the process Y at time s (which is $\{Y_u : u \geq s\}$). Then the utilities associated with $T_1, \ldots, T_n$ on the respective time intervals $(0, t]$ and $\mathbb{R}_+$ are

$$U(n,t) = \sum_{k=1}^{n} U_k(t), \qquad \hat{U}(n) = \sum_{k=1}^{n} U_k(\hat{T}_n). \tag{5.6}$$

This is an *additive utility process*. Note that the utility process for waiting times in the queueing system is an additive utility process. □

We will see in Sections 6.5 and 6.6 that sojourn and travel times for processes can be formulated by one-parameter uility processes defined as follows.

Example 5.14. *One-Parameter Utilities*. Suppose that $U(n, t)$ is a utility process in which $\hat{T}_n = T_n$, for each n. We call this a *one-parameter utility process based only on time* if $U(n, t) \equiv Z(t)$, independent of n, for some process $Z(t)$. In this case, $U(t) = Z(t)$ and $\hat{U}(n) = Z(T_n)$. Then the Little law for this utility process, which is $Z = \lambda \hat{Z}$, can often be established by Theorem 5.10.

Similarly, we say that $U(n, t)$ is a *one-parameter utility process based only on cycles* if it is of the form $U(n, t) \equiv Y_n$, independent of t, for some sequence Y_n. In this case, $U(t) = Y_{N(t)}$ and $\hat{U}(n) = Y_n$. Then the Little law for this utility process can often be established by the discrete analogue of Theorem 5.10 in Exercise 1. □

5.5 Omnibus Little Laws

In this section, we present limit theorems for the utility processes defined above. Then we apply these results to obtain the Little laws in Sections 5.1 and 5.2 for queueing systems.

Using the notation in the preceding section, we consider the utilities

$$U(t) = U(N(t), t), \qquad \hat{U}(n) = U(n, \hat{T}_n)$$

associated with the time interval $(0, t]$ and the times $T_1, \ldots, T_n$, respectively. The first result concerns the existence of the limits U, λ, and $\hat{U}$, as well as the relation $U = \lambda \hat{U}$, when only two of the limits exist.

Theorem 5.15. *Suppose the process $U(n, t)$ is nonnegative and nondecreasing in (n, t), and $\hat{T}_n \sim T_n$. Then the following statements are equivalent.*
(a) *The limits U, λ, and $\hat{U}$ exist, and $U = \lambda \hat{U}$.*
(b) *The limits λ and U exist.*
(c) *The limits λ and $\hat{U}$ exist.*
(d) *The limits U and $\hat{U}$ exist, and $\hat{U}(n) \sim U(\hat{T}_n)$.*

Proof. Clearly, (a) implies (b). We next show that (b) implies (c). To determine the existence of the limit $\hat{U}$, first note that

$$U(T_n-) \leq U(n, T_n) \leq U(n, \hat{T}_n) = \hat{U}(n) \leq U(\hat{N}(\hat{T}_n), \hat{T}_n) \leq U(\hat{T}_n). \tag{5.7}$$

These inequalities follow by the monotonicity of the utility process and

$$N(T_n-) < n \leq N(T_n), \quad T_n \leq \hat{T}_n, \quad \hat{N}(t) \leq N(t). \tag{5.8}$$

We will prove that (b) implies $\hat{U}(n) \sim n\lambda^{-1}U$. But this will follow by (5.7) upon showing that

$$U(T_n-) \sim n\lambda^{-1}U \sim U(\hat{T}_n). \tag{5.9}$$

To this end, note that the existence of λ ensures that $n/\hat{T}_n \sim n/T_n \sim \lambda$, and so $\hat{N}(t) \sim \lambda t$. Also, $U(t) \sim tU$ implies that $U(T_n-) \sim T_nU \sim n\lambda^{-1}U$. Similarly, $U(T_n-) \sim T_nU \sim n\lambda^{-1}U$. This proves (5.9) and hence (b) implies (c).

Next suppose that (c) is true. We will show that (d) holds. To obtain the existence of the limit U, we will use the inequalities

$$\hat{U}(\hat{N}(t)) \leq U(n,t) = U(t) \leq U(N(t)+1, T_{N(t)+1}) \leq \hat{U}(N(t)+1). \tag{5.10}$$

These relations are based on (5.8) and $\hat{T}_{\hat{N}(t)} \leq t < T_{N(t)+1}$. Now, arguing as in the last paragraph, we have $\hat{N}(t) \sim \lambda t$ and

$$\hat{U}(\hat{N}(t)) \sim \hat{N}(t)\hat{U} \sim t\lambda\hat{U}.$$

Similarly, $\hat{U}(N(t)+1) \sim t\lambda\hat{U}$. Applying these observations to (5.10) shows that $U(t) \sim t\lambda\hat{U}$. Furthermore, $U(T_n) \sim T_n\lambda\hat{U} \sim \hat{U}(n)$. This proves (d).

Finally, if (d) holds, then (a) follows since

$$T_n \sim U(T_n)U^{-1} \sim \hat{U}U^{-1},$$

which ensures that λ exists and $\lambda = U\hat{U}^{-1}$. □

We now consider systems that are recurrently empty or have regular termination times. The utility termination times $\hat{T}_n$ are said to be *regular* with respect to strictly increasing random times $\tau_n \uparrow \infty$ if $\tau_{n+1} \sim \tau_n$ and $\tau_n \sim \zeta_n$, where

$$\zeta_n \equiv \max\{\hat{T}_k : \tau_{n-1} \leq T_k < \tau_n\}.$$

This notion is consistent with regular departure times in a queueing system.

We will show that Theorem 5.15 reduces considerably for systems with regular termination times. The proof is based on the following result.

Lemma 5.16. *The following statements are equivalent.*
(a) λ *exists and the termination times are regular.*
(b) λ *and* $\hat{\lambda}$ *exist and* $\lambda = \hat{\lambda}$.
(c) *Either* λ *or* $\hat{\lambda}$ *exists and* $\hat{T}_n \sim T_n$.

PROOF. We first show that (b) is equivalent to (c). If (b) holds, then (c) follows since

$$\hat{T}_n \sim n/\hat{\lambda} = n/\lambda \sim T_n.$$

Now, suppose (c) holds. Then (b) follows since the existence of λ implies $\hat{T}_n \sim T_n \sim n/\lambda$, which yields $\hat{\lambda} = \lambda$; and the existence of $\hat{\lambda}$ implies $T_n \sim \hat{T}_n \sim n/\hat{\lambda}$, which yields $\lambda = \hat{\lambda}$.

Next, we show that (a) is equivalent to (b). If (b) holds, then the termination times are regular with respect to $\tau_n = n$ since

$$\begin{aligned}\tau_n \sim n - 1 \le \zeta_n &= \max\{\hat{T}_k : n - 1 \le T_k < n\} \\ &\le \hat{T}_{N(n)} \sim N(n)/\lambda \sim \tau_n.\end{aligned}$$

Thus (b) implies (a).

Now, suppose (a) holds, where the termination times are regular with respect to τ_n. To prove (b), it suffices to show $\hat{T}_n \sim n\lambda^{-1}$. The regularity assumption $\zeta_n \sim \tau_n$ and Lemma 5.11 imply that $\max_{k \le n} \zeta_k \sim \tau_n$. Using this along with $T_n \sim n/\lambda$, $T_n \le \hat{T}_n$ and $N(t) \sim \lambda t$, we have

$$N(\tau_n)/\lambda \sim T_{N(\tau_n)} \le \hat{T}_{N(\tau_n)} \le \max_{k \le n} \zeta_k \sim \tau_n \sim N(\tau_n)/\lambda. \tag{5.11}$$

This proves $\hat{T}_{N(\tau_n)} \sim N(\tau_n)\lambda^{-1}$. Also, $N(t) \sim \lambda t$ implies

$$N(\tau_{n+1})/N(\tau_n) \sim \tau_{n+1}/\tau_n \sim 1.$$

Then $\hat{T}_n \sim n\lambda^{-1}$ follows by an application of a discrete-time version of Lemma 5.9 with the pair $Z(t)$, T_n equal to $\hat{T}_n$, $N(\tau_n)$. □

We are now ready for the main result for systems with regular termination times.

Theorem 5.17. *Suppose $U(n, t)$ is nonnegative and nondecreasing in (n, t), and the utility termination times are regular. If any two of the limits U, λ, and $\hat{U}$ exist, then the other limit exists and $U = \lambda \hat{U}$.*

PROOF. Under the assumption that λ exists, the assertions follow by Theorem 5.15, since Lemma 5.16 ensures that $\hat{T}_n \sim T_n$. Now, consider the remaining case in which U and $\hat{U}$ exist. Let τ_n denote the times with respect to which the termination times are regular. To establish the existence of λ, we first show that $U(\tau_n) \sim \hat{U}(N(\tau_n))$.

Let $\eta_n = \max_{k \le n} \zeta_k$. We saw in (5.11) that $\hat{T}_{N(\tau_n)} \le \eta_n \sim \tau_n$. Using these relations along with

$$t < T_{N(t)+1}, \quad T_n \le \hat{T}_n, \quad \tau_n \le \eta_{n+1}, \quad \hat{U}(n) \sim n\hat{U},$$

and $U(t) \sim tU$, we have

$$\begin{aligned}U(\tau_n) &\le U(N(\tau_n), T_{N(\tau_n)+1}) \le U(N(\tau_n) + 1, \hat{T}_{N(\tau_n)+1}) \\ &= \hat{U}(N(\tau_n) + 1) \sim \hat{U}(N(\tau_n)) \le U(N(\eta_{n+1}), \eta_n) \\ &\le U(\eta_{n+1}) \sim \eta_{n+1} U \sim \tau_{n+1} U \sim \tau_n U \sim U(\tau_n).\end{aligned}$$

Thus $U(\tau_n) \sim \hat{U}(N(\tau_n))$.

This property and $\tau_{n+1} \sim \tau_n$ imply $U(t) \sim \hat{U}(N(t))$ by an application of Lemma 5.9 with the pair $Z(t)$, T_n equal to $U(t)$, τ_n. Therefore,

$$N(t) \sim \hat{U}(N(t))\hat{U}^{-1} \sim U(t)\hat{U}^{-1} \sim tU\hat{U}^{-1}.$$

In other words, λ exists and $U = \lambda \hat{U}$. □

Remark 5.18. *(Results for Nonmonotone Utility Processes).* Consider a utility process of the form $U(n,t) = U_1(n,t) - U_2(n,t)$, where $U_i(n,t)$ is nondecreasing in (n,t) for $i = 1, 2$. Define $\hat{U}_i(n) = U_i(n, \hat{T}_n)$, and let U_i and $\hat{U}_i$ denote limits of the respective averages $t^{-1}U_i(t)$ and $n^{-1}\hat{U}_i(n)$. Then by obvious applications of Theorems 5.15 and 5.17, it follows that their assertions also apply to this more general utility process with the modifications that "U exists" is replaced throughout by "U_1 and U_2 exist", and "$\hat{U}$ exists" is replaced throughout by "$\hat{U}_1$ and $\hat{U}_2$ exist".

We are now ready to apply the results above to prove the Little laws stated in Section 5.2 for queueing systems. Recall that, for a queueing system with regular departure times, Theorem 5.7 says that the existence of any two of the limits L, λ, and W ensures that the other limit exists and $L = \lambda W$. This result is an immediate corollary of the following result, which we stated previously as Theorem 5.4.

Theorem 5.19. *The following statements are equivalent.*
(a) *The limits L, λ, and W exist, and $L = \lambda W$.*
(b) *L exists, $\hat{T}_n \sim T_n$, and either λ or $\hat{\lambda}$ exists.*
(c) *L and λ exist, and $n^{-1}W_n \to 0$.*
(d) *L, λ, and $\hat{\lambda}$ exist, and $\lambda = \hat{\lambda}$.*
(e) *λ and W exist.*
(f) *L and W exit, and $\int_0^{\hat{T}_n} X_s\,ds \sim \sum_{k=1}^n W_k$.*

PROOF. We observed in Section 5.4 that waiting times in the queueing system under study are represented by the utility process

$$U(n,t) = \sum_{k=1}^{n} \int_0^t 1(T_k \le s \le T_k + W_k)\,ds.$$

Clearly $U(n,t)$ is nonnegative and nondecreasing in (n,t), and

$$U(t) = \int_0^t X_s\,ds, \qquad \hat{U}(n) = \sum_{k=1}^{n} W_k.$$

First, note that statements (b), (c), and (d) are equivalent by Lemma 5.20 below. The proof of the equivalence of (e) and (f) is left as Exercise 2.

The proof will be complete upon showing (b) $\Rightarrow$ (e) $\Rightarrow$ (a) $\Rightarrow$ (c). Now, Theorem 5.15 ensures that (b) $\Rightarrow$ (e) $\Rightarrow$ (a). Also, if (a) holds, then the existence of W implies

$$n^{-1}W_n = n^{-1}\Big(\sum_{k=1}^{n} W_k - \sum_{k=1}^{n-1} W_k\Big) \to 0.$$

Thus, (a) $\Rightarrow$ (c). □

Lemma 5.20. *For the setting of waiting times in a queueing system, where $\hat{T}_n = \max_{k \le n}(T_k + W_k)$, the following statements are equivalent:*
(a) *λ and $\hat{\lambda}$ exist and $\lambda = \hat{\lambda}$.*
(b) *Either λ or $\hat{\lambda}$ exists and $\hat{T}_n \sim T_n$.*
(c) *λ exists and $n^{-1}W_n \to 0$.*

PROOF. The equivalence of (a) and (b) follows from Lemma 5.16.

We finish the proof by showing that (a) is equivalent to (c). If (a) holds, then (c) follows since using $T_n \sim n\lambda^{-1} \sim \hat{T}_n$, we have

$$0 \leq n^{-1} W_n \leq n^{-1}(\hat{T}_n - T_n) \to 0.$$

Now, suppose (c) holds. Then by $T_n \sim n\lambda^{-1}$, $n^{-1} W_n \to 0$ and Lemma 5.11, we have

$$\lim_{n\to\infty} n^{-1}\hat{T}_n = \lim_{n\to\infty} n^{-1} \max_{k\leq n}(T_k + W_k) = \lim_{n\to\infty} n^{-1}(T_n + W_n) = \lambda^{-1}.$$

This proves that $\hat{\lambda}$ exists and equals λ. Thus (c) implies (a). □

Remark 5.21. *(Results for Nondiscrete Quantities).* The preceding results also apply to service systems that process nondiscrete quantities like fluids or other infinitely divisible items, or a combination of discrete and continuous quantities. In particular, a random measure $M(t)$ that represents the total mass that has arrived in $(0, t]$ replaces $N(t)$. And $T(x) = \sup\{t : M(t) \leq x\}$ replaces T_n. Similarly, a random measure $\hat{M}(t)$, representing the mass that has thoroughly departed in $(0, t]$, replaces $\hat{N}(t)$; and $\hat{T}(x)$, defined in the obvious way, replaces $\hat{T}_n$. Then $U(x, t)$ denotes the utility (or waiting time) associated with the quantity x up to time t; and $U(t) = U(M(t), t)$ and $\hat{U}(x) = U(x, \hat{T}(x))$ denote the utilities (or waiting times) up time t and for quantity x, respectively. All the results herein apply with the same interpretations; the only difference is the minor change in notation that the "quantity parameter" is now x instead of n.

5.6 Little Laws for Regenerative Systems

There are a number of queueing systems, such as the $GI/G/s$ system we discuss shortly, whose dynamics are expressible by regenerative processes. In this section, we apply the general results above to obtain Little laws for such systems.

We begin with a few comments on regenerative processes.

Definition 5.22. Let $\{Y_t : t \geq 0\}$ be a continuous-time stochastic process with arbitrary state space, and let $0 = \tau_0 < \tau_1 < \ldots$ be random times associated with Y such that $\tau_n \uparrow \infty$. The *nth cycle* of Y consists of the information

$$\xi_n = (\tau_n - \tau_{n-1}, \{Y_t : \tau_{n-1} \leq t < \tau_n\}).$$

This represents the trajectory of Y in the random time interval $[\tau_{n-1}, \tau_n)$. The process Y is *regenerative over the* τ_n*'s* if $\xi_1, \xi_2, \ldots$ are independent and identically distributed. For simplicity, the regenerations are assumed to start at time 0; otherwise, ξ_1 might not have the same distribution as the other ξ_n's.

Suppose Y is a regenerative process over τ_n. Note that the τ_n's form a renewal process. By the key renewal theorem, we know that if τ_1 is not periodic and

$E\tau_1 < \infty$, then the limiting distribution of Y is

$$\pi(A) = \lim_{t\to\infty} P\{Y_t \in A\} = \frac{1}{E\tau_1} E \int_0^{\tau_1} 1(Y_t \in A)\, dt. \tag{5.12}$$

There is an analogous limit when τ_1 is periodic. Many limit laws for functionals of Y are special cases of the following result.

Proposition 5.23. *If $Z(t)$ is a nonnegative, nondecreasing process on the same probability space as Y such that $Z(\tau_n) - Z(\tau_{n-1})$, $n \geq 1$, are independent and identically distributed, then $t^{-1}Z(t) \to EZ(\tau_1)/E\tau_1$ as $n \to \infty$.*

PROOF. The assertion follows by Theorem 5.10, since the classical law of large numbers ensures that $n^{-1}\tau_n \to E\tau_1$ and $n^{-1}Z(\tau_n) \to EZ(\tau_1)$. □

We are now ready to describe the average waiting times in regenerative systems.

Theorem 5.24. *Suppose the queue length process X is regenerative over τ_n and $E\tau_1 < \infty$. Let π denote the limiting distribution of X. Then the average queue length L and the arrival rate λ, which may be infinite, are*

$$L = \sum_x x\pi(x), \qquad \lambda = EN(\tau_1)/E\tau_1. \tag{5.13}$$

If, in addition, L and λ are finite and $\pi(0) > 0$, then W exists and $L = \lambda W$.

PROOF. Since X is regenerative, it follows by two applications of Proposition 5.23 with $Z(t) = \int_0^t X_s\, ds$ and $Z(t) = N(t)$, that the averages L and λ are as in (5.13). To prove W exists and $L = \lambda W$, it suffices by Theorem 5.7 to show that the system is recurrently empty.

To this end, let ν_n denote the nth cycle in which the system is empty: $\nu_0 = 0$ and

$$\nu_n = \min\{k > \nu_{n-1} : X_t = 0 \text{ for some } t \in [\tau_{k-1}, \tau_k)\}, \quad n \geq 1.$$

Set $\tau'_n = \tau_{\nu_n}$. Then by the regenerative property of X, the $\tau'_n - \tau'_{n-1}$, $n \geq 1$ are independent and identically distributed with mean $E\tau'_1 = E\nu_1 E\tau_1$. Furthermore, ν_1 is a geometric random variable with mean $1/\pi(0)$. Then by the classical strong law of large numbers, $\tau'_{n+1}/\tau'_n \sim (n+1)/n \sim 1$. Therefore the system is recurrently empty with respect to τ'_n. □

The preceding result for regenerative systems yields Theorem 5.1 for Markovian systems. Note that the assumption $\pi(0) > 0$ in Theorem 5.24 is used in the proof only to ensure that the system is recurrently empty. This assumption can clearly be replaced by any condition that implies the departures are regular or $n^{-1}W_n \to 0$.

In Theorem 5.24, the X is regenerative, but the W_n's do not have a standard probabilistic structure. Consequently, the limits L and λ also have interpretations as expected values, but W does not. The next result concerns the reverse situation in which the waiting times have a special structure, but the queue length process does not.

Consider a service system whose dynamics are represented by a process Y that is regenerative over τ_n. When a unit arrives in the time interval $[\tau_n, \infty)$, a natural assumption is that its waiting time is a function of the *system dynamics*

$$\eta_n = \{Y_{\tau_n+t} : t \geq 0\} \cup \{\tau_{n+k} - \tau_n : k \geq 1\} \tag{5.14}$$

in the time interval $[\tau_n, \infty)$. Formally, we say the sequence $\{W_n\}$ is a *τ_n-shift of Y* if $\{W_k : T_k \geq \tau_n\}$ is a function of η_n, for each n. This definition also applies when Y is not regenerative. For instance, such random time shifts arise naturally in stationary systems that we discuss in the next chapter. The arrival process and queue lengths may also have analogous links with the system dynamics η_n. We say that N (or X) is a *τ_n-shift of Y* if $\{N(\tau_n + t) - N(\tau_n) : t \geq 0\}$ (or $\{X_{\tau_n+t} : t \geq 0\}$) is a function of η_n, for each n.

The following result concerns two types of regenerative service systems. In part (i), the waiting times are a time-shift of the dynamics. In part (ii), the queue length process, and hence the arrival process, is a time-shift of the dynamics (this is a generalization of the system in Theorem 5.24).

Theorem 5.25. *Suppose Y is regenerative over τ_n and $E\tau_1 < \infty$.*
(i) *If $\{W_n\}$ is a τ_n-shift of Y and the limit λ exists and is finite, then the limits L and W exist and are*

$$W = \frac{1}{\lambda E\tau_1} E \sum_{n=0}^{N(\tau_1)-1} W_n, \qquad L = \lambda W. \tag{5.15}$$

(ii) *If X is a τ_n-shift of Y, then so is N and the limits L and λ are*

$$L = \frac{1}{\lambda E\tau_1} E \int_0^{\tau_1} X_t \, dt, \qquad \lambda = EN(\tau_1)/E\tau_1. \tag{5.16}$$

If, in addition, L and λ are finite and either $\{W_n\}$ is a τ_n-shift of Y or

$$P\{X_t = 0 \text{ for some } t \in [0, \tau_1)\} > 0,$$

then the limit W is given by $L = \lambda W$. The W is also given by (5.15) in case $\{W_n\}$ is a τ_n-shift of Y.

Proof. (i) Since $\{W_n\}$ is a τ_n-shift of Y, we can write

$$\sum_k W_k 1(\tau_n \leq T_k < \tau_{n+1}) = h(\eta_n),$$

for some function h, where η_n is defined by (5.14). The regenerative property of Y implies that the sequence η_n is stationary and ergodic—these concepts are defined in the next chapter. Furthermore, the sequence $h(\eta_n)$ is also stationary and ergodic. Then by the strong law of large numbers in Theorem 6.1 for stationary ergodic sequences,

$$n^{-1} \sum_{k=0}^{N(\tau_n)-1} W_k = n^{-1} \sum_{k=0}^{n-1} h(\eta_k) \to E \sum_{k=0}^{N(\tau_1)-1} W_k.$$

Since τ_n is a renewal process, we know that $\tau_n \sim nE\tau_1$. Also, the existence of λ implies that $N(\tau_n) \sim \lambda\tau_n \sim \lambda nE\tau_1$. From these observations and a discrete-time version of Theorem 5.10 with $Z(t) = \sum_{k=0}^{t-1} W_k$ and $T_n = N(\tau_n)$ we obtain expression (5.15) for W. Furthermore, since λ and W exist, Theorem 5.7 yields $L = \lambda W$.

(ii) Assuming X is a τ_n-shift of Y, it follows by the representation

$$N(t) = \sum_{s \le t} \max\{0, X_s - X_{s-}\}$$

that N is also a τ_n-shift of Y. To prove (5.16), first argue as in part (i) with $h(\eta_n) = N(\tau_{n+1}) - N(\tau_n)$ and obtain

$$n^{-1}N(\tau_n) = n^{-1}\sum_{k=0}^{n-1} h(\eta_k) \to EN(\tau_1).$$

In light of $\tau_n \sim nE\tau_1$ and $\tau_{n+1} \sim \tau_n$, Theorem 5.10 with $Z(t) = N(t)$ and $T_n = \tau_n$ yields expression (5.16) for λ. To prove expression (5.16) for L, argue as in the preceding case with

$$\int_{\tau_n}^{\tau_{n+1}} X_t \, dt = h(\eta_n).$$

Finally, the last assertion in part (ii) follows by part (i) and an argument like the proof of Theorem 5.24. □

Example 5.26. *GI/G/s and Regn/G/s Systems.* Consider a $GI/G/s$ system in which the arrival times form a renewal process and the service times of the s servers are independent and identically distributed and independent of the arrivals. We shall describe the system by the process $Y_t = (X_t, W_t^1, \ldots, W_t^s)$, where X_t is the number of customers in the system at time t and W_t^i is the remaining service time for the unit being served at server i at time t. Suppose the system parameters and rule for assigning units to servers are such that the process Y is regenerative over some τ_n's that are stopping times of Y. A typical rule is to route an arrival to one of the idle servers arbitrarily, or to the lowest numbered idle server. If the system empties out, it is natural to let τ_n denote the nth time that an arrival finds the system empty. Assume the waiting time of any arrival at time $\tau_n + t$ is a function of $Y_{\tau_n + t}$ (e.g., there are no delays caused by other factors). Then clearly $\{W_n\}$ is a τ_n-shift of Y. Note also that X is regenerative over τ_n. Consequently, Theorems 5.1 and 5.24 apply to this system. Special cases of this system are the classical $M/G/s$ and $GI/M/s$ systems.

A similar argument justifies that Theorems 5.1 and 5.24 apply to $Regn/G/s$ systems defined analogously, where the arrival process is a regenerative process. Special cases include the following systems:

- $M^X/G/s$ (compound Poisson arrivals).
- $GI^X/G/s$ (compound renewal arrival process).
- $SM/G/s$ (semi-Markov arrival process).

$Regn/G/s$ systems also arise naturally in tandem networks. For instance, in the two-station tandem network $GI/G/s \to \cdot/G/s$, the first station is a regenerative system, and so the second station is a $Regn/G/s$ system. Similarly, for the more general tandem system $Regn/G/s \to \cdot/G/s \to \ldots \cdot/G/s$, or an analogous tree-like network with one-way flows, one can obtain average waiting times at each station by Theorems 5.1 and 5.24. □

5.7 Exercises

1. *Discrete Analogue of Theorem 5.10.* Suppose that Y_n is an increasing sequence of random variables associated with a point process N on $\mathbb{R}_+$. Consider the averages

$$Y \equiv \lim_{n\to\infty} n^{-1} Y_n, \qquad \hat{Y} \equiv \lim_{t\to\infty} t^{-1} Y_{N(t)}.$$

Show that if any two of the limits Y, λ, and $\hat{Y}$ exist, then the other one exits and $\hat{Y} = \lambda Y$.
2. In Theorem 5.4 concerning a Little law, show that statement (e) is equivalent to statement (f). Hint: Use Theorem 5.15 and Lemma 5.20.
3. In the context of Theorem 5.4, show that if the limits L and W exit and $\int_0^{\hat{T}_n} X_s\, ds \sim \sum_{k=1}^n W_k$, then λ exists and $L = \lambda W$.
4. In Theorem 5.1, the assumption that the queueing process may equal 0 can be replaced by the weaker assumption that the departures are regular, which is implied by any one of the following conditions:
(i) The $\{W_n\}$ are functions of Y and there exist stopping times τ_n of Y such that Y is regenerative over τ_n (e.g., the τ_n are entrance times to a fixed set) and, for each $n \geq 1$, the $\{W_k : T_k \geq \tau_n\}$ are conditionally independent of $\{Y_s : s < \tau_n\}$ given Y_{τ_n}.
(ii) There are stopping times τ_n of Y such that Y is regenerative over τ_n and, for each $n \geq 1$, the $\{W_k : T_k \geq \tau_n\}$ is a function of $\{Y_{\tau_n+t} : t \geq 0\}$.
(iii) Y is regenerative over τ_n and $\{W_k : T_k \geq \tau_n\}$ is a function of Y in the time interval $[\tau_n, \infty)$ for each n.
Show that (i) implies (ii), and (ii) implies (iii).

5.8 Bibliographical Notes

The first references on Little laws are Morse (1958) and Little (1961). The extensive literature and history of these laws are reviewed in Whitt (1991), Serfozo (1994), and in the comprehensive monograph on sample path properties by El-Taha and Stidham (1999). Sample articles in the literature are Stidham (1972, 1974) (containing parts of Theorem 5.4); Heyman and Stidham (1980) and Miyazawa (1995) (covering $H = \lambda G$ related to Example 6.15); Rolski and Stidham (1983) and Miyazawa (1994) (fluid models); Glynn and Whitt (1988) (functional limit laws);

and Stidham and El-Taha (1989) (paths with embedded point processes). The material in Chapter 5 comes from Serfozo (1994), which is a distillation and extension of ideas from earlier articles.

6

Stationary Systems

This chapter is an introduction to the basics of stationary processes and Palm probabilities that are used in queueing theory. This includes Palm calculus and Campbell–Mecke formulas for functionals of stationary systems. This material is the foundation for modeling networks and queueing systems with stationary dynamics, and for obtaining Little laws for such systems.

6.1 Preliminaries on Stationary Processes

This section reviews ergodic theorems for stationary processes and the heredity property of stationarity.

We will use the following terminology. A stochastic process $X = \{X_t : t \in \mathbb{R}\}$ with values in a space $\mathbb{E}$ is *stationary* if the distribution of the time-shifted process $S_t X \equiv \{X_{s+t} : s \in \mathbb{R}\}$ is independent of t. A stationary process X is *ergodic* if $P\{X \in A\} = 0$ or 1 for each set A that satisfies $\{X \in A\} = \{S_t X \in A\}$, for $t \in \mathbb{R}$ (A is a time-shift-invariant set of X). Stationarity and ergodicity of sequences are defined similarly—the parameter t in these cases would simply be an integer.

The following are strong laws of large numbers (or ergodic statements) for stationary processes. Assume the processes here are real valued and all the expected values are finite.

Theorem 6.1. (i) *If* $\{X_n : n \in \mathbb{Z}\}$ *is a stationary, ergodic sequence, then*

$$\lim_{n\to\infty} n^{-1} \sum_{k=1}^{n} X_k = EX_0 \quad \textit{w.p.1.}$$

(ii) *If* $\{X_t : t \in \mathbb{R}\}$ *a is stationary ergodic process, then*

$$\lim_{t\to\infty} t^{-1} \int_0^t X_s\, ds = EX_0 \quad w.p.1.$$

(iii) *If* $\{Z_t : t \in \mathbb{R}_+\}$ *is a nondecreasing process and* τ_n *are increasing times such that the sequence* $\{(\tau_{n+1} - \tau_n, Z_{\tau_{n+1}} - Z_{\tau_n}) : n \in \mathbb{Z}_+\}$ *is stationary and ergodic, then*

$$\lim_{t\to\infty} t^{-1} Z_t = \frac{E(Z_{\tau_1} - Z_{\tau_0})}{E(\tau_1 - \tau_0)} \quad w.p.1.$$

PROOF. Statement (i) is proved in standard texts that cover ergodic theory or laws of large numbers. Statement (iii) is for processes with stationary increments at special embedded times τ_n; it follows from (i) and Theorem 5.10. Statement (ii) follows from (iii) with $Z_t = \int_0^t X_s\, ds$ and $\tau_n = n$, where $EZ_1 = \int_0^1 EX_t\, dt = EX_0$ by Fubini's theorem. □

A distinctive feature of stationarity is that many functions of stationary processes are also stationary. To discuss this heredity property, suppose that X is a stationary process. If $Y_t = f(X_t)$, where f is a function on the state space of X, then Y is stationary and also ergodic if X is. This assertion is an elementary example of the following result, which follows immediately from the definitions of stationarity and ergodicity.

Proposition 6.2. *Suppose*

$$Y_t = f(S_t X), \quad t \in \mathbb{R}, \tag{6.1}$$

where f *is a function on the (measurable) space of sample paths of* X*. Then* $\{Y_t : t \in \mathbb{R}\}$ *and the joint process* $\{(X_t, Y_t) : t \in \mathbb{R}\}$ *are stationary. These processes are also ergodic when* X *is.*

We will call the process Y defined by (6.1) a *stationary functional of* X. Note that Y is a time-shift invariant function of X in that if X is shifted in time by some value, then so is Y. The preceding proposition also applies to multiple processes. For instance, if Y and $\tilde{Y}$ are stationary functionals of X, then so are the multidimensional processes $(Y, \tilde{Y})$ and $(X, Y, \tilde{Y})$. This and the following transitivity property are useful for establishing joint stationarity of multidimensional processes.

Proposition 6.3. *If* Y *is a stationary functional of* X*, and* $\tilde{Y}$ *is a stationary functional of* (X, Y)*, then* $\tilde{Y}$ *is a stationary functional of* X*.*

PROOF. This follows since we can write $\tilde{Y}_t = \tilde{f}(S_t X, f(S_t X))$, where $Y_t = f(S_t X)$, and $\tilde{Y}_t = \tilde{f}(S_t X, S_t Y)$. □

Analyses of networks and systems with stationary dynamics often involve stationary point processes on the real line. We will use the point process terminology in the first section of Chapter 4. Suppose N is a point process on $\mathbb{R}$ with points at the locations

$$\ldots \le T_{-2} \le T_{-1} \le T_0 \le 0 < T_1 \le T_2 \ldots .$$

That is,

$$N(A) = \sum_n 1(T_n \in A), \quad A \subset \mathbb{R}.$$

The point process N is *stationary* if the time-shifted process $\{N(A+t) : A \subset \mathbb{R}\}$ is equal in distribution to N, for each $t \in \mathbb{R}$. The stationarity implies that $EN(A) = \lambda|A|$, where $|A|$ denotes the Lebesgue measure of A and $\lambda \equiv EN(0, 1]$ is the *intensity of* N. With no loss in generality, we assume the intensity is positive and finite.

Ergodicity of a stationary point process is defined in the same way as it is for a continuous-time stationary process. If the point process N is stationary and ergodic, then it satisfies the two strong laws of large numbers

$$t^{-1}N(t) \to \lambda \quad \text{w.p.1 as } t \to \infty. \tag{6.2}$$

$$n^{-1}T_n \to \lambda^{-1} \quad \text{w.p.1 as } t \to \infty. \tag{6.3}$$

The first strong law follows from (iii) in Theorem 6.1 just as (ii) in Theorem 6.1 does. The second law is equivalent to the first one by Theorem 5.8.

Stationary point processes arise naturally as shift-invariant functions of stationary processes. Namely, suppose X is a continous-time stationary process, and N is a point process such that $N(A + t) = f(S_t X)(A)$, for each $A \subset \mathbb{R}$, where f is a function from the space of sample paths of X to the counting measures. Then N is stationary, and it is also ergodic if X is. We call N a *stationary functional of* X. This terminology is consistent with that used for (6.1).

Example 6.4. *Stationary Functionals of Networks*. Suppose X is a stationary process that represents the numbers of units in an m-node network. The process that represents the number of units in the sector J is $X_t(J) = f(X_t)$, where $f(x) = \sum_{j \in J} x_j$. Clearly X_J is a stationary functional of X. Now, the point process N_J of arrival times of units in J is given by

$$N_J(A) = \sum_{t \in A} \max\{0, X_t(J) - X_{t-}(J)\}.$$

This covers the possibility of batch arrivals, in which case N_J is not a simple point process. Clearly $N(A + t) = \sum_{s \in A} \max\{0, X_{s+t} - X_{(s+t)-}\}$, and so N_J is a stationary functional of X_J. Furthermore, by the propositions above, N_J and (X_J, N_J) are stationary functionals of X.

6.2 Palm Probabilities

In Chapter 4, we discussed special types of Palm probabilities for stationary Markov processes. We now present a more comprehensive study of Palm probabilities for general stationary systems. This section covers elementary properties, and the next section covers a variety of formulas involving Palm probabilities.

In modeling a stationary system, the standard approach is to start with primitive information in terms of a stationary process that contains all the essential information of the system, and then express properties of the system as functionals of the primitive process. For our development, we will consider a system in which the primitive information is the stationary process θ defined as follows.

Definition 6.5. Let $(\Omega, \mathcal{F}, P)$ denote a probability space. For each $t \in \mathbb{R}$, suppose $\theta_t : \Omega \to \Omega$ is a bijection such that the map $(t, \omega) \to \theta_t(\omega)$ is measurable and

$$\theta_s(\theta_t(\omega)) = \theta_{s+t}(\omega), \quad \omega \in \Omega, \ s, t \in \mathbb{R}.$$

In particular, $\theta_0(\omega) = \omega$ and $\theta_{-t} = \theta_t^{-1}$. Assume the probability measure P on Ω is *invariant under* θ in the sense that

$$P\{\theta_t \in A\} = P\{A\}, \quad A \in \mathcal{F}, \ t \in \mathbb{R}.$$

Then $\theta \equiv \{\theta_t : t \in \mathbb{R}\}$ is a measurable stationary process on $(\Omega, \mathcal{F}, P)$ with values in Ω. The process θ is a *stationary flow* on $(\Omega, \mathcal{F}, P)$.

Stationary functionals of θ are sometimes called *compatible with the flow* θ. Unless specified otherwise, we assume that all stationary processes introduced below are stationary functionals of the flow θ, and hence they all reside on the single probability space $(\Omega, \mathcal{F}, P)$. Recall that any collection of stationary functionals of θ is jointly stationary.

To model a system in this framework, one takes θ as a stationary process that contains all the system information. Without loss of generality, one may assume that any stationary process $\{X_t : t \in \mathbb{R}\}$ is a stationary flow, since one can construct a flow that is equal in distribution to X. For instance, θ could be a stationary network process X as in Example 6.4 above, where the stationary functionals of X are the number of units in a sector of the network and the point process of arrival times to the sector.

We now consider a point process N on $\mathbb{R}$ that is a stationary functional of θ. Assume N is simple and its intensity $\lambda = EN(0, 1]$ is positive and finite. A typical problem is to compute the probability of some event under the condition that N has a point at time 0. This would be the conditional probability $P\{A|N(\{0\}) = 1\}$, provided it is well defined. This conditional probability does not exist in the usual sense, however, when the event $N(\{0\}) = 1$ has zero probability. In general, the desired "conditional probability" is represented by the Palm probability defined as follows.

Definition 6.6. The *Palm probability measure* P_N associated with N is defined on the underlying probability space $(\Omega, \mathcal{F}, P)$ by

$$P_N\{A\} = \frac{1}{\lambda|B|} E \int_B 1(\theta_t \in A) N(dt), \quad A \in \mathcal{F}, \tag{6.4}$$

where $B \subset \mathbb{R}$ is a set whose Lebesgue measure $|B|$ is positive and finite. We let E_N denote the expectation under P_N.

Since N is a stationary functional of θ, the right side of (6.4) does not depend on the choice of the set B, and so P_N is well defined. Note that (6.4) can also be written as

$$P_N\{A\} = \frac{1}{EN(B)} EN\{t \in B : \theta_t \in A\}.$$

Another representation based on the times T_n is

$$P_N\{A\} = \frac{1}{EN(B)} E \sum_n 1(\theta_{T_n} \in A, T_n \in B). \tag{6.5}$$

Since θ_{T_n} is what an observer "sees" of θ on all of $\mathbb{R}$ viewing it at T_n, one can say, loosely speaking, that $P_N\{A\}$ is the portion of the times T_n that an observer sees θ in A.

In the next result, (6.8) says that the Palm probability P_N is concentrated on the subspace $\{N(\{0\}) = 1\} \subset \Omega$. Consequently, P_N describes the probabilities of any event in this subspace "given that N has a point at 0." Accordingly, (6.6) below is the distribution of the process X *given that N has a point at* 0.

Proposition 6.7. *If X is a stationary functional of θ, then*

$$P_N\{X \in C\} = \lambda^{-1} E \int_{(0,1]} 1(S_t X \in C) N(dt), \tag{6.6}$$

where C is a set in the space of sample paths of X. In particular, this formula applies to $X = N$, and hence

$$P_N\{N(B) = n\} = \lambda^{-1} E \int_{(0,1]} 1(N(B+t) = n) N(dt), \tag{6.7}$$

$$P_N\{N(\{0\}) = 1\} = 1 = P_N\{T_0 = 0\}. \tag{6.8}$$

PROOF. By (6.4) with $B = (0, 1]$, we have

$$P_N\{X \in C\} = \lambda^{-1} E \int_{(0,1]} 1(\theta_t \in \{X \in C\}) N(dt). \tag{6.9}$$

Assuming X takes values in a space $\mathbb{E}$, we can write $X_t = g(\theta_t)$, for some $g : \Omega \to \mathbb{E}$. Since θ is a flow, it follows that, for any t and C,

$$\theta_t(\omega) \in \{\{g(\theta_s) : s \in \mathbb{R}\} \in C\} \text{ if and only if } \{g(\theta_{s+t}(\omega)) : s \in \mathbb{R}\} \in C.$$

That is, $1(\theta_t \in \{X \in C\}) = 1(S_t X \in C)$. Applying this to (6.9) yields (6.6).

Next, note that (6.7) is a special case of (6.6). Finally, (6.8) follows since $\{N(\{0\}) = 1\} = \{T_0 = 0\}$ and, by (6.7),

$$\begin{aligned} P_N\{N(\{0\}) = 1\} &= \lambda^{-1} E \int_{(0,1]} 1(N(\{t\}) = 1) N(dt) \\ &= \lambda^{-1} EN(0, 1] = 1. \end{aligned}$$

□

The next example shows that the Palm probabilities we discussed in Chapter 4 for $\mathcal{T}$-transitions of Markov processes are special cases of the Palm probability defined by (6.4).

Example 6.8. *Palm Probabilities of $\mathcal{T}$-transitions.* Let X be a stationary functional of θ. Suppose the sample paths of X are contained in the set of all functions from $\mathbb{R}$ to $\mathbb{E}$ that are piecewise constant and right-continuous. Let $\mathcal{T}$ be a subset of these functions. We say that a $\mathcal{T}$-transition of X occurs at time t if $X_t \neq X_{t-}$ and $S_t X \in \mathcal{T}$. Assume there a finite number of such transitions in any finite time interval. Then the times of these transitions form a simple point process N that is a functional of θ. Assume N has a positive, finite intensity λ. Then according to (6.4),

$$P_N\{X \in \mathcal{T}'\} = \lambda_{\mathcal{T}'}/\lambda, \tag{6.10}$$

where $\lambda_{\mathcal{T}'} = E\int_{(0,1]} 1(X_t \neq X_{t-}, S_t X \in \mathcal{T}')N(dt)$ is the rate of $\mathcal{T}'$-transitions. This example shows that P_N can be defined by the ratio-of-rates formula (6.10) if one is only interested in Palm probabilities for X under $\mathcal{T}$-transitions. □

We will now show that certain sequences of events with respect to the times T_n are stationary under P_N, even though they are not stationary under P. For example, the sequence of interpoint distances $\{T_{n+1} - T_n : n \in \mathbb{Z}\}$ is stationary under P_N, but it is not stationary under P. Also, if X is a stationary functional of θ, then the embedded sequence $\{X_{T_n} : n \in \mathbb{Z}\}$ is stationary under P_N, but it is not stationary under P. These properties are consequences of the following important result for Palm probabilities.

Theorem 6.9. *The sequence $\{\theta_{T_n} : n \in \mathbb{Z}\}$ is stationary under P_N. Moreover, this sequence is ergodic under P_N if and only if θ is ergodic under P.*

Proof. Consider the map $\hat{\theta} \equiv \theta_{T_1}$ on $\{T_0 = 0\} \subset \Omega$. Since $P_N\{T_0 = 0\} = 1$ and θ is a flow, we can write

$$\theta_{T_n} = \hat{\theta}^n \quad \text{w.p.1 under } P_N, \text{ for each } n \in \mathbb{Z}. \tag{6.11}$$

For instance, $\theta_{T_3} = \theta_{T_3 - T_2}(\theta_{T_2 - T_1}(\theta_{T_1 - T_0})) = \hat{\theta}^3$ w.p.1 under P_N. In light of (6.11), the first assertion of the theorem is equivalent to saying that P_N is invariant under the map $\hat{\theta}$ in the sense that

$$P_N\{\hat{\theta}^{-1}(A)\} = P_N\{A\}, \quad A \subset \mathbb{R}. \tag{6.12}$$

To see this, note that by (6.5) with $B = (0, t]$ and $N(t) = N(0, t]$, we have

$$\begin{aligned} P_N\{\hat{\theta}^{-1}(A)\} &= \frac{1}{\lambda t} E \sum_{n=1}^{N(t)} 1(\theta_{T_n} \in \hat{\theta}^{-1}(A)) = \frac{1}{\lambda t} E \sum_{n=1}^{N(t)} 1(\theta_{T_{n+1}} \in A) \\ &= P_N\{A\} + \frac{1}{\lambda t}[P\{\theta_{T_{N(t)+1}} \in A\} - P\{\theta_{T_1} \in A, N(t) \geq 1\}]. \end{aligned}$$

The last term converges to 0 as $t \to \infty$, and so (6.12) is true. The second assertion of the theorem follows by the definition of ergodicity. □

Theorem 6.9 applies as follows to sequences generated by stationary functionals of θ. Examples of this were the lead-in to Theorem 6.9.

Corollary 6.10. *If X is a stationary functional of θ, then the sequence $\{S_{T_n}X : n \in \mathbb{Z}\}$ of what one sees of X on the entire time axis at the times T_n is stationary under P_N. This sequence is ergodic under P_N if and only if θ is ergodic under P.*

PROOF. It suffices to show that $S_{T_n}X$ is a stationary functional of θ_{T_n} under P_N. But this follows, since we can write $X_t = g(\theta_t)$ for some $g : \Omega \to \mathbb{E}$, and hence

$$S_{T_n}X = \{X_{t+T_n} : t \in \mathbb{R}\} = \{g(\theta_t(\theta_{T_n})) : t \in \mathbb{R}\} = h(\theta_{T_n}),$$

where $h(\omega) = \{\{g(\theta_t(\omega)) : t \in \mathbb{R}\}$. □

The next result says that laws of large numbers for a stationary process X are valid under P_N as well as under P. Similarly, laws of large numbers for $\{\theta_{T_n} : n \in \mathbb{Z}\}$ are valid under P as well as under P_N. We use these "cross" ergodic theorems to link Little laws for expectations to those for limiting averages. For a proof, see the references at the end of this chapter.

Theorem 6.11. *If X is a stationary process and EX_0 exists, then*

$$t^{-1}\int_0^t X_s\,ds \to EX_0 \quad \text{w.p.1 under } P_N.$$

If $\{Y_n : n \in \mathbb{Z}\}$ is a stationary functional of $\{\theta_{T_n} : n \in \mathbb{Z}\}$ and EY_1 exists, then

$$n^{-1}\sum_{k=1}^{n} Y_k \to EY_1 \quad \text{w.p.1 under } P.$$

6.3 Campbell–Mecke Formulas for Palm Probabilities

This section covers further properties of the Palm probability P_N defined by (6.4), which is associated with the stationary point process N on $\mathbb{R}$. Keep in mind that N is simple and has a finite positive intensity λ. The focus is on the Campbell–Mecke formula, which is a framework for analysis involving Palm probabilities (sometimes called Palm calculus). We present this formula and several ostensibly different, but equivalent versions of it.

Many probabilities and expectations under P have natural representations in terms of expectations or probabilities under P_N. They can be obtained by the following formula, which is an example of Fubini's theorem. Here dt denotes Lebesgue measure.

Theorem 6.12. **(Campbell–Mecke Formula)** *For any $f : \mathbb{R} \times \Omega \to \mathbb{R}_+$,*

$$E\int_{\mathbb{R}} f(t, \theta_t)N(dt) = \lambda E_N \int_{\mathbb{R}} f(t, \theta_0)\,dt. \tag{6.13}$$

PROOF. This can be proved first for indicator functions f by the definition of P_N, then for linear combinations of indicators, and finally for general functions by monotone convergence. Another approach is to apply Fubini's theorem as follows. We can write

$$\int_{\mathbb{R}} f(t, \theta_t) N(dt) = \int_{\mathbb{R}\times\Omega} f(t, \omega') M(d(t, \omega')), \tag{6.14}$$

where M is the random measure defined by $M(B \times C) = \int_B 1(\theta_t \in C) N(dt)$. The dummy variable ω' is different from the suppressed ω in θ and N. By the definition (6.4) of P_N, we know that $EM(B \times C) = \lambda P_N\{C\} \int_B dt$. Then by (6.14) and Fubini's theorem, the left side of (6.13) equals

$$\begin{aligned} E\int_{\mathbb{R}\times\Omega} f(t, \omega') M(d(t, \omega')) &= \int_{\mathbb{R}\times\Omega} f(t, \omega') EM(d(t, \omega')) \\ &= \lambda \int_{\mathbb{R}\times\Omega} f(t, \omega') P_N\{d\omega'\}\, dt, \end{aligned}$$

which equals the right side of (6.13). □

The Campbell–Mecke formula (6.13) is sometimes called Campbell's formula or Mecke's formula, and so we combine the names. Some studies refer to (6.16) below as Campbell's formula.

In applications where one suppresses the formalism of the θ process, the Campbell–Mecke formula is as follows. Let $\{X_t : t \in \mathbb{R}\}$ denote a stochastic process with state space $\mathbb{E}$ and assume its sample paths are in the set D of all functions from $\mathbb{R}$ to $\mathbb{E}$ that are right continuous and have left-hand limits. Let N be a point process on $\mathbb{R}$ such that X, N are jointly stationary. Then for any $g : \mathbb{R} \times D \to \mathbb{R}$,

$$E\int_{\mathbb{R}} g(t, S_t X) N(dt) = \lambda E_N \int_{\mathbb{R}} g(t, X)\, dt, \tag{6.15}$$

provided the expectation exists. Note the resemblance of this formula to the extended Levy formula (4.3).

We now present several formulas that are ostensibly different from the Campbell–Mecke formula, but are actually equivalent to it (two formulas are equivalent if each one implies the other). The first theme concerns functionals of stationary marked point processes defined as follows.

Definition 6.13. Let $\{\xi_n : n \in \mathbb{Z}\}$ be random elements of some space $\mathbb{E}$ such that $\xi_n = h(\theta_{T_n})$, for some $h : \Omega \to \mathbb{E}$. The space–time point process

$$M(\cdot) = \sum_n 1\big((T_n, \xi_n) \in \cdot\big)$$

on $\mathbb{R} \times \mathbb{E}$, or its point locations $\{(T_n, \xi_n) : n \in \mathbb{Z}\}$, is a *stationary marked point process*. The ξ_n are *marks* of $N(\cdot) = M(\cdot \times \mathbb{E})$.

In the preceding definition, the stationarity of N and θ ensure that the space–time process M is stationary in time: The distribution of $\{(T_n - t, \xi_n) : n \in \mathbb{Z}\}$ is independent of t. Also, we know from Theorem 6.9 that $\{\xi_n : n \in \mathbb{Z}\}$ is a stationary sequence under P_N, but it is not stationary under P.

The following version of the Campbell–Mecke formula is useful for deriving Little laws for queueing systems. This motivates the inclusion of Little in the name.

Theorem 6.14. (Campbell–Little–Mecke Formula) *Suppose* $\{(T_n, \xi_n) : n \in \mathbb{Z}\}$ *is a stationary marked point process. For any* $f : \mathbb{R} \times \mathbb{E} \to \mathbb{R}$,

$$E \sum_n f(T_n, \xi_n) = \lambda E_N \int_{\mathbb{R}} f(t, \xi_0)\, dt, \tag{6.16}$$

provided the expectation exists. This formula is equivalent to the Campbell–Mecke formula (6.13).

PROOF. It suffices to prove (6.16) for nonnegative f. But this follows since (6.13) implies

$$E \sum_n f(T_n, \xi_n) = E \int_{\mathbb{R}} f(t, h(\theta_t))N(dt) = E_N \int_{\mathbb{R}} f(t, \xi_0)\, dt,$$

where $\xi_n = h(\theta_{T_n})$. Next, note that formula (6.16) with $\xi_n = \theta_{T_n}$ is (6.13). This and the preceding sentence prove that (6.16) is equivalent to (6.13). □

Example 6.15. *Functionals of Marked Point Processes.* Suppose $\{(T_n, \xi_n) : n \in \mathbb{Z}\}$ is a stationary marked point process. Consider the process

$$X_t = \sum_n f(t - T_n, \xi_n), \quad t \in \mathbb{R},$$

where $f : \mathbb{R} \times \mathbb{E} \to \mathbb{R}$. The X is a stationary functional of θ, since by the change of variable $u = s - t$

$$X_t = \int_{\mathbb{R}} f(t - s, h(\theta_s))N(ds) = \int_{\mathbb{R}} f(-u, h(\theta_u(\theta_t)))g(\theta_t)(du), \tag{6.17}$$

where $\xi_n = h(\theta_{T_n})$ and $N(A + t) = g(\theta_t)(A)$. By the Campbell–Little–Mecke formula (6.16), the mean of this process is

$$EX_0 = \lambda E_N \int_{\mathbb{R}} f(t, \xi_0)\, dt, \tag{6.18}$$

provided the expectation exists (the integral also equals $\int_{\mathbb{R}} f(-t, \xi_0)\, dt$). This formula is sometimes referred to in queueing applications as $H = \lambda G$.

A common form of the process X is

$$X_t = \sum_n \tilde{f}(t - T_n, \xi_n)1(T_n + \alpha_n \le t < T_n + \beta_n),$$

where $(\xi_n, \alpha_n, \beta_n)$ are marks of N such that $\beta_n \ge \alpha_n$. In this case,

$$EX_0 = \lambda E_N \int_{\alpha_0}^{\beta_0} \tilde{f}(t, \xi_0)\, dt. \qquad \square \tag{6.19}$$

The Campbell–Mecke formula allows us to express the probability P in terms of P_N as follows.

Corollary 6.16. (Inversion Formula) *If X is a real-valued stationary functional of θ, then*

$$EX_0 = \lambda E_N \int_0^{T_1} X_t \, dt, \tag{6.20}$$

provided the last expectation exists. Hence,

$$P\{A\} = \lambda E_N \int_0^{T_1} 1(\theta_t \in A) \, dt, \quad A \in \mathcal{F}. \tag{6.21}$$

PROOF. Expression (6.20) follows from (6.19) with $\beta_n = T_{n+1} - T_n$, since

$$X_t = \sum_n X_t 1(T_n < t \le T_{n+1}) = \sum_n X_{t-T_n}(\theta_{T_n}) 1(T_n \le t < T_n + \beta_n).$$

Expression (6.21) follows by applying (6.20) to $X_t = 1(\theta_t \in A)$ since $P\{A\} = EX_0$. □

Most of the material we have covered on stationary point processes automatically extends to random measures. A random measure M on $\mathbb{R}$ is a mapping from a probability space to the space of all measures on $\mathbb{R}$ that are finite on compact sets. The M is a point process if it is an integer-valued measure. A Palm probability P_M of a random measure M on $\mathbb{R}$ is also defined by (6.4). From the preceding proofs, it is clear that the Campbell–Mecke and inversion formulas above also apply to Palm probabilities of random measures. The following are two more equivalent versions of the Campbell–Mecke formula. We express these new formulas in terms of random measures instead of point processes to avoid technical differences between counting measures and general measures.

Theorem 6.17. (Integrals of Product Measures) *Suppose M and M' are random measures on $\mathbb{R}$ that are stationary functionals of θ and have respective intensities λ and λ' that are positive and finite. Then for any $g : \mathbb{E}^2 \times \Omega \to \mathbb{R}_+$,*

$$E \int_{\mathbb{E}^2} g(t, s, \theta_t) M'(ds) M(dt) = \lambda E_M \int_{\mathbb{E}^2} g(t, s+t, \theta_0) M'(ds) \, dt. \tag{6.22}$$

This formula is equivalent to the Campbell–Mecke formula (6.13) for random measures.

PROOF. Expression (6.22) follows from (6.13) for random measures with

$$f(t, \theta_t) = \int_{\mathbb{E}} g(t, s+t, \theta_t) M'(ds + t) = \int_{\mathbb{E}} g(t, s+t, \theta_t) h(\theta_t)(ds).$$

Here $M'(\cdot + t) = h(\theta_t)(\cdot)$. Conversely, (6.13) for random measures follows from (6.22) with $g(t, s, \omega) = f(t, \omega) 1(s \in (0, 1])$ and M' as the Lebesgue measure. □

One can view (6.22) as a "conditional" Campbell–Mecke formula for the bivariate random measure $\tilde{M}(ds \times dt) = M'(ds) M(dt)$, where the right side of (6.22) is like "conditioning" on the M part of $\tilde{M}$. The M and M' may be dependent. Expression (6.22) also extends to g that may be negative as well as nonnegative and the measure

M' may be a signed random measure: $M'(A) = M_1'(A) - M_2'(A)$, where M_1' and M_2' are nonnegative random measures. In this case, one applies (6.22) separately to the integrals of the positive and negative parts of g under the measures M_1' and M_2', provided the sum of the expectations is well defined (possibly infinite).

As an example of (6.22), consider the stationary random measure

$$\tilde{M}(B) = \int_{B\times\mathbb{R}} g(t,s,\theta_t)M'(ds)M(dt).$$

Its intensity, according to (6.22) is

$$E\tilde{M}(0,1] = \lambda E_M \int_{B\times\mathbb{R}} g(t,s+t,\theta_0)M'(ds)\,dt, \tag{6.23}$$

where $|B| = 1$. This is called the *Swiss Army formula.*

Here is another useful formula.

Theorem 6.18. (Neveu's Exchange Formula) *Suppose M and M' are random measures on $\mathbb{R}$ that are stationary functionals of θ and have respective intensities λ and λ' that are positive and finite. Then for any $f : \mathbb{R}\times\Omega \to \mathbb{R}_+$,*

$$\lambda E_M \int_{\mathbb{R}} f(t,\theta_t)M'(dt) = \lambda' E_{M'} \int_{\mathbb{R}} f(t,\theta_0)M(dt). \tag{6.24}$$

This formula is equivalent to the Campbell–Mecke formula (6.13) for random measures.

PROOF. To see this, fix $B \subset \mathbb{R}$ such that $|B| = 1$. Then applying (6.22) to M and then to M', we have

$$\begin{aligned}\text{Left side of (6.24)} &= \lambda E_M \int_{\mathbb{R}^2} f(t,\theta_t)1(s+t\in B)M'(dt)ds\\ &= E\int_{\mathbb{R}^2} f(t-s,\theta_t)1(s\in B)M'(dt)M(ds)\\ &= \lambda' E_{M'}\int_{\mathbb{R}^2} f(-s,\theta_0)1(t\in B)M(ds)\,dt\\ &= \text{Right side of (6.24)}.\end{aligned}$$

This proof also justifies that the Campbell–Mecke formula (6.13) for random measures implies (6.24), since (6.22) is equivalent to (6.13) by Theorem 6.17. Conversely, (6.13) for random measures follows from (6.24) when M is the Lebesgue measure (in this case, $P_M = P$ and $E_M = \mathbb{E}$). Thus, (6.24) is equivalent to (6.13) for random measures. □

We end this section with a rate conservation law that is implied by the Campbell–Mecke formula but is not equivalent to it.

Example 6.19. *Rate Conservation Law.* Suppose $\{X_t : t \in \mathbb{R}_+\}$ is a real-valued stochastic process of the form

$$X_t = X_0 + \sum_{i=1}^{n} \int_{(0,t]} X_s^i M_i(ds), \tag{6.25}$$

where X^i's are real-valued processes and M_i's are point processes or random measures on $\mathbb{R}_+$.

Corollary 6.20. *Suppose $X, X_1, \ldots, X_n, M_1, \ldots, M_n$ are jointly stationary, and let $\lambda_i = EM_i(1)$. Then*

$$\sum_{i=1}^{n} \lambda_i E_{M_i}[X_0^i] = 0,$$

provided the expectations exist.

This *rate conservation law* says that the expected rate of change of X under its n types of changes is 0. This is what one would anticipate for a stationary process.

PROOF. The assertion follows by taking the expectation of the terms in (6.25) with $t = 1$ and using the Campbell–Mecke formula (6.15) for each term in the sum. □

The preceding result applies if the processes X and $M = M_1 + \cdots + M_n$ are jointly stationary, the M_i's have disjoint supports, and each X^i is the Radon–Nikodym derivative of X with respect to M_i.

In particular, suppose X is a stationary process whose sample paths are of bounded variation and the number of its discontinuities forms a point process N on $\mathbb{R}_+$. Clearly N is a stationary functional of X. Since each sample path of X is absolutely continuous except at its discontinuity points, it follows that

$$X_t = X_0 + \int_0^t X_s' \, ds + \int_{(0,t]} (X_s - X_{s-}) N(ds), \tag{6.26}$$

where X_s' is the Radon–Nikodym derivative of X with respect to the Lebesgue measure ds. Then by Corollary 6.20 with M_1 as Lebesgue measure and $M_2 = N$, we have

$$EX_0' + \lambda E_N(X_0 - X_{0-}) = 0,$$

where λ is the intensity of N. When X is not stationary, there is an analogous rate conservation law in which the preceding expectations are limiting averages of the increments; see Exercise 4. □

6.4 Little Laws for Stationary Systems

We are now ready to describe average waiting times and the law $L = \lambda W$ for service systems with stationary characteristics.

Throughout this section, we assume the process $\{X_t : t \in \mathbb{R}\}$ represents the number of units in a service system over the entire time axis $\mathbb{R}$. The point process of arrivals N can be expressed as $N(A) = \sum_{t \in A} \max\{0, X_t - X_{t-}\}$ and batch arrivals are allowed. We assume that the waiting times W_n are well defined on the underlying probability space for X, but we do not assume any special functional relation between W_n and X. Additional assumptions on the structure of these processes are imposed in the theorem statements. Recall that L, λ, and W denote the average queue length, arrival rate, and waiting time, respectively. These limits may be infinite, and the phrase w.p.1 will be with respect to the underlying probability P, unless specified otherwise.

Our first result is for a system in which the process X is stationary.

Theorem 6.21. *Suppose the queueing process X is stationary and ergodic. Then the arrival process N is a stationary functional of X. Hence, the limits L and λ exist and are $L = EX_0$ and $\lambda = EN(1)$. If, in addition, λ is finite and $P\{X_0 = 0\} > 0$, then the limit W exists and $L = \lambda W$.*

PROOF. We noted that the arrival process $N(A) = \sum_{t \in A} \max\{0, X_t - X_{t-}\}$ is a stationary functional of X and hence it is stationary and ergodic. Now, the ergodic theorems for X and N yield $L = EX_0$ and $\lambda = EN(1)$.

The rest of the theorem will follow by Theorem 5.7 upon showing that X is recurrently empty with respect to the times τ_n at which X hits state 0. We first show that these times exist. Let

$$N_0(A) = \sum_{t \in A} 1(0 = X_t < X_{t-}), \quad A \subset \mathbb{R},$$

which is the number of times that X hits 0 in the time set A. Clearly N_0 is finite on finite time intervals since X takes at most a finite number of jumps in such an interval. Therefore, N_0 is a point process on $\mathbb{R}$. Also, it is clear by its definition that N_0 is a stationary functional of X. Consequently, N_0 is stationary and ergodic, and so $N_0(t)/t \to \lambda_0 \equiv EN_0(1)$. Note that λ_0 is finite since $N_0(1) \le 1 + N(1)$ and $\lambda = EN(1)$ is finite. Also, note that $\lambda_0 > 0$, since

$$P\{N_0(1) \ge 1\} \ge P\{N(-\infty, t) \ge 1, X_t = 0\} = P\{X_t = 0\} > 0,$$

for any fixed $t \in [0, 1]$. Here $N(-\infty, t) = \infty$ w.p.1 since the rate λ is positive. Now $\lambda_0 > 0$ implies that $N_0(t) \uparrow \infty$, and so the nth time $\tau_n = \min\{t : N_0(t) = n\}$ at which X hits state 0 is well defined and is finite. Clearly $X_{\tau_n} = 0$, and $\tau_{n+1} \sim \tau_n$ since $\tau_n/n \to 1/\lambda_0$. Thus, the process X is recurrently empty with respect to τ_n, which completes the proof. □

In Theorem 6.21, the assumption $P\{X_0 = 0\} > 0$ is needed only to imply that X is recurrently empty. Note that the stationarity of X ensures that the limiting averages L and λ are also the expected queue length and expected arrival rate. The stationarity of X, however, is not enough to guarantee that W is the expected waiting time of a customer. Stronger assumptions are needed as we will now describe.

Keep in mind that the arrival process N may have several arrivals at one time. We will also refer to the point process of *distinct arrival times* or *batch arrival times* given by

$$\bar{N}(A) = \sum_{t \in A} 1(N(\{t\}) \ge 1).$$

Note that $\bar{N} = N$ when units arrive one at a time (i.e., N is simple). Suppose the service system we are discussing has dynamics represented by the flow process θ. Then customer n, which arrives at time T_n, sees the system dynamics as θ_{T_n}. It would therefore be logical that this customer's waiting time W_n is a function of this information (i.e., a mark of N).

Theorem 6.22. *Suppose the arrival process N is a stationary functional of θ and the waiting times W_n are marks of N. Then $\bar{N}$ and X are stationary functionals of θ. Suppose, in addition, that the intensity λ of N is finite and positive. Then the intensity of $\bar{N}$ is $\bar{\lambda} = \lambda / E_{\bar{N}}[N(\{0\})]$ and*

$$EX_0 = \bar{\lambda} E_{\bar{N}} \sum_{n=0}^{N(\{0\})-1} W_n. \tag{6.27}$$

For the particular case in which N is simple, the last formula reduces to

$$EX_0 = \lambda E_N(W_0).$$

Furthermore, if θ is ergodic, then the limits L, λ, and W exist and satisfy $L = \lambda W$. In this case,

$$L = EX_0, \quad \lambda = EN(1),$$

$$W = \frac{1}{E_{\bar{N}}[N(\{0\})]} E_{\bar{N}} \sum_{n=0}^{N(\{0\})-1} W_n. \tag{6.28}$$

Proof. The point process $\bar{N}$ is a stationary functional of θ since N is. And $\bar{N}(1) \leq N(1)$ ensures that $\bar{\lambda}$ is finite. An application of the Campbell–Mecke formula yields

$$\begin{aligned} \lambda &= E \int_{\mathbb{R}} N(\{t\}) 1(0 < t \leq 1) \bar{N}(dt) \\ &= \bar{\lambda} E_{\bar{N}} \int_{\mathbb{R}} N(\{0\}) 1(0 \leq -u < 1)\, du = \bar{\lambda} E_{\bar{N}}[N(\{0\})]. \end{aligned}$$

Now, since the waiting times W_n's are marks of N, they are also marks of $\bar{N}$. Also, the batch sizes $N(\{\bar{T}_n\})$ are marks of $\bar{N}$. Then we can write

$$\begin{aligned} X_t &= \sum_n 1(T_n \leq t < T_n + W_n) \\ &= \sum_n \sum_{k=0}^{N(\{\bar{T}_n\})-1} 1(\bar{T}_n \leq t < \bar{T}_n + f_k(\theta_{\bar{T}_n})). \end{aligned}$$

The first indicator function is of the event that customer n is in the system at time t. The second indicator function is of the event that the kth customer in the nth batch is in the system at time t, where $\bar{T}_n$ denotes the nth batch arrival time associated with $\bar{N}$ and $f_k(\theta_t)$ denotes the waiting time of the kth unit in a batch that arrives at time t. Then, by Example 6.15, we know that X is a stationary process and that (6.27) follows from (6.19).

We now show that the expected values λ and EX_0 are also limiting averages under the assumption that θ is ergodic. Since X is a stationary functional of θ, the limit L exists and it equals EX_0. Also, the ergodic theorem for N justifies that the limit λ equals $EN(1)$. Next, note that since the batch sizes $N(\{\bar{T}_n\})$ are marks of $\bar{N}$, by the cross ergodic theorem (Theorem 6.11), we have w.p.1 under P,

$$n^{-1}N(\bar{T}_n) = n^{-1}\sum_{k=1}^{n} N(\{\bar{T}_k\}) \to E_{\bar{N}}[N(\{0\})].$$

By a similar argument, w.p.1 under P,

$$n^{-1}\sum_{k=1}^{N(\{\bar{T}_n\})} W_k = n^{-1}\sum_{k=1}^{n}\sum_{i=0}^{N(\{\bar{T}_k\})-1} f_i(\theta_{\bar{T}_k}) \to E_{\bar{N}}\sum_{i=0}^{N(\{0\})-1} W_i.$$

By these observations and a discrete-time version of Theorem 5.10 with $Z(t) = \sum_{k=0}^{t-1} W_k$ and $T_n = N(\bar{T}_n)$, it follows that the limit W is given by (6.28) as asserted. Lastly, since λ and W exist, it follows by Theorem 5.4 that $L = \lambda W$. □

Theorem 6.22 applies to many types of systems including the $G/G/m$ queue and systems that are parts or functions of stationary, Markovian, or regenerative phenomena. A typical example for a stationary network (as in Example 5.3 for Markovian networks) applies to the number of units X_t in a sector J (set of nodes) of the network at time t and the total sojourn time W_n in sector J of the nth unit to enter J. Another example is the slightly different situation in which X_t denotes the number of units in sector J "waiting in queues for service" and W_n is the total time the nth unit visiting J waits in queues for services during its sojourn in J. These and many other examples follow, without further analysis, simply by defining the process X and times W_n appropriately. The following is another example.

Example 6.23. *Customers within a Batch.* Consider the system as described in Theorem 6.22. For fixed $j \le k$, consider the waiting time of a unit that is the jth one in a batch of size k. The arrival times of these units are given by the point process

$$N_{jk}(A) = \sum_n 1(T_n \in A,\, T_{n-j-1} < T_{n-j} = \ldots = T_{n+k-j} < T_{n+k-j+1}).$$

This point process is a stationary functional of θ since N is. Let $\{T_n(j,k) : n \in \mathbb{Z}\}$ denote the times associated with N_{jk}, and let $W_n(j,k)$ denote the waiting time of the unit that arrives at time $T_n(j,k)$. Assume these waiting times are marks of N_{jk}. Now, the number of the jth units in a batch of size k that are in the system at time is

$$X_{jk}(t) = \sum_n 1(T_n(j,k) \le t \le T_n(j,k) + W_n(j,k)).$$

Then by Theorem 6.22,

$$E[X_{jk}(0)] = \lambda_{jk} E_{N_{jk}}[W_0(j,k)].$$

where λ_{jk} is the intensity of N_{jk}. □

Example 6.24. *Workloads for Service Systems.* Suppose the queueing system described in Theorem 6.22 is work conserving in that it cannot be idle when customers are present. Then the *workload process* representing the sum of the remaining service times of the units in the system at time t is given by

$$W(t) = \sum_n [S_n 1(T_n \leq t \leq T_n + W_n)$$
$$+(T_n + W_n + S_n - t)1(T_n + W_n \leq t \leq T_n + W_n + S_n)],$$

where W_n is the duration of time the nth unit waits in the queue before its service, and S_n is the unit's service time. The first part of the sum is the workload of those units still waiting in the queue at time t, and the other part of the sum is the workload of those units that have already entered service. Assume that the process θ contains enough information such that (W_n, S_n) are marks of the arrival process N. Then applying (6.19) for marked point processes to the two parts of the sum, we have

$$\begin{aligned} EW(0) &= \lambda E_N[S_0 W_0 + \int_{W_0}^{W_0+S_0} (W_0 + S_0 - s)\, ds] \\ &= \lambda E_N[S_0 W_0 + S_0^2/2]. \end{aligned}$$

□

The preceding results are for systems that contain some underlying stationarity, but the sequence of waiting times is not stationary. Here is a result for a system with a stationary waiting time sequence.

Theorem 6.25. *Suppose $\{(T_{n+1} - T_n, W_n) : n \in \mathbb{Z}\}$ is a stationary ergodic sequence and $E(T_1 - T_0)$ is finite. Then the limits L, λ, and W exist and $L = \lambda W$. In this case, $W = EW_0$ and $\lambda = 1/E(T_1 - T_0)$.*

PROOF. By the ergodic theorem for sequences, we have $W = EW_0$ and by Theorem 5.8,

$$\lambda = 1/\lim_{n\to\infty} T_n = 1/E(T_1 - T_0).$$

Then the existence of λ and W imply by Theorem 5.17 that L exists and $L = \lambda W$. □

In the preceding result, the process X is not stationary and $L \neq EX_0$. However, one can construct a stationary queueing process $\tilde{X}$ on a probability space $(\tilde{\Omega}, \tilde{\mathcal{F}}, \tilde{P})$ such that the limit L is equal to $\tilde{E}\tilde{X}(0)$. Furthermore, the sequence $\{(\tilde{T}_{n+1} - \tilde{T}_n, \tilde{W}_n) : n \in \mathbb{Z}\}$ for this new process under the Palm probability $\tilde{P}_{\tilde{N}}$ is equal in distribution to $\{(T_{n+1} - T_n, W_n) : n \in \mathbb{Z}\}$ under P. This construction is a correspondence between certain stationary processes and embedded sequences.

6.5 Sojourn Times and Related Functionals

In this section, we present a Little law for determining the average of an integral of a stochastic process. We use this result to describe the average sojourn time of a stochastic process in a subset of its state space.

The following is a framework that encompasses a variety of examples. Let $\{X_t : t \in \mathbb{R}\}$ be a stochastic process on a space $\mathbb{E}$ that is countable or is a complete separable metric space. Assume, for convenience, that the sample paths of X are in the space D of all functions from $\mathbb{R}$ to $\mathbb{E}$ that are right continuous and have left-hand limits. Suppose that N is a simple point process on $\mathbb{R}$, defined on the same probability space as X, and $N(\mathbb{R}_+) = \infty$. For $f : \mathbb{E} \to \mathbb{R}_+$, suppose that the process

$$Z(t) \equiv \int_0^t f(X_s)\, ds, \qquad t \geq 0,$$

exists. We will consider the existence of the time and interval averages

$$Z \equiv \lim_{t\to\infty} t^{-1} Z(t), \qquad \hat{Z} \equiv \lim_{n\to\infty} n^{-1} Z(T_n) \quad \text{w.p.1},$$

and the average $\lambda \equiv \lim_{t\to\infty} t^{-1} N(t)$ of the time points. We use the term "a limit exists" to also include that the limit is not zero or infinite.

Corollary 6.26. (a) *If any two of the limits Z, λ, $\hat{Z}$ exist, then the other one also exists and $Z = \lambda \hat{Z}$.*
(b) *Suppose X, N are jointly stationary and ergodic with $EN(1) < \infty$. Then the limits Z, λ, and $\hat{Z}$ exist and $Z = \lambda\hat{Z}$, where*

$$Z = Ef(X_0), \qquad \lambda = EN(1). \tag{6.29}$$

Furthermore, the sequence $\{\int_{T_{n-1}}^{T_n} f(X_s)\, ds : n \geq 1\}$ is stationary with respect to the Palm probability P_N, and

$$Ef(X_0) = \lambda E_N \int_0^{T_1} f(X_s)\, ds. \tag{6.30}$$

Proof. Part (a) follows by Theorem 5.10. For part (b), the limits Z and λ exist by the ergodic theorems (recall Theorem 6.1) for the processes X and N. Then part (a) ensures that the limit $\hat{Z}$ exists and $Z = \lambda\hat{Z}$. By Corollary 6.10, the sequence $\{S_{T_n} X\}$ is stationary and ergodic with respect to the Palm probability P_N, and hence $\{\int_{T_{n-1}}^{T_n} f(X_s)\, ds : n \geq 1\}$ also has this property since it is a stationary functional of $\{S_{T_n} X\}$. Finally, (6.30) follows by the inversion formula in Corollary 6.16. □

We will now consider sojourn times of the process X in a fixed subset $B \in \mathcal{E}$ of its state space. Suppose that the point process N on $\mathbb{R}$ represents the times $\{T_n\}$ at which the process X enters B, and assume $N(\mathbb{R}_+) = \infty$. The average amount of time that X spends in B is

$$\pi(B) \equiv \lim_{t\to\infty} t^{-1} \int_0^t 1(X_s \in B)\, ds \quad \text{w.p.1}.$$

provided the limit exists (which also means that it is not 0 or 1). In many instances, $\pi(B)$ would also be the limiting probability $\lim_{t\to\infty} P\{X_t \in B\}$. Let W_n denote the duration of the nth visit or sojourn of X in B that begins at time T_n. The next result gives an expression for the average sojourn time

$$W \equiv \lim_{n\to\infty} n^{-1} \sum_{k=0}^{n-1} W_k \quad \text{w.p.1.}$$

Corollary 6.27. (a) *If any two of the limits $\pi(B)$, λ, W exist, then the other one also exists and $\pi(B) = \lambda W$.*
(b) *Suppose that X is stationary and ergodic, and $B \in \mathcal{E}$ is such that $0 < P\{X_0 \in B\} < 1$. Then the limits $\pi(B)$, λ, W exist and $\pi(B) = \lambda W$, where $\pi(B) = P\{X_0 \in B\}$ and $\lambda = EN(1)$. Furthermore, the sequence of sojourn times $\{W_n\}$ in B is stationary and ergodic with respect to the Palm probability P_N, and hence $W = E_N[W_0]$ and*

$$P\{X_0 \in B\} = \lambda E_N[W_0].$$

Proof. The assertions follow by Corollary 6.26. Part (b) also uses the fact that N is a stationary functional of X. □

Note that a special case of assertion (b) for Markov processes is Theorem 1.3.

How do the preceding result apply to networks? Suppose that X is a process that represents the numbers of customers in a network. Sojourn times of X that may be of interest are time periods during which the following events occur.

- A node, sector or the entire network is idle.
- The maximum number of units in a certain sector exceeds a certain value.
- The total number of units in a certain sector exceeds a certain value.
- The number of units in a certain sector exceeds that of another sector.

By Corollary 6.26, we know that if X is stationary and ergodic, then the expected value of such a sojourn time is $E_N[W_0] = \pi(B)/\lambda$, where the expectation is with respect to the Palm probability P_N that such a sojourn is beginning. Assuming the stationary distribution π is known, the expected sojourn time would be determined by evaluating λ. In some cases, λ can be determined directly from π and the dynamics of the network process.

6.6 Travel Times for Stochastic Processes

A travel time of a stochastic process, loosely speaking, refers to the time it takes for the process to follow a certain trajectory or route in the state space. Similarly, a travel time of a unit in a network is the time it takes for the unit to traverse a certain route in the network. An example is the time it takes a unit in a network to travel n times from one sector to another sector (recall Corollary 4.33). Another example is the time it takes a unit in a network to visit each node in a certain sector at least once. In this section, we describe general travel times for processes and networks and give Little-type formulas for their limiting averages or expected values.

How do travel times compare with sojourn times, which were the focus of the last section? A sojourn time of a process in a certain subset of its state space is characterized by an entrance and exit time of the set, which are stopping times of the process. The beginning and end of a travel time, however, need not be stopping times; they may depend on the future of the process as well as the past and present. For instance, the travel time of a process from one set B to another set B' begins at an exit time from B with the additional property that the process in the future enters B' before it returns to B.

Although travel times are more complicated than sojourn times, the framework for analyzing average sojourn times in the last section also applies to travel times. The key idea is that one can analyze travel times by using stopping times for "subsets of sample paths" in the same way that one uses stopping times for sets to analyze sojourn times.

We will use the following notation throughout this section. As in the preceding section, we assume that $\{X_t : t \in \mathbb{R}\}$ is a stochastic process on a space $\mathbb{E}$ and its sample paths are in the function space D. If an observer of X at time t can see the entire time-shifted process

$$\tilde{X}_t \equiv S_t X, \quad t \in \mathbb{R}, \tag{6.31}$$

then the observer should be able to tell if X is traveling on a special route at time t. We will associate routes with subsets of D as follows.

Definition 6.28. The process $\tilde{X}$ defined by (6.31) is the *sample-path process* of X. A subset $\mathcal{R} \subset D$ of sample paths of X is a *route of* X if the times at which $\tilde{X}$ enters $\mathcal{R}$ is a point process on $\mathbb{R}$. This point process N of entrance times $\{T_n\}$ is defined by

$$N(A) \equiv \sum_{t \in A} 1(\tilde{X}_{t-} \notin \mathcal{R}, \tilde{X}_t \in \mathcal{R}) = \sum_n 1(T_n \in A), \quad A \subset \mathbb{R}.$$

The process X *enters the route* at each time T_n, and then the *travel time on the route* is

$$W_n \equiv \inf\{t > 0 : \tilde{X}_{T_n + t} \notin \mathcal{R}\}.$$

The process X *exits the route* at time $T_n + W_n$.

Our interest is in the average travel time of X on the route $\mathcal{R}$:

$$\pi(\mathcal{R}) \equiv \lim_{t \to \infty} t^{-1} \int_0^t 1(S_u X \in \mathcal{R})\, du \quad \text{w.p.1.}$$

Since $\pi(\mathcal{R})$ is the average sojourn time of the process $\tilde{X}_t = S_t X$ in the subset $\mathcal{R}$ of D, we have the following result, which is simply a restatement of Corollary 6.27 for this setting in which $\tilde{X}_0 = X$.

Corollary 6.29. (a) *If any two of the limits* $\pi(\mathcal{R})$, λ, W *exist, then the other one also exists and* $\pi(\mathcal{R}) = \lambda W$.
(b) *Suppose that* X *is stationary and ergodic, and* $\mathcal{R} \subset D$ *is such that* $0 < P\{X \in \mathcal{R}\} < 1$. *Then the limits* $\pi(\mathcal{R})$, λ, W *exist and* $\pi(\mathcal{R}) = \lambda W$, *where*

$\pi(\mathcal{R}) = P\{X \in \mathcal{R}\}$ and $\lambda = EN(1)$. Furthermore, the sequence of travel times $\{W_n\}$ on the route $\mathcal{R}$ is stationary and ergodic with respect to the Palm probability P_N, and hence $W = E_N[W_0]$ and

$$P\{X \in \mathcal{R}\} = \lambda E_N[W_0].$$

The preceding result covers most travel times that one might imagine. Stopping times of processes generate a vast family of travel times. To analyze a travel time, the first steps are to formulate it in terms of a subset $\mathcal{R}$ of sample paths, and then verify that the number of entrance times of $\tilde{X}$ into that subset in any finite time interval is finite. A fundamental example is as follows.

Example 6.30. *Travel Time between Two Sets.* Consider the travel time of the process X from some set $B \in \mathcal{E}$ to another set $B' \in \mathcal{E}$. For each sample path $x \in D$, the first time the path enters (or hits) the set B after time 0 is

$$H_B(x) \equiv \inf\{t > 0 : x(t) \in B\},$$

and the last exit time of the path from B prior to time 0 is

$$L_B(x) \equiv \sup\{t < 0 : x(t) \in B\}.$$

Then the process X is on the route from B to B' at time t if $S_t X \in \mathcal{R}$, where

$$\mathcal{R} \equiv \{x \in D : x(0) \notin B \cup B',\ L_{B'}(x) < L_B(x),\ H_{B'}(x) < H_B(x)\}.$$

Now, suppose that X is stationary and ergodic. Assume that X_0 has a nonzero probability of being in each of the sets B, B' and $B \cup B'$. This implies that X enters each of these sets infinitely often, which ensures that the times at which the process X begins a traverse from B to B' form a point process. Then by Corollary 6.29, the average or expected travel time between B and B' is

$$E_N[W_0] = \lambda^{-1} P\{X_0 \notin B \cup B',\ L_{B'}(X) < L_B(X),\ H_{B'}(X) < H_B(X)\}.$$

For the case in which X is a Markov process, this expression has the tractable form shown in Corollary 4.33.

As a variation of this travel time, suppose one is interested in the travel time of X in some subset $C \subset B^c \cap B'^c$ during a traverse of X between B and B'. This new travel time is defined by the route $\mathcal{R}' \equiv \{x \in \mathcal{R} : x(0) \in C\}$. Assume that $P\{X_0 \in C\} > 0$, which implies that X visits C infinitely often. Then the expectation of the new travel time is given as above with $X_0 \notin B \cup B'$ replaced by $X_0 \in C$. □

6.7 Sojourn and Travel Times in Networks

The preceding two sections focused on characterizing the average of a sequence $\{W_n\}$ of sojourn or travel times of a stochastic process. One distinguishing feature of such times is that their associated time intervals $(T_n, T_n + W_n]$ do not overlap. In this section, we characterize average sojourn and travel times of units in a network.

Here a sequence $\{W_n\}$ of sojourn or travel times of units is associated with time intervals $(T_n, T_n + W_n]$ that typically overlap. Although the discussion will be be in terms of networks, the results also apply to sojourn and travel times for general multivariate processes.

We will first analyze travel times for networks that can be represented by the locations of its units. Consider a m-node network that is either closed with ν units or open with capacity ν. In case the network is closed, we label the units as $1, \ldots, \nu$. In case the network is open, we assume that the indices $1, \ldots, \nu$ are labels or tokens that the units in the network carry as follows. Whenever there are $n < \nu$ units in the network, a unit entering the network selects one of the $\nu - n$ unused labels with equal probability. The unit retains the label until it exits the network, and then the label becomes available for another unit. The unit carrying the label i is called unit i.

We will represent the network by the stochastic process $Y(t) \equiv (Y_1(t), \ldots, Y_\nu(t))$, where $Y_i(t)$ denotes the node location of unit i at time t. A typical state of the process Y is a vector $y = (y_1, \ldots, y_\nu)$ in M^ν, where $M \equiv \{1, \ldots, m\}$ or $M \equiv \{0, 1, \ldots, m\}$ according to whether the network is closed or open. Assume that the location process Y is stationary and ergodic. For instance, this may represent customer locations in a Whittle network.

We associate each unit i with a route $\mathcal{R}_i \subset D$ that satisfies the property $O < P\{Y_i \in \mathcal{R}_i\} < 1$. Then the time-shifted process $S_t Y_i$, which is a stationary functional of Y, enters $\mathcal{R}_i$ and $\mathcal{R}_i^c$ infinitely often. This ensures that the times at which Y_i enters the route $\mathcal{R}_i$ form a stationary, ergodic point process N_i on $\mathbb{R}$. Assume that N_i has a finite intensity λ_i. Then by Corollary 6.29, the sequence of travel times $\{W_n^i\}$ on the route $\mathcal{R}_i$ are stationary with respect to the Palm probability P_{N_i}, and their average is

$$E_{N_i}[W_0^i] = \lambda_i^{-1} P\{Y_i \in \mathcal{R}_i\}.$$

Our focus will be on the average travel time of an arbitrary unit on its route. The times $\{T_n\}$ at which the units enter their routes are described by the point process $N \equiv N_1 + \cdots + N_\nu$. Clearly N is a stationary functional of Y and its intensity is $\lambda = \lambda_1 + \cdots + \lambda_\nu$. For simplicity, assume that only one unit moves at a time. Then the point process N is simple. Let γ_n denote the index i on the process Y_i that enters the route $\mathcal{R}_i$ at time T_n; that is, $S_{T_n-} Y_{\gamma_n} \notin \mathcal{R}_{\gamma_n}$, and $S_{T_n} Y_{\gamma_n} \in \mathcal{R}_{\gamma_n}$. Consider the time

$$W_n = \inf\{t > 0 : S_{T_n+t} Y_{\gamma_n} \notin \mathcal{R}_{\gamma_n}\},$$

which is the travel time of the process Y_{γ_n} on the route $\mathcal{R}_{\gamma_n}$. We call W_n a *travel time of an arbitrary unit* on its route. Clearly γ_n and W_n are marks of N.

Corollary 6.31. *Under the preceding assumptions, the sequence of travel times $\{W_n\}$ is stationary under the Palm probability P_N and*

$$E_N[W_0] = \lambda^{-1} \sum_{i=1}^{\nu} P\{Y_i \in \mathcal{R}_i\} = \lambda^{-1} \sum_{i=1}^{\nu} \lambda_i E_{N_i}[W_0^i].$$

In particular, if the distributions of $Y_1, \ldots, Y_\nu$ *are identical and* $\mathcal{R}_i = \mathcal{R}$ *for each* i*, then* $E_N[W_0] = \lambda_1^{-1} P\{Y_1 \in \mathcal{R}\}$.

PROOF. The process

$$X_t \equiv \sum_{i=1}^{\nu} \mathbf{1}(S_t Y_i \in \mathcal{R}_i), \quad t \in \mathbb{R} \tag{6.32}$$

records the number of the processes Y_i that are traversing their routes at time t. Viewing X as a stationary queueing process, the first assertion follows by the Little law for stationary systems given in Theorem 6.22. The second assertion is an obvious special case of the first one. □

The preceding result is useful when one knows the stationary distributions of the location processes. We now discuss another approach for modeling travel times in terms of the quantities of units at the nodes rather than the locations of specific units.

Consider an m-node network represented by a process $\{X_t : t \in \mathcal{R}\}$ with states $x \equiv (x_1, \ldots, x_m)$, where x_j denotes the number of units at node j. The network may be closed or open, with a finite or unlimited capacity. Assume that the process X is stationary and ergodic. Suppose that the units in the network move one at a time, and that the times at which units move from node j to node ℓ form a point process $N_{j\ell}$ on $\mathbb{R}$. Here j and ℓ are in the node set $M \equiv \{1, \ldots, m\}$ or $M \equiv \{0, 1, \ldots, m\}$ according as the network is closed or open. Assume that $N_{j\ell}$ is a stationary functional of X. Its intensity $\rho_{j\ell} \equiv E[N_{j\ell}(1)]$ is the throughput from j to ℓ.

Since the process X does not include the entire information about the sample paths of each of its units, it is natural to consider only special routes that can be described by X and its routing process. Accordingly, we assume the routes of the units are independent, and independent of the quantities at the nodes. This assumption is satisfied by Jackson and Whittle networks when the service discipline at each node is processor sharing and each unit is treated equally.

We will consider a route $\mathcal{R}$ that satisfies the following properties:

Traversing Assumptions. At any time t, the event that a unit at node j is traversing the route is independent of the disposition of the other units in the network at that time, and the probability of this event is γ_j, independent of t. The process $\{X'_t : t \in \mathbb{R}\}$ that denotes the number of units that are traversing the route $\mathcal{R}$ at time t is a stationary functional of X.

Entry Rate Assumptions. The times $\{T_n\}$ at which units begin traversing the route $\mathcal{R}$ form a point process N on R, and N is a stationary functional of X. If a unit moves from a node j to some node ℓ at time t, the event that the unit begins a traverse of the route $\mathcal{R}$ is independent of the disposition of the other units in the network at the transition, and the probability of this event is $b_{j\ell}$, independent of t.

Let W_n denote the travel time on the route $\mathcal{R}$ that begins at time T_n. An expression for the average of these times is as follows. Here we use $L_j \equiv \sum_x x_j \pi(x)$, which is the expected number of units at node j.

Corollary 6.32. *Under the preceding assumptions, the sequence* $\{W_n\}$ *of travel times on* $\mathcal{R}$ *is stationary under the Palm probability* P_N*, and*

$$E_N[W_0] = \frac{\sum_j L_j \gamma_j}{\sum_{j,\ell} \rho_{j\ell} b_{j\ell}}, \tag{6.33}$$

provided that these sums are positive and finite.

PROOF. Consider the stationary process X' representing the number of units traversing $\mathcal{R}$ as a queueing process. Then by the Little law for stationary systems given in Theorem 6.22, the sequence $\{W_n\}$ is stationary under the Palm probability P_N, and

$$E_N[W_0] = E[X'_0]/E[N(1)]. \tag{6.34}$$

Now, under the traversing assumptions,

$$E[X'_0] = \sum_j E[\sum_{n=1}^{X_0^j} U_n^j],$$

where X_0^j is the number of units at node j at time 0, and U_n^j is 1 or 0 according to whether or not the nth unit at node j is traversing the route $\mathcal{R}$. Furthermore, $U_1^j, U_2^j, \ldots$ are independent random variables that are independent of X_t^0 and $E[U_n^j] = \gamma_j$. Therefore,

$$E[X'_0] = \sum_j L_j \gamma_j. \tag{6.35}$$

Next, under the entry assumptions for the route,

$$E[N(1)] = E[\sum_{j,\ell} \sum_{n=1}^{N_{j\ell}(1)} \eta_n(j,\ell)],$$

where $\eta_n(j,\ell)$ is 1 or 0 according to whether or not the nth unit moving from node j to node ℓ begins a traverse of the route $\mathcal{R}$ at that transition. Furthermore, $\eta_1(j,\ell), \eta_2(j,\ell), \ldots$ are independent variables that are independent of $N_{j\ell}(1)$, and $E[\eta_n(j,\ell)] = b_{j\ell}$. Therefore,

$$E[N(1)] = \sum_{j,\ell} \rho_{j\ell} b_{j\ell}. \tag{6.36}$$

Then substituting (6.35) and (6.36) in (6.34) yields (6.33). □

To use the average travel time expression (6.33), one only has to evaluate the throughputs $\rho_{j\ell}$ of units moving from j to ℓ, the probabilities γ_j that a unit at node j in equilibrium is traversing the route, and the probabilites $b_{j\ell}$ that a unit moving from j to ℓ in equilibrium begin traversing the route. The following is a basic example.

Example 6.33. *Travel Times in Whittle Networks.* Suppose the network we are discussing is a Whittle network in which the services at each node are under a processor-sharing discipline, where each unit at a node receives the same service treatment. We will consider the average time it takes an arbitrary unit to travel from one sector J to another sector K. The J and K may overlap, but assume their union is not M.

A little thought justifies that all the assumptions above are satisfied for this route from J to K. Hence Corollary 6.32 applies, and the average travel time from J to K is

$$E_N[W_0] = \frac{\sum_{\ell \notin J \cup K} L_\ell \gamma_\ell}{\sum_{j \in J} \sum_{\ell \notin J} \rho_{j\ell} b_{j\ell}}. \tag{6.37}$$

Note that the throughput rates $\rho_{j\ell}$ and expected numbers of units L_j at the nodes can be obtained as in Chapter 1 from the stationary distribution of a Whittle process. The following discussion describes the other probabilities in the preceding expression.

We will represent a typical unit's path among the nodes by an ergodic Markov chain $\{\xi_n : n \in \mathbb{Z}\}$ with transition probabilities $p_{j\ell}$. Consider the hitting and last exit times

$$H_J(\xi) \equiv \inf\{n > 0 : \xi_n \in J\},$$

$$L_J(\xi) \equiv \sup\{n < 0 : \xi_n \in J\}.$$

Now, the probability that starting at node j, the routing chain ξ enters K before it enters J is

$$\alpha_j \equiv P\{H_K(\xi) < H_J(\xi) \mid \xi_0 = j\}.$$

These probabilities are solutions to the following equations: $\alpha_j = 1$ or 0 according to whether j is in K or is in J, and

$$\alpha_j = \sum_{k \in K} p_{jk} + \sum_{\ell \notin K} p_{j\ell}\alpha_\ell, \quad j \notin J \cup K. \tag{6.38}$$

Analogously, looking backward in time, the probability that conditioned on being in state j, the routing chain exited J more recently than it exited K is

$$\bar{\alpha}_j \equiv P\{L_K(\xi) < L_J(\xi) \mid \xi_0 = j\}.$$

These probabilities are the solution to the equations (6.38), where J and K are interchanged and $p_{j\ell}$ is replaced by the "time-reversed" routing probabilities

$$\bar{p}_{j\ell} \equiv w_j^{-1} w_\ell p_{\ell j}.$$

These observations, which are similar to those in Proposition 4.32, are well-known properties of Markov chains.

We now complete our description of expression (6.37) for the average travel time on the route from J to K. Clearly, the equilibrium probability that a unit moving from $j \in J$ to $\ell \notin J$ begins traversing the route is

$$b_{j\ell} = \alpha_\ell.$$

Finally, the equilibrium probability that a unit at node $\ell \notin J \cup K$ is traversing the route is

$$\gamma_\ell = P\{H_K(\xi) < H_J(\xi),\ L_K(\xi) < L_J(\xi) \mid \xi_0 = \ell\} = \alpha_\ell \tilde{\alpha}_\ell.$$

The last equality follows from the property that the past and future of a Markov chain are conditionally independent given its present state. □

6.8 Exercises

1. *Marked Random Measures*. Suppose M is a random measure on $\mathbb{R}$ that is a stationary functional of θ. Assume each t in the support of M has associated with it a quantity $\xi_t = h(\theta_t)$, where $h : \Omega \to \mathbb{E}'$. We call ξ_t a *mark of M* at the location t. Consider the process $X_t = \int_{\mathbb{E}} f(t-s, \xi_s)M(ds)$. Justify that this is a stationary functional of θ, whose mean is $EX_0 = \lambda E_N \int_{\mathbb{R}} f(t, \xi_0)\, dt$.

2. *Little Laws for Semi-Stationary Systems*. According to Definition 5.22, the process Y is regenerative if its cycle variables ξ_n are i.i.d. More generally, if the sequence $\{\xi_n\}$ is stationary, then we say that Y is a *semi-stationary* process over τ_n (sometimes called a *synchronous* process, or a process with *stationary cycles*). Furthermore, we say that Y is ergodic if ξ_n is. These notions also apply when the process Y is defined on the entire time axis, which we assume here. Proofs of the following statements are minor modifications of the referenced theorems. Specify the needed modifications in the proofs that would justify the statements.
(i) If the queue length process X is semi-stationary and ergodic over τ_n, then the assertions of Theorem 5.24 are true.
(ii) Suppose the service system is as in Theorem 5.25, but the underlying process Y, instead of being regenerative, is semi-stationary and ergodic over τ_n. Then the assertions of Theorem 5.25 are true for this more general system.

3. *Campbell–Mecke and Exchange Formulas for Random Kernels*. Suppose the function $K : \mathbb{R} \times \Omega \times \mathbb{R} \to \mathbb{R}$ is such that $K(t, \omega, \cdot)$ is a measure on $\mathbb{R}$ for each t, ω. The K is a *random kernel from* $\mathbb{R}$ *to* $\mathbb{R}$. Show that the following formula is equivalent to the Campbell–Mecke formula. For $f : \mathbb{R} \times \mathbb{R} \times \Omega \to \mathbb{R}_+$,

$$E \int_{\mathbb{R}} \left[\int_{\mathbb{R}} f(t, s, \theta_t) K(t, \theta_t, ds) \right] M(dt)$$
$$= \lambda_M E_M \int_{\mathbb{R}} \left[\int_{\mathbb{R}} f(t, s, \theta_0) K(t, \theta_0, ds) \right] dt.$$

Next, suppose K and K' are random kernels from $\mathbb{R}$ to $\mathbb{R}$ that are stationary functionals of θ and satisfy

$$M(dt)K(t, ds) = M'(dt)K'(t, ds)$$

(the omegas in the kernels are now suppressed), where M and M' are random measures that are stationary functionals of θ with respective intensities λ and λ'.

Show that the following variation of the exchange formula (6.24) is equivalent to the Campbell–Mecke formula. For $f : \mathbb{R} \times \mathbb{R} \times \Omega \to \mathbb{R}_+$,

$$\lambda E_M \int_{\mathbb{R}} f(s, \theta_s) K'(0, ds) = \lambda' E_{M'} \int_{\mathbb{R}} f(s, \theta_0) K(0, ds).$$

4. *Rate Conservation Law for Averages.* Consider the stochastic process X defined by (6.26), where N is the point process of discontinuity points T_n of X. Assume that $t^{-1}X_t \sim 0$ and that the limit $\lambda = \lim_{t\to\infty} t^{-1}N(t)$ exists and is positive. Show that if either one of the following limits exists, then the other one does, and they are related as indicated:

$$\lim_{t\to\infty} t^{-1} \int_0^t X_s'\, ds + \lambda \lim_{n\to\infty} n^{-1} \sum_{k=1}^{n} (X_{T_k} - X_{T_k-}) = 0.$$

This conservation law says the average derivative of X plus the rate of jumps times the average jump size of X equals 0; one would expect this relation for a stable process. This result is analogous to the conservation law in Corollary 6.20 for expected values. Hint: Write (6.26) as $X_t - X_0 = U(t) - \hat{U}(N(t))$, where U and $\hat{U}$ are defined in the obvious way. Then apply Theorem 5.15.

6.9 Bibliographical Notes

For more background on stationary processes, see for instance Cramér and Leadbetter (1967) and Rozanov (1967); and for ergodic theory and laws of large numbers, see Krengel (1985) and Révész (1968). Monographs on stationary queueing systems are Rolski (1981), Franken et al. (1981), Brandt et al. (1990), Baccelli and Brémaud (1994), and Sigman (1995). The basic results on Palm probabilities in this chapter can be found in these references, with some nuances on Campbell's formula coming from Schmidt and Serfozo (1995). Brandt et al. (1990) and Baccelli and Brémaud (1994) discuss many examples of stationary queueing systems. The Little laws for stationary systems come from these references, with the example on workloads coming from Brumelle (1971). The rate conservation laws are due to Miyazawa (1994), and Exercise 4 is from Sigman (1991).

7

Networks with String Transitions

Chapter 2 discussed network models with batch or concurrent movements of units under reversibility assumptions. Are there comparable models of Whittle networks with batch movements? More generally, are there tractable network models in which a transition may involve a series of simultaneous single- or multiple-unit changes? This chapter describes networks with such characteristics called *networks with string transitions*, or *string-nets* for short. In a string-net, a transition consists of a string of instantaneous subtractions or additions of units at the nodes, where the string is randomly selected from a family of variable-length strings. Invariant measures for string-nets resemble those of Whittle networks, but now key parameters in the measures are obtained as solutions to more complicated nonlinear traffic equations.

7.1 Definition of a String-Net

Throughout this chapter, we will consider an m-node network that operates as follows. As in Chapter 1, we will represent the network by a stochastic process $X = \{X_t : t \geq 0\}$ that represents the numbers of units at the respective nodes. The state space is a set $\mathbb{E}$ of m-dimensional vectors $x = (x_1, \ldots, x_m)$, where x_j denotes the number of units at node j. We place no further assumptions on the form of $\mathbb{E}$, and so our results apply to a variety of network types, including the standard ones that are closed, or open with finite or unlimited capacity.

Whenever the process is in state x, a typical transition will be to some state of the form $x - (s^1 + \cdots + s^\ell) + a$ or $x - (s^1 + \cdots + s^i)$, $1 \leq i < \ell$, where the increment vectors a and s^i are in a set A, and the string $s = (s^1, \ldots, s^\ell)$ is in a set $\mathcal{S}$. The A is a finite set of m-dimensional vectors with negative or nonnegative integer entries and A contains the zero vector 0. The $\mathcal{S}$ is a countable set of *strings* $s = (s^1, \ldots, s^\ell)$, where $s^i \in A\backslash\{0\}$ and $\ell \equiv \ell(s)$ denotes the string length. Let $L \leq \infty$ denote the supremum of these string lengths. Assume that $\mathcal{S}$ contains the empty or zero string, denoted by 0, whose length is zero.

Associated with each string $s \in \mathcal{S}$ are its ith partial sum vectors

$$s(i) = s^1 + \ldots + s^i, \quad 0 \leq i \leq \ell,$$

where $s(0) = 0$ for the zero string. Denote the set of all partial sums of the strings by $S = \{s(i) : 1 \leq i \leq \ell, s \in \mathcal{S}\}$. Think of A as the set of allowable increment vectors and $\mathcal{S}$ as the set of feasible strings of vectors from A that can be subtracted in a transition. Then S (which contains A) is the entire set of network increment vectors. For each $x \in \mathbb{E}$ and $d, a \in S$, we define the vector $T_{da}x = x - d + a$, which may or may not be in $\mathbb{E}$. A transition $x \to T_{da}x$ means that the vectors a and d are added and subtracted from x.

In terms of this notation, the transitions of the process X are as follows. Whenever the process X is in state x, a transition is determined by a pair sa in $\mathcal{S} \times A$ that results in one of the following ℓ possibilities:

- A *complete sa-transition*: $x \to T_{s(\ell)a}x = x - (s^1 + \ldots + s^\ell) + a$.
- An *ith partial sa-transition*: $x \to T_{s(i)0}x = x - (s^1 + \ldots + s^i)$, where $0 \leq i < \ell$.

Keep in mind that ℓ, with s suppressed, is the length of the string s. Note that the complete sa-transition uses a as well as the whole string s, but the ith partial sa-transition uses only the part $s^1, \ldots, s^i$ of s. Some of these transitions may be infeasible as discussed below. Under the preceding assumptions, each state $x \in \mathbb{E}$ is a linear combination of vectors in A. Assume that the standard m-dimensional unit vectors form a basis that generates the vectors in A and $\mathbb{E}$. This is not a restriction since one can always represent these vectors by a basis and the form of the basis is not important here.

We assume the rates of these string transitions are as follows:

Type of Transition	Rate
Complete sa-transition $x \to T_{s(\ell)a}x$	$\lambda_{sa}\phi_{s(\ell)}(x)$,
A ith partial sa-transition $x \to T_{s(i)0}x$	$\lambda_{sa}(\phi_{s(i)}(x) - \phi_{s(i+1)}(x))$.

These transition rates can be viewed as the compounding of two rates as follows. The nonnegative λ_{sa} is the rate (or probability) at which an sa-transition occurs, where $\lambda_{00} = 0$. A typical example is a product of probabilities $\lambda_{sa} = p(s^1)p(s^1, s^2) \cdots p(s^{\ell-1}, s^\ell)$ of Markovian selections of the vectors s^i. Within an sa-transition, $\phi_{s(\ell)}(x)$ is the nonnegative rate of subtracting the complete vector $s(\ell)$ from x and adding a; and $\phi_{s(i)}(x) - \phi_{s(i+1)}(x)$ is the rate of subtracting exactly $s(i)$ (the ith partial of s) from x, where $0 \leq i < \ell$. A compounding of these two rates yields the transition rates above.

We assume the ϕ_d's are Φ-balanced as in earlier chapters, where Φ is a positive function on $\mathbb{E}$ such that, for any $x \in \mathbb{E}$, $d \in S$, and $a \in A$ with $\sum_s \lambda_{sa} > 0$,

$$\Phi(x)\phi_a(x) = \Phi(T_{ad}x)\phi_d(T_{ad}x). \tag{7.1}$$

For convenience, we extend the definition of ϕ_d to all integer-valued, m-dimensional vectors by setting $\phi_d(x) = 0$, for $x \notin \mathbb{E}$.

The preceding description says that whenever the Markov process X is in state x, the times to the next complete sa-transition and ith partial sa-transition are independent, exponentially distributed with rates shown in the table above. Then the time to the next $x \to y$ transition is exponentially distributed, and its rate $q(x, y)$ is the sum of appropriate sa-transition rates. That is, the transition rates of the process X are

$$q(x, y) = \sum_{s,a} \lambda_{sa} r_{sa}(x, y), \qquad y \neq x \text{ in } \mathbb{E}, \tag{7.2}$$

where

$$\begin{aligned} r_{sa}(x, y) &= \phi_{s(\ell)}(x) 1(y = T_{s(\ell)a}x) \\ &\quad + \sum_{i=0}^{\ell-1} [\phi_{s(i)}(x) - \phi_{s(i+1)}(x)] 1(y = T_{s(i)0}x), \qquad x, y \in \mathbb{E}. \end{aligned} \tag{7.3}$$

All sums on s, a herein are for $s \in \mathcal{S}$ and $a \in A$, unless specified otherwise, and $\sum_{i=0}^{-1} = 0$. Since a transition $x \to y$ is possible under several combinations of subtractions and additions, its rate $q(x, y)$ is a sum of rates, some of which may be 0 due to the λ_{sa}, ϕ_d or the indicator functions being 0. The rate functions λ_{sa} and ϕ_d as well as the sets A and $\mathcal{S}$ can have a variety of forms depending on the routing and service rules of the network. For instance, for a closed network, the rate λ_{sa} can be positive only if $|s(i)| = |a - s(\ell)| = 0$, for $1 \leq i < \ell$.

Note that the rate of the exponential sojourn time in state x is

$$\sum_{y \neq x} q(x, y) = \sum_{s,a} \lambda_{sa} [\phi_0(x) - r_{sa}(x, x)]. \tag{7.4}$$

This follows since

$$\sum_{y \in \mathbb{E}} r_{sa}(x, y) = \phi_0(x), \tag{7.5}$$

which is due to the telescoping series in (7.3), and the fact that the sum of the indicators over y is 1.

To complete the definition of the process, a few more technical assumptions are in order. We assume that $\sum_s \lambda_{sa} < \infty$ for each $a \in A$. This and the finiteness of A ensure that the rate (7.4) is finite. Next, we assume the following condition.

Dominance of ϕ_0: If $L \geq 2$, then $\phi_0(x) \geq \phi_a(x)$, for each $x \in \mathbb{E}$ and $a \in A$.

Finally, we adopt the standard assumption that the process is irreducible on the space $\mathbb{E}$ (otherwise, we could let $\mathbb{E}$ denote a closed communicating class).

Definition 7.1. The process X defined above with rates (7.2) that satisfy the preceding assumptions is a *Markov Network process with string transitions* or a *string-net*.

The data for the string-net process X are $\mathbb{E}$, A, $\mathcal{S}$, L, $\{\lambda_{sa} : s \in \mathcal{S}, a \in A\}$, and $\{\phi_d(\cdot) : d \in S\}$. To model an actual network with this process, one would specify this data from the operational features of the network. Note that Jackson and Whittle networks with single-unit movements are examples of string-nets.

The following are a few more observations about the definition. From the Φ-balance and ϕ_0-dominance, it follows that

$$\begin{aligned}\Phi(T_{0d}x)\phi_{d+a}(T_{0d}x) &= \Phi(x)\phi_a(x)\\ &\le \Phi(x)\phi_0(x) = \Phi(T_{0d}x)\phi_d(T_{0d}x), \quad d \in S,\ a \in A.\end{aligned}$$

Thus $\phi_d(x) \ge \phi_{d+a}(x)$. This ensures that the rates in the second sum in (7.2) are not negative.

Note that (7.1) implies $\phi_a(x) = 0$ when $T_{ad}x \notin \mathbb{E}$ for some $d \in S$, because $\phi_d(x') = 0$ when $x' \notin \mathbb{E}$. This says that an sa-transition in state x is not feasible or is blocked if any one of the possible new states resulting from a complete or partial transition is not in $\mathbb{E}$. Recall that the Φ-balance of the ϕ_d's is equivalent to their being of the form

$$\phi_d(x) = \Psi(x-d)/\Phi(x), \quad d \in S, x \in \mathbb{E}, \tag{7.6}$$

for some function Ψ that is nonnegative on $\{x - d : d \in S, x \in \mathbb{E}$ and is 0 outside of $\mathbb{E}$. These ϕ_d's also satisfy the ϕ_0-dominance assumption when Ψ is nonincreasing and each vector $a \in A$ is nonnegative.

7.2 Invariant Measures of String-Nets

In this section, we characterize invariant measures of the process X under the assumption that certain polynomial "traffic equations" have a solution. Conditions for the existence of solutions are given in the next section.

In addition to the notation above, we denote the rate of all string transitions with s as the initial segment by

$$\Lambda_s = \sum_{s',a} \lambda_{(ss')a}, \quad s \in \mathcal{S}, \tag{7.7}$$

where the string (ss') denotes the concatenation of the strings s and s'. We sometimes use $\Lambda_{(sa)}$ for $sa \in \mathcal{S}$, where $\Lambda_{(0a)} \equiv \Lambda_a$. The following is the main result.

Theorem 7.2. *Suppose there exist positive numbers $w_1, \ldots, w_m$ that satisfy*

$$\eta(a)\sum_s \prod_{i=1}^{\ell} \eta(s^i)\Lambda_{(sa)} = \sum_s \prod_{i=1}^{\ell} \eta(s^i)\lambda_{sa}, \quad a \in A_0 \equiv A\backslash\{0\}, \tag{7.8}$$

where $\eta(x) = \prod_{j=1}^{m} w_j^{x_j}$ *and the sums in (7.8) are finite. Then an invariant measure for the string-net process* X *is*

$$\pi(x) = \Phi(x)\eta(x), \quad x \in \mathbb{E}.$$

Furthermore, a necessary and sufficient condition for the process to have an invariant measure of this form is

$$\sum_{a \in A_0} D(a)[\phi_0(x) - \phi_a(x)\eta(a)^{-1}] = 0, \quad x \in \mathbb{E}, \tag{7.9}$$

where $D(a)$ *denotes the right side of (7.8) minus its left side.*

The invariant measure $\pi(x) = \Phi(x)\eta(x)$ resembles the invariant measure in Theorem 1.15 for Whittle processes. In particular, π is a weak coupling of Φ determined only by the ϕ_a's, and η determined only by the λ_{sa}'s. The traffic equations (7.8) are what is left of the balance equations upon substituting the measure π into the equations and cancelling the ϕ functions. Consequently, the existence of solutions for the traffic equations is a necessary condition for invariant measures of the form as shown. The second statement in the theorem gives more precise information in this regard.

From a key identity (7.11) in the proof below, it follows that the summation in (7.9) times $\pi(x)$ is the difference between the two sides of the balance equations for the process X (this should be 0 for the balance equations to be satisfied). Note that the summation is a weighted average of the differences $D(a)$ of the two sides of the traffic equations (7.8). The weights $\phi_0(x) - \phi_a(x)\eta(a)^{-1}$, which arise in (7.11), don't seem to have any special meaning.

PROOF. The balance equations that an invariant measure π must satisfy are

$$\pi(x) \sum_{y \in \mathbb{E}} q(x, y) = \sum_{y \in \mathbb{E}} \pi(y) q(y, x), \quad x \in \mathbb{E}. \tag{7.10}$$

The usual convention is that $q(x, x) = 0$, but here we define $q(x, x) = \sum_{s,a} \lambda_{sa} r_{sa}(x, x)$. This does not affect the equality, and it simplifies some expressions.

Let $L(x)$ and $R(x)$ denote the left and right sides of (7.10), respectively, and suppose $\pi(x) = \Phi(x)\eta(x)$. The proof will proceed as follows. A short calculation yields $L(x) = \pi_0(x)\Lambda_0$, and a more complicated analysis of $R(x)$ yields the identity

$$L(x) = R(x) + \pi(x) \sum_{a \neq 0} D(a)[\phi_0(x) - \phi_a(x)\eta(a)^{-1}]. \tag{7.11}$$

From this it follows that if $D(a) = 0$, $a \in A_0$, then $L(x) = R(x)$, $x \in \mathbb{E}$, and hence π is an invariant measure of the process. This proves the first assertion of the theorem. Also, π is an invariant measure if and only if the last summation in (7.11) is 0. This proves the second assertion of the theorem.

It remains to prove (7.11). Using the transition rate formulas (7.2), (7.3) and the property $x = T_{dd'}y$ if and only if $y = T_{d'd}x$, it follows that the right side of (7.10) is

$$R(x) = \sum_{y\in\mathbb{E}} \pi(y) \sum_{s,a} \lambda_{sa} r_{sa}(y,x) = \sum_{s,a} \pi(T_{as(\ell)}x)\lambda_{sa} r_{sa}(T_{as(\ell)}x, x)$$

$$= \sum_{s,a} \pi(T_{as(\ell)}x)\lambda_{sa}\phi_{s(\ell)}(T_{as(\ell)}x) + \sum_{s,a}\sum_{i=0}^{\ell-1} \pi(T_{0s(i)}x)\lambda_{sa}$$
$$\times [\phi_{s(i)}(T_{0s(i)}x) - \phi_{s(i+1)}(T_{s^{i+1}s(i+1)}x)]. \tag{7.12}$$

Here we also use our convention that the functions π and ϕ_d are defined to be zero outside of $\mathbb{E}$ and that $T_{0s(i)}x = T_{s^{i+1},s(i+1)}x$. Now, the Φ-balance assumption and $\pi(x) = \Phi(x)\eta(x)$ and $\eta(x+y) = \eta(x)\eta(y)$, ensure that, for $x \in \mathbb{E}$, $a \in A$, and $d \in S$,

$$\pi(T_{ad}x)\phi_d(T_{ad}x) = \pi(x)\phi_a(x)\eta(d)\eta(a)^{-1}. \tag{7.13}$$

Applying this to (7.12), we obtain

$$R(x) = \pi(x)\sum_{s,a} \eta(s(\ell))\eta(a)^{-1}\lambda_{sa}\phi_a(x)$$
$$+ \pi(x)\sum_{s,a}\lambda_{sa}\sum_{i=0}^{\ell-1}\eta(s(i))[\phi_0(x) - \phi_{s^{i+1}}(x)]. \tag{7.14}$$

To proceed, we need a convenient expression for the last sum on s, a, i. Note that $s = 0$ has no contribution to the sum, and hence we ignore it. Also, any $s \neq 0$ can be written as the concatenation $s = (s'as'')$ for some $s', s'' \in \mathcal{S}$ and $a \in A_0$. Now, make the change-of-variables $s^{i+1} = a$ and $sa = (s'as'')a'$ and reverse the order of the summations and recall the definition of Λ_s. Then the last sum in (7.14) becomes

$$\sum_{s',a\neq 0} \eta(s'(\ell'))\sum_{s'',a'}\lambda_{(s'as'')a'}[\phi_0(x) - \phi_a(x)]$$
$$= \sum_{s}\eta(s(\ell))\sum_{a\neq 0}\Lambda_{(sa)}[\phi_0(x) - \phi_a(x)]. \tag{7.15}$$

Substituting this in (7.14) and recalling that $D(a)$ equals the right side of (7.8) minus its left side, we arrive at

$$R(x) = \pi(x)\sum_{a\neq 0}\phi_a(x)\eta(a)^{-1}D(a) + \pi(x)\phi_0(x)$$
$$\times \sum_{s}\eta(s(\ell))\sum_{s}\eta(s(\ell))[\lambda_{s0} + \sum_{a\neq 0}\Lambda_{(sa)}]. \tag{7.16}$$

Next, note that the left side of the balance equation (7.10), in light of (7.5), is

$$L(x) = \pi(x)\sum_{s,a}\lambda_{sa}\sum_{y\in\mathbb{E}} r_{sa}(x,y)$$
$$= \pi(x)\sum_{s,a}\lambda_{sa}\phi_0(x) = \pi(x)\phi_0(x)\Lambda_0. \tag{7.17}$$

Now, using the fact that $s' \neq 0$ can be expressed as $s' = (sas'')$ for some $s, s'' \in S$ and $a \in A_0$, we have the identity

$$\sum_s \eta(s(\ell))\Lambda_s = \Lambda_0 + \sum_{s \neq 0} \eta(s(\ell))\Lambda_s = \Lambda_0 + \sum_{s, a \neq 0} \eta(s(\ell))\eta(a)\Lambda_{(sa)}. \tag{7.18}$$

Also, by its definition, $\Lambda_s = \lambda_{s0} + \sum_{a \neq 0}(\lambda_{sa} + \Lambda_{(sa)})$. Substituting this in the left side of (7.18) and using terms from its right side yields

$$\Lambda_0 = \sum_{a \neq 0} D(a) + \sum_s \eta(s(\ell))[\lambda_{s0} + \sum_{a \neq 0} \Lambda_{(sa)}]. \tag{7.19}$$

Finally, substituting this in (7.17) and using (7.16) yields the identity (7.11). □

7.3 Traffic Equations, Partial Balance, and Throughputs

We begin this section with insights into the existence of solutions to the traffic equations (7.8). Next, we show that the traffic equations are equalities of certain average flows in the network (a partial balance property). We end the section with an expression for throughputs at the nodes.

Note that the hypothesis (the first sentence) of Theorem 7.2 is actually two hypotheses:
(i) There are positive γ_a, $a \in A_0$, that satisfy the traffic equations

$$\gamma_a \sum_s \prod_{i=1}^{\ell} \gamma_{s^i} \Lambda_{(sa)} = \sum_s \prod_{i=1}^{\ell} \gamma_{s^i} \lambda_{sa}, \quad a \in A_0, \tag{7.20}$$

where $\gamma_0 = 1$, and these sums are finite.
(ii) A solution to the preceding equations is of the form

$$\gamma_a = \prod_{j=1}^{m} w_j^{a_j}, \quad a \in A_0,$$

for some positive numbers $w_1, \ldots, w_m$.

Let's first consider hypothesis (i). With a slight abuse of notation, interpret A_0 as an ordered set and view $\gamma \equiv (\gamma_a, a \in A_0)$ as a vector. Write (7.20) as $\gamma_a g_a(\gamma) = h_a(\gamma)$, where $g_a(\gamma)$ and $h_a(\gamma)$ denote the summations on the left and right sides of (7.20) as functions of γ. In other words, (7.20) is the same as $\gamma = f(\gamma)$, where $f(\gamma) = \{f_a(\gamma) : a \in A_0\}$ is the vector-valued function defined by $f_a(\gamma) = h_a(\gamma)/g_a(\gamma)$ for γ in the region where the numerator is finite and the denominator is not zero. Here a vector inequality $\gamma \leq \overline{\gamma}$ means $\gamma_a \leq \overline{\gamma}_a$, for each $a \in A_0$, and 0 and 1 are the vectors of all zeros and all ones.

From the preceding observations, it follows that the set of solutions to (7.20) is equal to the set of fixed points of f. Here is a general criterion for the existence of a solution to (7.20) (i.e., a sufficient condition for hypothesis (i)).

Theorem 7.3. *Suppose there there are vectors* $0 \le \underline{\gamma} < \overline{\gamma}$ *such that* $0 < g_a(\underline{\gamma}) < \infty$ *and*

$$\underline{\gamma}_a g_a(\overline{\gamma}) < h_a(\underline{\gamma}), \quad \text{and} \quad h_a(\overline{\gamma}) < \overline{\gamma}_a g_a(\underline{\gamma}), \quad a \in A_0. \tag{7.21}$$

Then there exists a vector γ *that satisfies (7.20) and* $\underline{\gamma} < \gamma < \overline{\gamma}$.

PROOF. Let $C \equiv \{\gamma : \underline{\gamma} \le \gamma \le \overline{\gamma}\}$. Since $g_a(\gamma)$ and $h_a(\gamma)$ are increasing in γ, it follows that all the terms in (7.21) are finite and

$$\underline{\gamma}_a < h_a(\underline{\gamma})/g_a(\overline{\gamma}) \le f_a(\gamma) \le h_a(\overline{\gamma})/g_a(\underline{\gamma}) < \overline{\gamma}_a, \quad \gamma \in C, \ a \in A_0.$$

Thus, f maps C into C. Also, f is clearly continuous. Then f has a fixed point $\gamma \in C$ by Brouwer's fixed-point theorem. Furthermore, $\underline{\gamma} < \gamma < \overline{\gamma}$, because of the strict inequalities in the preceding display. □

Theorem 7.3 is a framework for obtaining specific conditions for a solution to (7.20) in terms of the structure of the λ_{sa}'s. Examples are in the following sections. The next result is a simpler version of Theorem 7.3. It assumes that $\underline{\gamma}$ exists and $g_a(0) > 0$. Assumption (a) is typically satisfied by open networks, and $\overline{\gamma} = 1$ is often adequate for (b).

Corollary 7.4. *There exists a positive solution to the traffic equations (7.20) if the following conditions hold.*
(a) *The set* $A^* = \{a \in A_0 : \lambda_{0a} > 0\}$ *is not empty and, for each* $a \in A_0 \backslash A^*$, *there is an* $s \in S$ *such that* $\lambda_{sa} > 0$ *and* $s^i \in A^*$, *for* $1 \le i \le \ell$.
(b) *There is a positive vector* $\overline{\gamma}$ *such that* $g_a(\overline{\gamma}) < \infty$ *and* $h_a(\overline{\gamma}) < \overline{\gamma}_a g_a(0)$.

PROOF. Let $\underline{\gamma} = \{\underline{\gamma}_a : a \in A_0\}$ be a vector in $(0, \overline{\gamma})$ such that $\underline{\gamma}_a < \lambda_{0a}/g_a(\overline{\gamma})$, for $a \in A^*$ and

$$\underline{\gamma}_a < \sum_s \prod_{i=1}^{\ell} \underline{\gamma}_{s^i} \lambda_{sa} 1(s^i \in A^*, 1 \le i \le \ell)/g_a(\overline{\gamma}), \quad a \in A_0 \backslash A^*.$$

Assumptions (a) and (b) ensure that $0 < g_a(\underline{\gamma}) < \infty$. Since $g_a(\gamma)$ and $h_a(\gamma)$ are increasing in γ, it follows that (7.21) holds. Thus, the assertion follows by Theorem 7.3. □

Now, consider the hypothesis (ii) that a solution γ to (7.20) has the geometric form $\gamma_a = \prod_{j=1}^m w_j^{a_j}$, $a \in A_0$, for some positive $w_1, \ldots, w_m$. This hypothesis is satisfied for the large class of networks discussed in the next sections on strings composed of unit vectors. For the general case, we have the following observation. The problem is to determine when there are positive $w_1, \ldots, w_m$ that satisfy the linear equations

$$\log \gamma_a = \sum_{j=1}^m a_j \log w_j, \quad a \in A_0, \tag{7.22}$$

for known γ_a's. From a standard property of linear algebra, we have the following result.

Remark 7.5. *(Geometric Solutions).* Let M denote the matrix, whose rows are vectors in A_0, and let M' denote the matrix M augmented by the column $(\log \gamma_a)_{a\in A_0}$. Then there is a solution $w_1, \dots, w_m$ to (7.22) if and only if M and M' have the same rank, which is at most m. If they have the same rank and it is less than m, then there are an infinite number of solutions. Uniqueness is not important for our purposes. However, the solution is unique if M and M' have the same rank m, which is true when A_0 consists of m linearly independent vectors.

We now justify why equations (7.8) are traffic equations. Throughout the rest of this section, we assume that the network process X is ergodic with stationary distribution $\pi(x) = c\Phi(x)\eta(x)$, where $\eta(x) = \prod_{j=1}^m w_j^{x_j}$ and the w_j's satisfy (7.8). We first note that the traffic equation (7.8) for $a = 0$ is

$$\sum_{a\neq 0} \lambda_{0a} = \sum_{s\neq 0} \eta(s(\ell))[\lambda_{s0} + \sum_{a\neq 0} \Lambda_{(sa)}], \qquad x \in \mathbb{E}. \tag{7.23}$$

This follows from the identity (7.19) in which $D(a) = 0$ follows from (7.8), for $a \neq 0$.

Now, recall that by the ergodic theorem for Markov processes, the quantity $\sum_{(x,y)\in\mathcal{T}_0} \pi(x)q(x, y)$ is the average number of $x \to y$ transitions of X per unit time, where (x, y) is in some subset $\mathcal{T}_0$ of $\mathbb{E}^2$. This average number of $\mathcal{T}_0$*-transitions*, which is a limiting average, is also the expected number of $\mathcal{T}_0$-transitions in a unit time interval when the process is stationary.

We shall consider two types of transitions related to the traffic equations. For $a \in A$ and $x \in \mathbb{E}$, let $\tilde{\lambda}_a(x)$ denote the average number of transitions of X per unit time in which the vector a is added to the state x such that the transition leads to the new state $x + a$. We call $\tilde{\lambda}_a(x)$ the *rate of exits from x via an a-addition.* Similarly, let $\bar{\lambda}_a(x)$ denote the *rate of entrances into x via an a-subtraction*: the average number of transitions of X per unit time in which the process enters state x (during a transition) from a subtraction of the vector a. Here are expressions for these rates.

Proposition 7.6. (Partial Balance) *For each $x \in \mathbb{E}$,*

$$\tilde{\lambda}_a(x) = \begin{cases} \pi(x)\phi_0(x)\sum_s \eta(s(\ell))\lambda_{sa} & \text{if } a \neq 0 \\ \pi(x)\phi_0(x)\sum_{s\neq 0} \eta(s(\ell))[\lambda_{s0} + \sum_{a\neq 0}\Lambda_{(sa)}] & \text{if } a = 0 \end{cases} \tag{7.24}$$

$$\bar{\lambda}_a(x) = \begin{cases} \pi(x)\phi_0(x)\eta(a)\sum_s \eta(s(\ell))\Lambda_{(sa)} & \text{if } a \neq 0 \\ \pi(x)\phi_0(x)\sum_{a\neq 0}\lambda_{0a} & \text{if } a = 0. \end{cases} \tag{7.25}$$

Hence, the traffic equations (7.8), (7.23) are equivalent to

$$\bar{\lambda}_a(x) = \tilde{\lambda}_a(x), \qquad a \in A, x \in \mathbb{E}. \tag{7.26}$$

Expression (7.26) says the average number of entrances into x via an a-subtraction is equal to (or balanced with) the average number of exits out of x via an a-addition. The equivalence between this balance (7.26) of traffic flows and the equations (7.8), (7.23) is the reason why we call the latter traffic equations. The equality (7.26) is a partial balance property for the process since it is the balance equations (7.10) for only a part of the summation. Note that (7.26) also implies that, for each fixed state x, the average number of entrances into x via any a-subtraction equals the average number of exits from x via any a-addition (namely $\sum_{a\in A}\bar{\lambda}_a(x) = \sum_{a\in A}\tilde{\lambda}_a(x)$). A similar sum on x, for a fixed a, says the average number of a-subtractions is equal to the average number of a-additions, regardless of the state x.

PROOF. First consider the case $a \neq 0$. A transition of X in which the vector a is added to a state x such that the transition leads to the new state $x+a$ is necessarily a complete sa-transition that starts from $x+s(\ell)$ and lands in $x+a$, for any $s \in \mathcal{S}$. Then by the comment above on the ergodic theorem for Markov processes,

$$\tilde{\lambda}_a(x) = \sum_s \pi(x+s(\ell))\lambda_{sa}\phi_{s(\ell)}(x+s(\ell)).$$

This reduces, in light of (7.13), to the first line in (7.24). Next, note that a transition of X in which it enters state x (during a transition) due to a subtraction of the vector a can only happen when the process is in state $x+s(\ell)+a$ and an $(sas')a'$-string transition occurs, causing the process to enter state x at the stage in which a is subtracted. Arguing as above,

$$\begin{aligned}\bar{\lambda}_a(x) &= \sum_s \pi(x+s(\ell)+a)\sum_{s',a'}\lambda_{(sas')a'}\phi_{s(\ell)+a}(x+s(\ell)+a)\\ &= \sum_s \pi(x+s(\ell)+a)\Lambda_{(sa)}\phi_{s(\ell)+a}(x+s(\ell)+a),\end{aligned}$$

and this reduces to the first line in (7.25).

Now, consider the case $a=0$. Since 0-additions involve complete $s0$-transitions and other partial transitions as well, we have

$$\tilde{\lambda}_0(x) = \sum_{s\neq 0}\pi(x+s(\ell))\phi_{s(\ell)}(x+s(\ell))[\lambda_{s0} + \sum_{a\neq 0}\sum_{s',a'}\lambda_{(sas')a'}],$$

and this reduces to the second line in (7.24). Also, the second line in (7.25) clearly follows since 0-subtractions only involve complete $0a$-transitions. Finally, a glance at (7.24), (7.25), and the traffic equations (7.8) and (7.23) verifies that the traffic equations are equivalent to (7.26). □

A network's performance is often measured by its throughputs $\tilde{\lambda}_1, \ldots, \tilde{\lambda}_m$, where $\tilde{\lambda}_j$ denotes the average number of units per unit time that enter node j. Since the process is ergodic, $\tilde{\lambda}_j$ is also the average number of departures per unit time from j.

Proposition 7.7. (Throughputs at Nodes) *The throughput at node j is*

$$\tilde{\lambda}_j = \sum_{s,a} \alpha_a \lambda_{sa} (a_j^+ + \sum_{i=1}^{\ell} (s_j^i)^-) + \sum_{s,a} \lambda_{sa} \sum_{i=0}^{\ell-1} (\alpha_0 - \alpha_{s^{i+1}}) \sum_{n=1}^{i} (s_j^n)^-, \qquad (7.27)$$

where $\alpha_a \equiv \sum_x \pi(x)\phi_a(x)$, $c^+ \equiv \max\{0, c\}$ and $c^- \equiv c^+ - c$.

PROOF. By the ergodic theorem for Markov processes, it follows that $\tilde{\lambda}_j = \sum_{x,y} \pi(x) q(x, y) f(x, y)$, where $f(x, y)$ describes the number of arrivals to j in a $x \to y$ transition. That is,

$$\begin{aligned}\tilde{\lambda}_j = &\sum_x \pi(x) \sum_{s,a} \lambda_{sa} \phi_{s(\ell)}(x) 1(x - s(\ell) + a \in \mathbb{E})[a_j^+ + \sum_{i=1}^{\ell} (s_j^i)^-] \\ &+ \sum_x \pi(x) \sum_{s,a} \lambda_{sa} \sum_{i=0}^{\ell-1} [\phi_{s(i)}(x) - \phi_{s(i+1)}(x)] \\ &\qquad\qquad \times 1(x - s(i) \in \mathbb{E}) \sum_{n=1}^{i} (s_j^n)^-.\end{aligned}$$

Now, by two uses of Φ-balance and the structure of π, we have

$$\begin{aligned}&\sum_x \pi(x) \phi_{s(i)}(x) 1(x - s(i) + a \in \mathbb{E}) \\ &\quad = \sum_x \pi(T_{as(i)} x) \phi_{s(i)}(T_{as(i)} x) \eta(a) \eta(s(i))^{-1} \\ &\quad = \sum_x \pi(x) \phi_a(x) = \alpha_a.\end{aligned}$$

Similarly,

$$\sum_x \pi(x) \phi_{s(i+1)}(x) 1(x - s(i+1) - s^{i+1} \in \mathbb{E}) = \alpha_{s^{i+1}}.$$

Then applying these equalities to the two sums on x in the preceding display yields (7.27). □

7.4 String-Nets with Unit-Vector Transitions

This section describes the results above when the allowable increment vectors $a \in A$ are unit vectors instead of general vectors.

Suppose the string-net process X represents an open network in which the allowable increment vectors consist of the m-dimensional unit vectors $e_1, \ldots, e_m$ and $e_0 = 0$. We say that the process has *unit-vector string transitions*. In this case, the unit vectors are associated with the node numbers: It is convenient to let $A = \{0, 1, \ldots, m\}$ denote the node numbers instead of the vectors. Accordingly, $s = (s^1, \ldots, s^\ell)$ is a string of node numbers, $s(i) = \sum_{n=1}^{i} e_{s^n}$, and the rates are

λ_{sj}, $j \in A$. Then the results above are the same, aside from the change in notation from vectors to node numbers. For instance, the traffic equations (7.8) are

$$w_j \sum_s \prod_{i=1}^{\ell} w_{s^i} \sum_{s',j'} \lambda_{(sjs')j'} = \sum_s \prod_{i=1}^{\ell} w_{s^i} \lambda_{sj}, \quad 1 \leq j \leq m. \tag{7.28}$$

The following is a combination of Theorems 7.2 and 7.3 for unit-vector string transitions. Consider (7.28) written as $w_j g_j(w) = h_j(w)$, where $g_j(w)$ and $h_j(w)$ denote the summations on the left and right sides of (7.28) as functions of $w = (w_1, \ldots, w_m)$.

Theorem 7.8. *Suppose there are vectors* $0 \leq \underline{w} < \overline{w}$ *such that* $0 < g_j(\overline{w}) < \infty$ *and*

$$\underline{w}_j g_j(\overline{w}) < h_j(\underline{w}), \quad \text{and} \quad h_j(\overline{w}) < \overline{w}_j g_j(\underline{w}), \quad 1 \leq j \leq m.$$

Then there is a vector w *that satisfies (7.28) and* $\underline{w} < w < \overline{w}$. *Moreover,* $\pi(x) = \Phi(x) \prod_{j=1}^{m} w_j^{x_j}$, $x \in \mathbb{E}$, *is an invariant measure for the network process* X.

Note that this result is simpler than Theorems 7.2 and 7.3 because w_j plays the role of γ_a in Theorem 7.3, and hence there is no issue of verifying that γ_a is a product.

For closed networks, unit-vector transitions make sense only for the case of one-stage transitions ($L = 1$). One can also define analogous unit-vector transitions when the set A of increment vectors consists of only the negative unit vectors, or when it consists of a combination of negative and positive unit vectors.

We now derive expressions for throughputs and service rates. For the rest of this section, assume that the network process X is ergodic and denote its stationary distribution by $\pi(x) = c\Phi(x) \prod_{j=1}^{m} w_j^{x_j}$. Let $\tilde{\mu}_j$ denote the average number of departures per unit time from node j when the node is not empty. The $\tilde{\lambda}_j$ and $\tilde{\mu}_j$ are often called the *effective arrival and service rates* for node j and the ratio $\tilde{\lambda}_j / \tilde{\mu}_j$ is the *traffic intensity*.

Proposition 7.9. *For the network process* X *with unit-vector string transitions,*

$$\tilde{\lambda}_j = \sum_x \pi(x) \phi_j(x) \sum_s \lambda_{sj}, \quad \text{and} \quad \tilde{\mu}_j = \tilde{\lambda}_j / \sum_x 1(e_j \leq x) \pi(x).$$

PROOF. The first expression is an obvious special case of (7.27). By the strong law of large numbers for Markov processes, the effective departure rate is

$$\tilde{\mu}_j = \sum_x \overline{\lambda}_j(x) / \sum_x 1(e_j \leq x) \pi(x),$$

which is the average number of departures from j per unit time divided by the portion of time j is nonempty. And $\sum_x \overline{\lambda}_j(x) = \tilde{\lambda}_j$ by Proposition 7.6. □

Another important performance measure of the network is the average sojourn or waiting time of a unit in a node j of the network. This average is defined by $W_j \equiv \lim_{n \to \infty} n^{-1} \sum_{\nu=1}^{n} W_j(\nu)$, where $W_j(\nu)$ is the waiting time in j of the νth unit to enter j. We will use the average number of units in j, which is $L_j \equiv \sum_x x_j \pi(x)$. The Little law for Markovian systems in Theorem 5.1 yields the following result.

Proposition 7.10. (Little Law for Waiting Times) *If the network has unit-vector string transitions and the state space contains a vector x with $x_j = 0$, then W_j exists and $L_j = \tilde{\lambda}_j W_j$. Here $\tilde{\lambda}_j$ is necessarily finite, but W_j and L_j may both be finite or infinite. This assertion for averages also holds for expected values when the system is stationary (or in equilibrium). In this case, L_j is the expected number of units in j; the $\tilde{\lambda}_j$ is the expected number of units that enter j in a unit time interval; and W_j is the expected sojourn time for an arbitrary unit in j under the Palm probability that a unit enters j at time* 0.

Similar Little laws apply to batch arrivals into j, but more information is needed on how the "order of units" in a batch affect their individual service times. In these cases, one can state a law for all units labeled as the kth unit within a batch arriving into j—the W_j and L_j would be the average waiting times and queue lengths for these kth arrivals, and $\tilde{\lambda}_j$ would be the arrival rate of batches into j of size k or more. The expected waiting time in a sector (subset of nodes) in a Jackson network is described in Chapter 1. To obtain similar results for vector-transitions, one would need more information on where each unit in a batch actually moves; the net number of movements is not adequate to describe waiting times as it is under single-unit movements.

The computation of throughputs, average waiting times, and other performance parameters—even for a Jackson network—is difficult for a moderate-size network. However, since there is a closed-form expression for the stationary distribution of the network, one can compute these parameters by Monte Carlo simulation as discussed in Chapter 1.

7.5 Networks with One-Stage Batch Transitions

A Whittle-type network with batch movements is a string-net in which all transitions are of the form $x \to T_{da}x$, for $d, a \in A$. The d and a are departure and arrival vectors. Units represented by d may form part or all of the vector a, in which case they are transferred within the network. In this string-net, all of the strings are exactly of length 1 (i.e., each transition involves only one pair of addition/subtraction vectors). This section describes the results above for these batch-movement networks.

Consider the string-net process X with strings exactly of length 1 as described in the preceding paragraph. The transition rates (7.2) for the network are

$$q(x, y) = \sum_{d,a \in A} \lambda_{da} \phi_d(x) 1(y = T_{da}x), \quad x \neq y \text{ in } \mathbb{E}. \tag{7.29}$$

In other words, whenever the process is in state x, the time to its next potential move to $T_{da}x$ via a da-transition is exponentially distributed with rate $\lambda_{da}\phi_d(x)$.

The Φ-balance assumption implies that $\phi_d(x) = 0$, if $T_{da}x \notin \mathbb{E}$ for some $a \in A$ with $\lambda_{da} > 0$. Note that the ϕ_0-dominance assumption is not relevant since $L = 1$.

We call X a network process with *one-stage batch transitions*—the vectors in A are the allowable batch increments in the process. Invariant measures for this process are given by the following special case of Theorem 7.2.

Theorem 7.11. *Suppose that $\gamma = (\gamma_a : a \in A)$ is a positive vector, with $\gamma_0 = 1$, that satisfies*

$$\gamma_a \sum_{d \in A_0} \lambda_{ad} = \sum_{d \in A_0} \gamma_d \lambda_{da}, \quad a \in A_0, \tag{7.30}$$

and $\gamma_a = \prod_{j=1}^m w_j^{a_j}$, $a \in A_0$, for some positive $w_1, \ldots, w_m$. Then an invariant measure for the process X is $\pi(x) = \Phi(x) \prod_{j=1}^m w_j^{x_j}$, $x \in \mathbb{E}$. Furthermore, a necessary and sufficient condition for an invariant measure of this form is

$$\sum_{a \in A_0} D(a)[\phi_0(x) - \phi_a(x) \prod_{j=1}^m w_j^{-a_j}] = 0, \quad x \in \mathbb{E},$$

where $D(a)$ denotes the right side of (7.30) minus its left side.

In this case, the traffic equations (7.8) reduce to (7.30) because in $\Lambda_{(sa)}$, the s must be 0 since $L = 1$, and $\Lambda_{(0a)} = \sum_{d \in A} \lambda_{ad}$. Note that (7.30) is a balance equation for a Markov process on the finite set A_0 with transition rates λ_{ad} and hence there exists a positive solution γ to the equation. The solution is a geometric product form under the criterion in Remark 7.5.

According to Proposition 7.6, the measure π in Theorem 7.11 satisfies the partial balance property

$$\pi(x) \sum_{d \in A_0} q(x, T_{ad}x) = \sum_{d \in A_0} \pi(T_{ad}x) q(T_{ad}x, x) 1(T_{ad}x \in \mathbb{E}), \quad a \in A_0, x \in \mathbb{E}. \tag{7.31}$$

The following is an example in which the units in a batch movement are independently transferred among nodes via Markovian routing probabilities.

Example 7.12. *Independent Concurrent Movements of Units.* Let X denote the network process described above with rates (7.29), where A denotes a set of m-dimensional vectors. For simplicity, assume the network is closed (the open case is similar). In a da-transition in state x, think of $\phi_d(x)$ as the rate at which the batch d is released from the network and λ_{da} as the rate in which d is changed into the addition batch a. To describe the units in these vectors by their node locations, we define

$$\mathcal{I}(d) = \{\mathbf{i} = (i_1, \ldots, i_{|d|}) : \sum_{n=1}^{|d|} 1(i_n = j) = d_j, 1 \le j \le m\},$$

which is the set of node indices that "represent" d.

Assume the units in the batch d move concurrently such that r_{jk} is the rate (probability or propensity) for a single unit in the batch to move from j to k in the node set $\{1, \dots, m\}$; and that $r_{j_1,k_1} \cdots r_{j_{|d|},k_{|d|}}$ is the rate that the released batch $\mathbf{j} \in \mathcal{I}(d)$ results in the batch addition $\mathbf{k} \in \mathcal{I}(a)$, where $|d| = |a|$. This rate is a compounding of the single-unit rates. Then the rate of a da-transition in state x is $\lambda_{da}\phi_d(x)$, where

$$\lambda_{da} = \sum_{\mathbf{j}\in\mathcal{I}(d)} \sum_{\mathbf{k}\in\mathcal{I}(a)} r_{j_1,k_1} \cdots r_{j_{|d|},k_{|d|}}, \quad d, a \in A, \text{ with } |d| = |a|.$$

Note that the probability of d and a being generated is $\lambda_{da}/\sum_a \lambda_{da}$; and if the r_{jk}'s are probabilities with $\sum_{k=1}^m r_{jk} = 1$ for each j, then $\sum_a \lambda_{da} = |d|!/d_1! \cdots d_m!$.

Assume the rates r_{jk} are irreducible and let $w_1, \dots, w_m$ be positive numbers that satisfy

$$w_j \sum_{k=1}^m r_{jk} = \sum_{k=1}^m w_k r_{kj}, \quad 1 \le j \le m.$$

Define $\eta(x) = \prod_{j=1}^m w_j^{x_j}$. Because the w_j's satisfy the preceding equations, we have

$$\begin{aligned}
\eta(a)\sum_{d\in A}\lambda_{ad} &= \sum_{\mathbf{j}\in\mathcal{I}(a)} \sum_{k_1\dots,k_{|a|}} w_{j_1} r_{j_1,k_1} \cdots w_{j_{|a|}} r_{j_{|a|},k_{|a|}} \\
&= \sum_{\mathbf{j}\in\mathcal{I}(a)} \sum_{k_1\dots,k_{|a|}} w_{k_1} r_{k_1,j_1} \cdots w_{k_{|a|}} r_{k_{|a|},j_{|a|}} \\
&= \sum_{d\in A} \eta(d)\lambda_{da}, \quad a \in A_0.
\end{aligned}$$

Therefore, by Theorem 7.11 it follows that $\pi(x) = \Phi(x)\eta(x)$, $x \in \mathbb{E}$, is an invariant measure for the process. □

7.6 Networks with Compound-Rate String Transitions

A large class of string-nets are those in which the rate λ_{sa} of an sa-string is a product or compounding of several rates representing micro features of the network. This section illustrates this class with an example of a network in which a string is generated by a Markov chain mechanism. The ideas here readily extend to a variety of networks with compound-rate string transitions.

Consider the m-node network that operates as follows. The network is open, and its state space consists of all m-dimensional vectors with nonnegative integer-valued entries. Units enter the network at the nodes according to independent Poisson processes with respective rates $\lambda_1, \dots, \lambda_m$; a zero rate for a node means it has no external arrivals. The services at each node j are independent and exponentially distributed with rate μ_j. The results below also apply, with minor modifications, to general Φ-balanced service rates and closed or open networks with other types of state spaces.

A transition of the network is triggered by the movement of a single unit. An external arrival to a node just adds one unit to the node and no other units move.

On the other hand, a service completion at a node may trigger a transition in which single units are successively deleted from a string of nodes $s_1, \ldots, s_\nu$ and, at the end, one unit might be added to some node k in $A \equiv \{0, 1, \ldots, m\}$. All of this occurs instantaneously and the number of deletions $\nu \le L$ is a stopping index that may be random.

The procedure for such a transition triggered by a service completion is as follows. Whenever a normal service completes at some node $s_1 \in A_0 \equiv A\backslash\{0\}$ (with rate μ_{s_1}), then with probability $Q_{s_1 k}$ one unit moves to some node $k \in A$ and the procedure stops; or with probability $P_{s_1 s_2}$, one unit exits the network from node s_1 and a signal goes to node $s_2 \in A_0$ to delete a unit there provided that node is not empty ($\sum_k (P_{jk} + Q_{jk}) = 1$ for each j). If node s_2 is empty or if $L = 1$, the procedure stops. Otherwise, the preceding events are repeated until stopping. That is, for each $i \ge 1$, the departure from node s_i, with probability $Q_{s_i k}$, adds one unit to node k and stops the procedure; or with probability $P_{s_i s_{i+1}}$, it triggers another departure from node s_{i+1} provided this node is nonempty and, if node s_{i+1} is empty or $i = L$, the procedure stops. Think of P_{jk} as probabilities of "propagating new departures" and Q_{jk} as probabilities of "quitting" (or stopping) the string deletions.

In summary of the preceding description, typical transitions of the network are as follows.

- An arrival into node k from outside the network: $x \to x + e_k$.
- String deletions stopped because node s_{i+1} is empty or $i = L$:
$$x \to x - e_{s_1} - \cdots - e_{s_i}.$$
- String deletions stopped by the quitting probability $Q_{s_i k}$:
$$x \to x - e_{s_1} - \cdots - e_{s_i} + e_k.$$

As in the previous sections, we let X denote the stochastic process representing the numbers of units at the nodes. The data for this process are the arrival rates $\lambda = (\lambda_1, \ldots, \lambda_m)$, service rates $\mu = (\mu_1, \ldots, \mu_m)$, maximum string length L, and propagating and quitting probabilities P_{jk}, Q_{jk}. Define $P = (P_{jk})$ and $Q = (Q_{jk})$ for $j, k \in A_0$. We also assume that the inverse of the matrix $I - Q$ exists, where I denotes the identity matrix. We will frequently use the vector

$$\tilde{\lambda} = \lambda(I - Q)^{-1},$$

whose entries are effective arrival rates, as we will soon see.

We first justify that this network is a string-net.

Proposition 7.13. *Under the preceding assumptions, X is a Markov network process with unit-vector string transitions and its associated traffic equation (7.28) in matrix form is*

$$\mu \sum_{n=0}^{L-1} (WP)^n W = \lambda + \mu \sum_{n=0}^{L-1} (WP)^n WQ, \tag{7.32}$$

where W is a diagonal matrix with diagonal entries $w_1, \ldots, w_m$.

Proof. Because of the Poisson arrivals and exponential service times, the network process is clearly Markovian. The rates of its sj-transitions are $\lambda_{0j} = \lambda_j$

and, for $s \neq 0$,

$$\lambda_{sj} = \Lambda_s Q_{s_\ell j}, \quad j \neq 0, \quad \text{and} \quad \lambda_{s0} = \Lambda_s[Q_{s_\ell 0} + 1(\ell = L)\sum_k P_{s_\ell k}],$$

where $\Lambda_s = \mu_{s_1} \prod_{i=1}^{\ell-1} P_{s_i s_{i+1}}$. Now, consistent with its definition in the last section, $\Lambda_0 = \sum_j (\lambda_j + \mu_j)$ and, for $s \neq 0$, $\Lambda_s = \sum_{s',j} \lambda_{(ss')j}$, which is the rate of all string transitions whose first part is s.

Next note that the departure rate functions must satisfy

$$\phi_{s(i)}(x) - \phi_{s(i+1)}(x) = 1(0 \leq x - s(i), 0 \nleq x - s(i+1)).$$

Consequently, they have the special form $\phi_j(x) = 1(e_j \leq x)$. Clearly, these ϕ_j's are Φ-balanced with $\Phi(\cdot) = 1$, and the ϕ_0-dominance assumption is satisfied. Under these specifications, X is a Markov network process with unit-vector string transitions.

In this setting, the traffic equations (7.28) reduce to (7.32) since

$$\begin{aligned} w_j \sum_s \eta(s(\ell))\Lambda_{(sj)} &= \sum_{s\neq 0} \mu_{s_1} \prod_{i=1}^{\ell-1} w_{s_i} P_{s_i s_{i+1}} w_j 1(s_\ell = j) \\ &= \left(\mu \sum_{n=0}^{L-1} (WP)^n W\right)_j, \end{aligned} \tag{7.33}$$

$$\sum_s \eta(s(\ell))\lambda_{sj} = \left(\lambda + \mu \sum_{n=0}^{L-1} (WP)^n WQ\right)_j. \qquad \square \tag{7.34}$$

The following characterization of solutions to the traffic equation is analogous to Theorem 7.3. We will use $\overline{w}_j = \tilde{\lambda}_j/\mu_j$, $1 \leq j \leq m$. Recall that $\tilde{\lambda} = \lambda(I - Q)^{-1}$.

Theorem 7.14. (a) *Suppose $L = \infty$ and $\sum_{n=0}^{\infty}(PW)^n < \infty$, where*

$$w_j = \tilde{\lambda}_j/(\mu + \tilde{\lambda}P)_j, \quad 1 \leq j \leq m.$$

Then $w_1, \ldots, w_m$ is the unique solution to the traffic equation (7.32).
(b) *Suppose $L < \infty$ and $\sum_{n=0}^{\infty}(P\overline{W})^n < \infty$. Then there exists a solution w to the traffic equation (7.32) in the open rectangle $(0, \overline{w})$. Furthermore, let $\mathbf{w}_n$ denote a sequence of vectors defined by $\mathbf{w}_0 = 0$ and $\mathbf{w}_{n+1} = h(\mathbf{w}_n)$, where $h(w) = (h_1(w), \ldots, h_m(w))$ and*

$$h_j(w) = (\tilde{\lambda} + \mu(WP)^L W)_j/(\mu + \tilde{\lambda}P)_j, \quad w \in C \equiv [0, \overline{w}], \ 1 \leq j \leq m.$$

Then $\mathbf{w}_n$ is a nondecreasing sequence whose limit is the minimal solution to the traffic equation (7.32) (any other solution is greater than or equal to this limit).

PROOF. First note that by subtracting the right side of (7.32) from its left and dividing by $I - Q$, this traffic equation can be written as

$$\mu \sum_{n=0}^{L-1} (WP)^n W = \tilde{\lambda}. \tag{7.35}$$

Now, assume the assumptions in part (a) hold. Multiplying both sides of (7.35) on the right by the matrix $(I - PW)$ yields $\mu W = \tilde{\lambda} - \tilde{\lambda} PW$. That is, $(\mu + \tilde{\lambda} P)W = \tilde{\lambda}$, for $w \in C$. This proves the assertion in part (a).

Next, assume $L < \infty$. Note that equation (7.35) is the same as $f(w) = w$, where $f(w) = (f_1(w), \ldots, f_m(w))$ is defined by

$$f_j(w) = \tilde{\lambda}_j / (\mu \sum_{n=0}^{L-1} (WP)^n)_j, \quad w \in C, \quad 1 \le j \le m.$$

Clearly f is positive, continuous, nonincreasing, and its range is contained in C since $0 < f(\overline{w}) < \overline{w}$. Then, by Brouwer's fixed point theorem, f has a fixed point in C and hence this point is a solution to the traffic equation (7.35). Furthermore, this solution is in the open rectangle $(0, \overline{w})$ since this set contains the range of f.

For the rest of the proof, we need another representation of the traffic equation (7.35). Multiplying both sides of it on the right by the matrix $(I - PW)$ yields

$$\mu[I - (WP)^L]W = \tilde{\lambda} - \tilde{\lambda} PW, \quad w \in C. \tag{7.36}$$

Writing this as $(\mu + \tilde{\lambda} P)W = \tilde{\lambda} + \mu(WP)^L W$ and recalling the definition of h in part (b), it is clear that the traffic equation (7.35) is equivalent to $w = h(w)$, for $w \ge 0$. Hence the solutions to (7.35) are the same as the fixed points of h. Since h is nondecreasing, it follows by induction that $\mathbf{w}_n$ is a nondecreasing sequence. Then the limit $w^* = \lim_{n\to\infty} \mathbf{w}_n$ exists. Now, as $n \to \infty$ in $\mathbf{w}_{n+1} = h(\mathbf{w}_n)$, the continuity of h ensures that $w^* = h(w^*)$. Thus w^* is a solution to the traffic equation (7.35). It remains to show that if w' is any solution to the equation, then $w^* \le w'$. To prove this, it suffices to show that $\mathbf{w}_n \le w'$ for each n. But this follows by induction, since $\mathbf{w}_0 = 0 \le w'$ and, assuming $\mathbf{w}_n \le w'$ for some n, then $\mathbf{w}_{n+1} = h(\mathbf{w}_n) \le h(w') = w'$ because h is nondecreasing. □

Obtaining a solution of the traffic equation by successively computing $\mathbf{w}_{n+1} = h(\mathbf{w}_n)$ is very efficient: $\mathbf{w}_n$ converges to its limit very fast.

The next result describes invariant measures for the network process and says the process is ergodic if the arrival rates to the nodes are less than the service capacities of the nodes.

Theorem 7.15. *Suppose the vector w is a solution to the traffic equation (7.32) and $0 < w < \overline{w}$. Then $\prod_{j=1}^m w_j^{x_j}$, $x \ge 0$, is an invariant measure for the process X. This process is ergodic if and only if $0 < w < 1$. In particular, the process is ergodic if $\tilde{\lambda} < \mu$, and, in case $L = \infty$, the process is ergodic if and only if $\tilde{\lambda} < \mu + \tilde{\lambda} P$. When the process is ergodic, its stationary distribution is*

$$\pi(x) = \prod_{j=1}^m (1 - w_j) w_j^{x_j}, \ x \ge 0.$$

In addition, $\tilde{\lambda}$ *is the throughput vector and the vector of average numbers of units that depart from the respective nodes per unit time when they are busy is*

$$\tilde{\mu} = \begin{cases} \mu + \tilde{\lambda} P & \text{if } L = \infty \\ \mu \sum_{n=0}^{L} (WP)^n & \text{if } L < \infty. \end{cases} \tag{7.37}$$

Also, $w_j = \tilde{\lambda}_j / \tilde{\mu}_j$*, which follows by the traffic equation, is the* traffic intensity *at node* j.

PROOF. The first two assertions follow by Theorems 7.2 and 7.14 (recall that $\Phi(\cdot) = 1$) and the fact that $\prod_{j=1}^{m} w_j^{x_j}$ is finite if and only if $w_j < 1$ for each j. If $\tilde{\lambda} < \mu$ (i.e., $\overline{w} < 1$), then by Theorem 7.14 we know that there is a solution w to the traffic equation that satisfies $0 < w < 1$; hence the process X is ergodic. The assertion for the case $L = \infty$ also follows by Theorem 7.14.

Now, assume that X is ergodic. By Proposition 7.6, we know that the effective arrival rate to j is $\sum_s \eta(s_\ell)\lambda_{sj}$. Then this rate equals $\tilde{\lambda}_j$, since the equality of (7.33) and (7.34) along with (7.35) yield

$$\sum_s \eta(s_\ell)\lambda_{sj} = \Big(\mu \sum_{n=0}^{L-1} (WP)^n W\Big)_j = \tilde{\lambda}_j.$$

Next, observe that Proposition 7.9 says that the effective departure rate from j is

$$\tilde{\lambda}_j / \sum_x 1(e_j \le x)\pi(x) = \tilde{\lambda}_j / w_j.$$

Then this average equals $\tilde{\mu}_j$ defined by (7.37) since $\tilde{\mu}_j = w_j^{-1}\tilde{\lambda}_j$ by the traffic equation (7.32). □

7.7 Networks with Multiple, Compound-Rate String Transitions

In this section, we discuss string-nets with compound-rate string transitions for multiple types of string initiations.

Consider the network process X in the previous section with the following generalizations:

(1) There are *multiple types of services* or string initiations indexed by $\iota \in \mathcal{I}$, and $\mu(\iota) = (\mu_1(\iota), \ldots, \mu_m(\iota))$ denotes the type ι service rates at the nodes.

(2) A type ι service completion generates a string of deletions and a possible addition as before, but now the *propagation and quitting probabilities depend on the stage* i *and index* ι. Specifically, an ith departure from node s_i may trigger a departure from node s_{i+1} with probability $P_{s_i s_{i+1}}(\iota, i)$ or, with probability $Q_{s_i k}(\iota, i)$, it may add one unit to k and then quit. With no loss in generality, the maximum string length L is independent of ι.

The following result is analogous to Theorem 7.15. The condition (7.40) for ergodicity is that the arrival rates $\tilde{\lambda}(I)$ into the nodes are less than the service rates $\tilde{\mu}(0)$.

Theorem 7.16. *The process X described above is a string-net and its traffic equation is $\tilde{\mu}(W)W = \tilde{\lambda}(W)$, where*

$$\tilde{\lambda}(W) = \lambda + \sum_{\iota\in\mathcal{I}} \mu(\iota) \sum_{n=0}^{L-1} [\prod_{i=1}^{n} WP(\iota, i)] WQ(\iota, n+1), \tag{7.38}$$

$$\tilde{\mu}(W) = \sum_{\iota\in\mathcal{I}} \mu(\iota) \sum_{n=0}^{L-1} [\prod_{i=1}^{n} WP(\iota, i)]. \tag{7.39}$$

If $\tilde{\lambda}(\overline{W}) < \tilde{\mu}(0)\overline{W} = \sum_{\iota\in\mathcal{I}} \mu(\iota)\overline{W}$ for some $m \times m$ diagonal matrix $\overline{W}$ with positive diagonal entries, then there is a solution w to the traffic equation in $(0, \overline{w})$. In particular, if

$$\tilde{\lambda}(I) < \sum_{\iota\in\mathcal{I}} \mu(\iota), \tag{7.40}$$

then there exists a solution w to the traffic equation in $(0, 1)$, and hence the process X is ergodic. In this case, its stationary distribution is

$$\pi(x) = \prod_{j=1}^{m} (1 - w_j) w_j^{x_j}, \quad x \geq 0,$$

and $\tilde{\lambda}(W)$ and $\tilde{\mu}(W)$ are the effective arrival and service rate vectors.

PROOF. The traffic equation $\tilde{\mu}(W)W = \tilde{\lambda}(W)$ is the obvious analogue of (7.32). The rest of the proof is similar to that of Theorems 7.14 and 7.15 because a solution of the traffic equation is a fixed point of the vector-valued function

$$f(w) = (\tilde{\lambda}(W)_1/\tilde{\mu}(W)_1, \ldots, \tilde{\lambda}(W)_m/\tilde{\mu}(W)_m), \quad w \in [0, \overline{w}]. \qquad \square$$

The rest of this section is devoted to examples of the preceding theorem. We start with a case that can be treated as in the last section.

Example 7.17. *Homogeneous Propagation and Quitting Probabilities.* Consider the process described above in which all the propagating and quitting probabilities are P_{jk} and Q_{jk}, respectively. Then the traffic equation is the same as that in Proposition 7.13 with $\mu = \sum_\iota \mu(\iota)$. Consequently, the assertions of Proposition 7.13 and Theorems 7.14 and 7.15 apply automatically. $\square$

Another special situation of interest is when the propagating and quitting matrices are homogeneous and equal to P and Q, respectively, after the first stage. Then (7.38) and (7.39) reduce to

$$\tilde{\lambda}(W) = \lambda + \sum_{\iota\in\mathcal{I}} \mu(\iota)[WQ(\iota, 1) + WP(\iota, 1) \sum_{n=0}^{L-2} (WP)^n WQ], \tag{7.41}$$

$$\tilde{\mu}(W) = \sum_{\iota \in \mathcal{I}} \mu(\iota)[I + WP(\iota, 1) \sum_{n=0}^{L-2} (WP)^n]. \tag{7.42}$$

The following are examples of this case.

Example 7.18. *Regular and Negative Units with Two-Stage Strings*. Consider the process described above in which there are regular and negative units (types 1 and 2) with two-stage strings ($L = 2$) that evolve as follows. Whenever a regular unit finishes a service at node j with rate $\mu(1)_j$, it either enters a node k with probability Q'_{jk} for another service, or becomes a negative unit and enters node k with probability P_{jk}. If this negative unit encounters no units at k, nothing more happens. Otherwise, one unit is deleted from k and one regular unit enters a node k' with probability $Q_{kk'}$ (entering $k' = 0$ means the unit exits the network). In addition, negative units from outside enter the nodes according to independent Poisson processes with rates $\mu(2) = (\mu(2)_1, \ldots, \mu(2)_m)$. If a negative unit entering k encounters no units there, then nothing more happens; otherwise, one unit is deleted from k and one regular unit enters a node k' with probability $Q_{kk'}$. In terms of the notation above,

$$P(1,1) = P, \quad Q(1,1) = Q', \quad Q(1,2) = Q = Q(2,1),$$

$$P(1,2) = P(2,1) = P(2,2) = Q(2,2) = 0.$$

Then Theorem 7.16 applies with (7.41), (7.42) reduced to

$$\tilde{\lambda}(W) = \lambda + \mu(1)[WQ' + WPWQ] + \mu(2)WQ,$$
$$\tilde{\mu}(W) = \mu(1) + \mu(2) + \mu(1)WP.$$

The sufficient condition for ergodicity is

$$\left(\lambda + \mu(1)[Q' + PQ] + \mu(2)Q\right)_j < (\mu(1) + \mu(2))_j, \quad 1 \leq j \leq m. \qquad \square$$

Example 7.19. *Regular and Negative Units with Infinite Strings*. Consider the process related to (7.41), (7.42) in which $L = \infty$ and all the propagation and quitting probability matrices are P, Q except for those at the first stage. Clearly $\sum_{n=0}^{\infty} (WP)^n = (I - WP)^{-1}$ exists for $W \leq I$, provided that $(I - P)^{-1}$ exists. In this case, Theorem 7.16 applies and the sufficient condition for ergodicity is

$$\left(\lambda + \sum_{\iota \in \mathcal{I}} \mu(\iota)[Q(\iota, 1) + P(\iota, 1)(I - P)^{-1}Q]\right)_j < \sum_{\iota \in \mathcal{I}} \mu(\iota)_j, \quad 1 \leq j \leq m.$$

Although we know the traffic equations have a solution, we cannot obtain a closed-form expression for it as we did in Example 7.17.

7.8 String-Nets with Two-Node Batch Transitions

In this section, we discuss string-nets in which a transition involves a batch deletion at a single node and a batch addition at another single node.

Assume that the string-net X is such that each element of A is of the form $a = ne_k$ for some $0 \le k \le m$ and $n \ge 1$. Also, assume that each nonzero string s in S consists of ℓ copies of of some e_j (i.e., $s = (e_j \dots e_j)$), where $\ell \le \infty$. This means that for such a pair sa, the complete transition is $x \to x - \ell e_j + ne_k$, and the ith partial transition is $x \to x - ie_j$. We express the rate of sa as $\lambda_{sa} = \lambda_{\ell j,nk}$. Under an sa-transition with $s = e_je_j\dots$ and $\ell = \infty$, all units from node j are cleared out, and we denote its rate simply by $\lambda_{\infty j}$. Such a "clearing" transition might represent a dispatching or assembly of units (or a catastrophe) that clears out all units at j. We say that X has *two-node string transitions* since exactly two nodes are affected in a transition.

Here we let γ_{nj} denote positive real numbers, for $1 \le j \le m$ and $n \ge 1$, where $\gamma_{0j} = 1$. Also, the summations on ℓ are the conventional ones that do not include a term for $\ell = \infty$.

Theorem 7.20. *For the network process X with two-node batch transitions, the traffic equations (7.8) are*

$$\gamma_{nj} \sum_{\ell=0}^{\infty} \gamma_{\ell j}[r_j(n+\ell) + \lambda_{\infty j}] = \sum_{\ell k} \gamma_{\ell k}\lambda_{\ell k,nj}, \qquad 1 \le j \le m,\ n \ge 1, \tag{7.43}$$

where $r_j(n) \equiv \sum_{\ell' \ge n} \sum_{n'j'} \lambda_{\ell' j,n'j'}$. *If these equations have a solution of the form* $\gamma_{nj} = w_j^n$, *for some positive* $w_1, \dots, w_m$, *then* $\pi(x) = \Phi(x) \prod_{j=1}^m w_j^{x_j}$, $x \in \mathbb{E}$, *is an invariant measure for the process.*

Proof. This follows by Theorem 7.2, where the traffic equations (7.8) reduce to (7.43) since, for any $sa = (\ell k, nj)$, the $\Lambda_{(sa)} = r_j(n+\ell)1(j = k)$ for $\ell < \infty$ and $\Lambda_{(sa)} = \lambda_{\infty j}1(j = k)$ for $\ell = \infty$. □

Here are a few examples.

Example 7.21. *Open Whittle Process with Periodic Clearing*. Suppose the process X with two-node batch transitions has strings of only length 1 or ∞, and $A = \{e_1, \dots, e_m\}$. Then all transitions are of the form $(x \to x - e_j + e_k)$, as in a Whittle network, or there is a clearing $(x \to x - x_je_j)$. The rates of these transitions are

$$q(x, x - e_j + e_k) = \phi_j(x)\lambda_{j,k}, \quad q(x, x - x_je_j) = \phi_j(x)\lambda_{\infty j}. \tag{7.44}$$

We call X a *Whittle network process with periodic clearing*. Without loss in generality, assume that $\lambda_{j,k}$ is an irreducible matrix.

Theorem 7.22. *Suppose the process X with transition rates (7.44) satisfies* $\sum_x \Phi(x) < \infty$. *Then it is ergodic and its stationary distribution is*

$$\pi(x) = c\Phi(x) \prod_{j=1}^m w_j^{x_j}, \qquad x \in \mathbb{E},$$

where $w_0 = 1$ *and* $w_1, \dots, w_m$ *in* $(0, 1)$ *satisfy the traffic equations*

$$w_j[\sum_k \lambda_{j,k} + \lambda_{\infty j} \sum_{\nu=0}^{\infty} w_j^{\nu}] = \sum_k w_k\lambda_{k,j}, \qquad 1 \le j \le m. \tag{7.45}$$

Furthermore, the effective arrival and service rates, $\tilde{\lambda}_j$ and $\tilde{\mu}_j$, for node j are given by the sums on the right and left sides of (7.45), respectively.

PROOF. First note that equations (7.45) are clearly a special case of the traffic equations (7.43). We will consider (7.45) written as $w_j g_j(w) = h_j(w)$ and apply Theorem 7.3 to justify that it has a solution. To this end, let $\overline{w}$ be a vector in (0, 1) that satisfies

$$\overline{w}_j \sum_k \lambda_{j,k} = \sum_k \overline{w}_k \lambda_{k,j}, \quad 1 \le j \le m.$$

Define $A^* = \{j : \lambda_{0,j} > 0\}$. This set is not empty because $\lambda_{j,k}$ is irreducible. Let $\underline{w}$ be a vector in $(0, \overline{w})$ such that

$$\underline{w}_j < \lambda_{0,j}/g_j(\overline{w}), \quad j \in A^*,$$

$$\underline{w}_j < \sum_{k \in A^*} \underline{w}_k \lambda_{kj}/g_j(\overline{w}), \quad j \in \{1, \ldots, m\} \backslash A^*.$$

From the definition of $\overline{w}$ and these inequalities, it follows that

$$\underline{w}_j g_j(\overline{w}) < h_j(\underline{w}), \qquad h_j(\overline{w}) = \overline{w}_j \sum_k \lambda_{j,k} < \overline{w}_j g_j(\underline{w}), \quad 1 \le j \le m.$$

From these inequalities and Theorem 7.3, it follows that (7.45) has a solution $w \in (0, 1)$. Then the first assertion of the theorem follows by Theorem 7.2. Also, Proposition 7.9 justifies, as in the proof of Theorem 7.15, that (7.45) is the same as $w_j \tilde{\mu}_j = \tilde{\lambda}_j$. □

Example 7.23. *Assembly Networks.* Consider the string-net described in Theorem 7.22 with the following features. Units arrive to the nodes by independent Poisson processes with rates $\lambda_1, \ldots, \lambda_m$. Services at node j are exponential with rate μ_i. When a service at j completes, K_j units, if available at j, are assembled into one unit and sent to node k with probability Q_{jk}. If there are less than K_j units at j, then all the units at j are assembled into one defective unit and discarded (sent to node 0). Then the process X that represents the numbers of units at the nodes has two-node batch transitions. Its traffic equations (7.43) are

$$w_j \mu_j \sum_{\ell=0}^{K_j - 1} w_j^\ell = \lambda_j + \sum_{k=1}^{m} \mu_k w_k^{K_k} Q_{kj}, \quad 1 \le j \le m.$$

Then Theorem 7.22 applies in this setting under the assumption that

$$\lambda_j + \sum_{k=1}^{m} \mu_k Q_{kj} < K_j \mu_j,$$

which says that the service capacity at node j is greater than the arrival rate. □

7.9 Single Service Station With String Transitions

Further understanding of string transitions can be obtained by considering a single node or service station, which we will now do. In addition to giving more insight into string transitions, these processes for single nodes can be used as building blocks for networks, comparable to quasi-reversible nodes that are coupled together to form networks.

Consider a Markov process $\{X_t : t \geq 0\}$ with state space $\mathbb{E} = \{0, 1, 2, \ldots\}$. Assume that it has the following type of string transitions, where $a = 0$ or 1 and s is a nonnegative integer.

- A complete sa-transition: $x \to x - s + a$.
- An ith partial sa-transition: $x \to x - i$.

Then the transition rates (7.2) of the process X are

$$\begin{aligned} q(x, x+1) &= \lambda_{01}\phi_0(x) \\ q(x, x-i) &= \sum_{a=0,1} \lambda_{i+a,a}[\phi_{i+a}(x) + \phi_i(x) - \phi_{i+1}(x)) \\ &= \lambda_{i,0}[2\phi_i(x) - \phi_{i+1}(x)] + \lambda_{i+1,1}\phi_i(x), \quad 0 < i < x. \end{aligned}$$

The following result is an immediate consequence of Theorem 7.2.

Corollary 7.24. *An invariant measure for the process defined above is $\pi(x) = \Phi(x)w^x$, $x \in \mathbb{E}$, where $w > 0$ satisfies*

$$\sum_{s=0}^{L-1} w^{s+1}\Lambda_{s+1} = \sum_{s=0}^{L} w^s \lambda_{s1}, \tag{7.46}$$

and $\Lambda_s = \sum_{s'=s}^{L}(\lambda_{s'0} + \lambda_{s'1})$. The process is positive recurrent if and only if $0 < w < 1$.

Clearly w is the unique positive solution to (7.46) since this equation is equivalent to $\sum_{s=1}^{L} w^s[\lambda_{s0} + \Lambda_{s+1}] = \lambda_{01}$, which has a unique solution. A special case of this model is as follows.

Example 7.25. *A Simple Production–Inventory System.* Consider a production system whose cumulative output over time is a Poisson process with rate λ. As the units are produced, they are put in inventory to satisfy random demands. Let X_t denote the quantity of units in inventory at time t. Whenever there are x units in inventory, the time to the next demand has an exponential distribution with rate μ and the demand is for i units with probability p_i, where $i \leq L$. Also, the probability that the demand can be satisfied is $P\{Z \leq x - i | Z \leq x\}$, where $1 \leq i \leq \min\{x, L\}$. Think of Z as a nonnegative integer-valued random variable that denotes a feasible inventory level. Then the process X is clearly a Markov process, and its nonzero transition rates are $q(x, x+1) = \lambda$ and

$$q(x, x-i) = \mu p_i P\{Z \leq x - i | Z \leq x\}, \quad 1 \leq i \leq \min\{x, L\}.$$

An easy check shows that this process is a special case of the preceding example in which

$$\lambda_{s0} = 0, \quad \lambda_{01} = \lambda, \quad \lambda_{i+1,1} = \mu p_i, \quad \phi_i(x) = P\{Z \leq x\}.$$

Therefore, $\pi(x) = \Phi(x)w^x$, $x \geq 0$, is a stationary distribution, where w is the unique solution to $\sum_{s=1}^{L} w^s(p_s + \ldots + p_{L-1}) = \lambda/\mu$.

7.10 Bibliographical Notes

This chapter is based on Serfozo and Yang (1998), which built on earlier works. Specifically, Gelenbe (1991, 1993) and Gelenbe and Schassberger (1992) introduced the model of negative customers in Example 7.18. Chao and Pinedo (1993) extended this model to more general services and deletion mechanisms described in Examples 7.17 and 7.19, and they gave insights on quasi-reversibility. Henderson et al. (1995) extended the model to one or two-stage "batch" deletions, and Chao (1995) and Chao and Miyazawa (1998) gave further insights into similar networks. Another extension to batch arrivals, as well as batch departures, is in Henderson et al. (1995). Chao Pinedo and Shaw (1996) introduced Example 7.23. Other related articles are Henderson (1993) and Henderson et al. (1994a). A good reference for Brower's fixed point theorem is Horn and Johnson (1994).

8

Quasi-Reversible Networks and Product Form Distributions

This chapter addresses the following question for a Markov network process. Under what conditions is the stationary distribution of the process a product of stationary distributions associated with the nodes? We consider a network in which the state of each node may contain more information than the number of units at the node, and a network transition may be triggered by an internal node change as well as by a unit moving from one node to another. The network process is viewed as a linkage of certain artificial Markov "node processes" that mimic the operation of the nodes as if they were operating in isolation. The main results are necessary and sufficient conditions under which the stationary distribution of the network is a product of the stationary distributions of the individual node processes.

An important example of a network with a product form distribution is a quasi-reversible network. Loosely speaking, a single queueing system is quasi-reversible if Poisson arrivals imply Poisson departures when the system is stationary. A network is called quasi-reversible if each of its nodes viewed in isolation is quasi-reversible. A major result of this chapter is that a network has a product form stationary distribution and is "biased locally balanced" if and only if the network is quasi-reversible and certain traffic equations are satisfied. We also characterize product form distributions for queueing networks in which the routing is reversible, but the entire process need not be reversible. The chapter ends with a discussion of how the results extend to networks with multiclass transitions.

8.1 Quasi-Reversibility

An open, unlimited-capacity Jackson network process is a quintessential process with a product form stationary distribution. The transition rates of the process are consistent with viewing the nodes in isolation as birth–death queueing processes linked together by customer routing rates. Then the stationary distribution of the network is a product of the stationary distributions of the birth–death processes. This approach of constructing or analyzing network processes is developed in the following sections.

As an introduction, this section describes such a construction of a classical quasi-reversible network. We begin by defining quasi-reversibility for a single service system, and then discuss a network of quasi-reversibile nodes.

Consider a service system with queueing whose state is represented by a Markov jump process $\{X_t : t \geq 0\}$ on a countable state space $\mathbb{E}$. A typical state $x \in \mathbb{E}$ includes all the relevant information about the system including the number of customers present, denoted by $n(x)$. Assume the system evolves as follows. A transition of the process is triggered by one of the three following events: arrival of one customer; departure of one customer, or internal change without the customer population changing. The transition rates of the process are defined as

$$q(x, y) = q^{\mathrm{a}}(x, y) + q^{\mathrm{d}}(x, y) + q^{\mathrm{i}}(x, y), \quad x \neq y \in \mathbb{E}.$$

The three parts are the transition rates associated with arrivals, departures, and internal changes. At most, one of these rates may be nonzero for each pair x, y, and so

$$q(x, y) = \begin{cases} q^{\mathrm{a}}(x, y) & \text{if } n(y) = n(x) + 1 \\ q^{\mathrm{d}}(x, y) & \text{if } n(y) = n(x) - 1 \\ q^{\mathrm{i}}(x, y) & \text{if } n(y) = n(x). \end{cases}$$

Assume the process X is ergodic, and let $\pi(x)$ denote its stationary distribution.

Definition 8.1. The process X defined above is *quasi-reversible* if

$$\alpha \equiv \sum_{y \neq x} q^{\mathrm{a}}(x, y) \quad \text{is independent of } x \in \mathbb{E}, \text{ and} \tag{8.1}$$

$$\pi(x)^{-1} \sum_{y \neq x} \pi(y) q^{\mathrm{d}}(y, x) = \alpha, \qquad \text{for each } x \in \mathbb{E}. \tag{8.2}$$

This definition is a special case of Definition 8.12 below for more general Markov processes. To see the meaning of the conditions above, let $N(t)$ denote the number of customer arrivals and $D(t)$ denote the number of departures in the time interval $[0, t]$. These point processes are functionals of the Markov process X, like the functionals discussed in Chapter 4. For instance,

$$N(t) = \sum_{s \leq t} 1(X_s \neq X_{s-},\ n(X_s) = n(X_{s-}) + 1), \quad t \geq 0.$$

Now, by Theorem 4.11, it follows that condition (8.1) is equivalent to N being a Poisson process with rate α and $N_+ \perp X_-$ (i.e., N on (t, ∞) is independent of X

on $[0, t]$ for each t). Similarly, by Theorem 4.12, we know that, when the process X is stationary, then condition (8.2) is equivalent to D being a Poisson process with rate α and $D_- \perp X_+$.

Because of these observations, a quasi-reversible system is said to have Poisson input and output processes, and the current state of the system is independent of prior departures and subsequent arrivals.

We now show how quasi-reversible nodes can be connected to form a quasi-reversible network with a product form distribution.

Example 8.2. *Classical Quasi-Reversible Network.* Consider a network consisting of m quasi-reversible nodes as defined above. Specifically, each node j "in isolation" operates as a quasi-reversible process on a countable state space $\mathbb{E}_j$, and its transition rates are

$$q_j(x_j, y_j) = q_j^{\mathrm{a}}(x_j, y_j) + q_j^{\mathrm{d}}(x_j, y_j) + q_j^{\mathrm{i}}(x_j, y_j), \quad x_j \neq y_j \in \mathbb{E}_j.$$

The process is ergodic with stationary distribution $\pi_j(x_j)$, and its Poisson arrival rate α_j is given by (8.1). Condition (8.2) is also satisfied for each node j.

We will link the nodes together to form a network as follows. The network may be closed or open or a mixed network with a combination of transient and permanent customers. For the last two cases, assume the outside, denoted by node 0, also operates in isolation as a quasi-reversible process on a space $\mathbb{E}_0$ (the number of units at node 0 can be taken to be infinite). Denote the node set M by $\{1, \dots, m\}$ or $\{0, 1, \dots, m\}$ according to whether the network is closed or not closed. A typical state of the network is a vector $x = (x_j : j \in M)$ in a space $\mathbb{E}$ that is the cartesian product of the sets $\{\mathbb{E}_j : j \in M\}$, or a subspace of this product space. Let $n_j(x)$ denote the number of customers at node j when the network is in state x.

A change in the network is triggered by one unit moving from one node to another in M, or by an internal change at a node. Such transitions will be described by the sets

$$\begin{aligned} \mathcal{T}_{jk}(x) \equiv \quad & \{y \in \mathbb{E} : y_\ell = x_\ell,\ \ell \neq j, k; \\ & \qquad n_j(y) = n_j(x) - 1, n_k(y) = n_k(x) + 1\} \\ \mathcal{T}_j(x) \equiv \quad & \{y \in \mathbb{E} : y_\ell = x_\ell, \ell \neq j;\ n_j(y) = n_j(x)\}. \end{aligned}$$

The set $\mathcal{T}_{jk}(x)$ consists of all states that can be reached in a transition from the state x due to a unit moving from j to k. The set $\mathcal{T}_j(x)$ consists of all states that can be reached in a transition from the state x due to an internal change at j.

We assume the transition rates of the network process X are

$$q(x, y) = \begin{cases} \lambda_{jk} q_j^{\mathrm{d}}(x_j, y_j) q_k^{\mathrm{a}}(x_k, y_k) & \text{if } y \in \mathcal{T}_{jk}(x) \\ q_j^{\mathrm{d}}(x_j, y_j) & \text{if } y \in \mathcal{T}_j(x) \\ 0 & \text{otherwise.} \end{cases} \tag{8.3}$$

The λ_{jk}, as in preceding chapters, denotes a rate of routing units from j to k, or the probability of such a movement. Without loss of generality, assume the routing rates λ_{jk} are irreducible. In addition, assume the process X is irreducible on $\mathbb{E}$.

The process X defined above is an example of a *quasi-reversible network process*. A product form invariant measure for it is as follows.

Theorem 8.3. *An invariant measure for the network process X is*

$$\pi(x) = \prod_{j \in M} \pi_j(x_j) w_j^{n_j(x)}, \quad x \in \mathbb{E},$$

where w_j are positive values that satisfy the traffic equations

$$w_j \sum_{k \in M} \lambda_{jk} \alpha_j \alpha_k = \sum_{k \in M} w_k \lambda_{kj} \alpha_k \alpha_j, \quad j \in M,$$

and $w_0 = 1$ if the network is not closed.

One can prove this result by showing that the specified π satisfies the total balance equations. This theorem is also a consequence of Theorem 8.14 below.

Note that the traffic equations in Theorem 8.3 have the following interpretation (Exercise 3): The arrival rate into each node j is equal to the sum of the arrival rates into node j from all other nodes. Keep in mind that although the input and output processes for a node in isolation are Poisson, the input and output processes for a node in the network may not be Poisson. Conditions for the latter can be obtained by the results in Chapter 4. □

For the network described above with transition rates (8.3), there are several definitions of the node transition rates q_j that yield different distributions π_j and traffic equations, but lead to the same invariant measure of the network. This idea is illustrated in the next example.

Example 8.4. *Alternative Formulation of Example 8.2*. Consider the network process X in Example 8.2 with q_j defined slightly differently as

$$q_j(x_j, y_j) = \beta_j^a q_j^{\mathrm{a}}(x_j, y_j) + \beta_j^d q_j^{\mathrm{d}}(x_j, y_j) + q_j^{\mathrm{i}}(x_j, y_j), \quad x_j \neq y_j \in \mathbb{E}_j.$$

Now there are coefficients on the first two terms. We assume that β_j^a is a dummy variable that will be determined. As above, we assume that $\alpha_j \equiv \sum_{y_j \neq x_j} q_j^{\mathrm{a}}(x_j, y_j)$ is independent of x_j. Motivated by Theorem 8.7 below, we set $\beta_j^d = \sum_{k \neq j} \lambda_{jk} \alpha_k$.

Assume q_j is an ergodic transition rate and let $\pi_j(x_j)$ denote an invariant measure for it. Note that π_j as well as q_j is a function of β_j^a. Finally, assume π_j is such that

$$\overline{\alpha}_j(\beta_j^a) \equiv \pi_j(x_j)^{-1} \sum_{y_j \neq x_j} \pi_j(y_j) q_j^{\mathrm{d}}(y_j, x_j), \quad \text{is independent of } x_j \in \mathbb{E}_j. \quad (8.4)$$

This quantity is a function of β_j^a because π_j is. In this setting, node j need not be quasi-reversible as in Example 8.2; it would be if $\overline{\alpha}_j(\beta_j^a) = \alpha_j$.

Theorem 8.5. *Suppose there exist positive β_j^a's that satisfy the traffic equations*

$$\beta_j^a = \sum_{k \neq j} \overline{\alpha}_k(\beta_k^a) \lambda_{kj}, \quad j \in M. \quad (8.5)$$

Let these β_j^a's be the coefficients in the q_j's. If each q_j has an invariant measure π_j, then an invariant measure for the network process X is

$$\pi(x) = \prod_{j \in M} \pi_j(x_j), \quad x \in E.$$

This result follows by showing that the specified π satisfies the total balance equations. This theorem is also a consequence of Theorem 8.14 below.

Since the network in this example is the same as the one in the preceding example, the invariant measures for it in Theorems 8.3 and 8.5 are essentially the same. The only difference are the q_j's, π_j's, and traffic equations. In comparing these theorems, note that the assumption (8.4) in Theorem 8.5 is weaker than the related assumption $\overline{\alpha}_j(\beta_j^a) = \alpha_j$ in Theorem 8.3. On the other hand, the traffic equations in Theorem 8.3 are linear and known to have a solution, whereas in Theorem 8.5, the traffic equations are more complex nonlinear equations that may be difficult to solve.

One can show, as we suggest in Exercise 5, that $\overline{\alpha}_j(\beta_j^a)$ is the rate at which customers depart from node j. Furthermore, the traffic equations (8.5) say that the arrival rate into each node j is equal to the sum of the arrival rates into node j from all other nodes. □

Quasi-reversibility also applies to multiclass customers as follows.

Example 8.6. *Multiclass Customers.* Consider the queueing system described in Definition 8.1, with the difference that it contains a countable number of customer classes. Define the state x so that it includes the numbers of customers of each class in the system. Let $n(x, c)$ denote the number of class c customers in the system when it is in state x. Assume a transition of the system from a state x to a state y is triggered by one of the following events: arrival of one of class-c customer ($n(y, c) = n(x, c)+1$); departure of one of class-c customer ($n(y, c) = n(x, c)-1$); or internal change ($n(y, c) = n(x, c)$ for each c). Under these possibilities, the transition rates of the process are

$$q(x, y) = \sum_c [q^{\mathrm{a,c}}(x, y) + q^{\mathrm{d,c}}(x, y) + q^{\mathrm{i,c}}(x, y)], \quad x \neq y \in \mathbb{E}.$$

Then the system is quasi-reversible if it satisfies the conditions

$$\alpha^c \equiv \sum_y q^{\mathrm{a,c}}(x, y) \quad \text{is independent of } x \in \mathbb{E} \text{ for each } c,$$

$$\pi(x)^{-1} \sum_y \pi(y) q^{\mathrm{d}}(y, x) = \alpha^c, \quad \text{for each } x \text{ and } c.$$ □

8.2 Network to be Studied

In this section, we define the network process that is the focus of the rest of this chapter.

We shall consider an open m-node stochastic network, where each node j in the set $M \equiv \{0, 1, \ldots, m\}$ is represented by a state x_j in a countable set $\mathbb{E}_j$. Later we comment on how the results apply to closed networks. The network is represented by a stochastic process $\{X_t : t \geq 0\}$ with values $x = (x_j : j \in M)$ in the cartesian product $\mathbb{E}$ of the spaces $\mathbb{E}_j$, $j \in M$. Note that this includes a value x_0 for node 0. The network need not contain customers that move among the nodes—it may just be a multidimensional system with interactions. Therefore, we do not refer to the numbers of customers at the nodes. The changes in the network are triggered by three types of transitions. To be more descriptive, however, we call these transitions arrivals, departures, and internal node changes.

The major assumption is that X is a Markov jump process with transition rates

$$q(x, y) = \sum_{j,k} q_{jk}(x, y), \quad x, y \in \mathbb{E}, \tag{8.6}$$

where

$$q_{jk}(x, y) = \begin{cases} q_j^{\mathrm{d}}(x_j, y_j)\lambda_{jk} q_k^{\mathrm{a}}(x_k, y_k) 1(y_\ell = x_\ell, \ell \neq j, k) & \text{if } j \neq k \\ q_j^{\mathrm{i}}(x_j, y_j) 1(y_\ell = x_\ell, \ell \neq j) & \text{if } j = k. \end{cases}$$

Think of q_j^{a}, q_j^{d} and q_j^{i} as state-dependent *rate components* associated with "arrival," "departure," and "internal" transitions, respectively, at node j. We call them rate components because they are only parts of a compound transition rate. The λ_{jk} is the rate component or tendency for a departure from node j to trigger an arrival at node k; it is often assumed to be a probability, with $1 - \sum_{k \neq j} \lambda_{jk}$ being the probability of an attempted internal change at node j. The $q_j^{\mathrm{i}}(x_j, y_j)$ can be augmented by multiplying it by a factor λ_{jj}, but we will assume that such a coefficient is already included in q_j^{i}.

The usual convention for a Markov process is to disregard bogus transitions from a state back to itself. For our analysis, however, it is convenient to include bogus jumps, and so we assume $q(x, x)$ are well-defined rates (possibly 0). We adopt this convention for all transition functions in this chapter.

For simplicity, we assume the network process X is irreducible and positive recurrent. The aim is to determine conditions under which the stationary distribution of X is a product form $\pi(x) = \prod_{j \in M} \pi_j(x_j)$, where π_j is the jth marginal distribution. The results also apply to closed queueing networks and other networks, where $\mathbb{E}$ is a subset of a product space and π, π_j are invariant measures instead of normalized distributions. The approach is to relate the marginal distributions of the network process X to stationary distributions of one-dimensional Markov processes defined as follows.

For each $j \in M$, consider a Markov jump process on $\mathbb{E}_j$ with transition rates

$$q_j(x_j, y_j) = \beta_j^{\mathrm{a}} q_j^{\mathrm{a}}(x_j, y_j) + \beta_j^{\mathrm{d}} q_j^{\mathrm{d}}(x_j, y_j) + q_j^{\mathrm{i}}(x_j, y_j), \quad x_j, y_j \in \mathbb{E}_j. \tag{8.7}$$

Think of this process as representing the state of node j as if it were operating in isolation. The three terms in the summation are transition rates associated respectively with an arrival into node j; a departure from node j; and an internal change. For a fixed x_j and y_j, any combination of these three terms may be positive. For instance, if all are positive for some x_j, y_j, then a transition from x_j to y_j might consist of a simultaneous occurrence of an arrival, a departure, and internal change.

Keep in mind that an arrival transition does not represent the arrival process at the node in the network; it only specifies a fictitious arrival environment for the isolated node. Similar statements apply to departure and internal-change transitions. The coefficients β_j^{a}, β_j^{d} at this point are dummy variables. Our results determine the form of these coefficients in order for the stationary distribution of q_j to be the jth marginal stationary distribution of the network process. There is no coefficient on q_j^{i}, which is consistent with it having no coefficient in (8.6). For simplicity, we assume the transition function q_j is irreducible and positive recurrent.

Throughout this chapter, each π_j will denote an arbitrary positive probability measure on $\mathbb{E}_j$. The role of π_j will be specified in the theorem statements. Associated with each transition rate component $q_j^{\mathrm{s}}(x_j, y_j)$, for s $=$ a, d, i, we define

$$\alpha_j^{\mathrm{s}}(x_j) = \sum_{y_j} q_j^{\mathrm{s}}(x_j, y_j), \tag{8.8}$$

$$\tilde{\alpha}_j^{\mathrm{s}}(x_j) = \pi_j(x_j)^{-1} \sum_{y_j} \pi_j(y_j) q_j^{\mathrm{s}}(y_j, x_j), \tag{8.9}$$

$$\overline{\alpha}_j^{\mathrm{s}} = \sum_{x_j} \sum_{y_j} \pi_j(x_j) q_j^{\mathrm{s}}(x_j, y_j). \tag{8.10}$$

Assume each $\overline{\alpha}_j^{\mathrm{s}}$ is finite. Keep in mind that $\tilde{\alpha}_j^{\mathrm{s}}(x_j)$ and $\overline{\alpha}_j^{\mathrm{s}}$ are functions of π_j. Also, note that

$$\sum_{x_j} \pi_j(x_j) \tilde{\alpha}_j^{\mathrm{s}}(x_j) = \overline{\alpha}_j^{\mathrm{s}}. \tag{8.11}$$

8.3 Characterization of Product Form Distributions

We now present necessary and sufficient conditions on the one-dimensional node processes defined by the q_j's above under which the network process X has a product form stationary distribution.

We begin by showing that if the network process has a product form stationary distribution, then the coefficients β_j^{a}, β_j^{d} of q_j must be of the form (8.12) below.

Theorem 8.7. *If $\pi(x) = \prod_{j \in M} \pi_j(x_j)$ is the stationary distribution of the network process X, then each π_j is the stationary distribution for q_j with coefficients*

$$\beta_j^{\mathrm{a}} = \sum_{k \neq j} \overline{\alpha}_k^{\mathrm{d}} \lambda_{kj}, \qquad \beta_j^{\mathrm{d}} = \sum_{k \neq j} \lambda_{jk} \overline{\alpha}_k^{\mathrm{a}}. \tag{8.12}$$

PROOF. The balance equations for q that π satisfies are

$$\pi(x) \sum_y q(x, y) = \sum_y \pi(y) q(y, x), \qquad x \in \mathbb{E}. \tag{8.13}$$

Since

$$\pi(y) = \pi(x)\pi_j(y_j)\pi_k(y_k)/(\pi_j(x_j)\pi_k(x_k)),$$

for y such that $x_\ell = y_\ell$ for all $\ell \neq j, k$, it follows by the definition of q that (8.13) is

$$\pi(x)\sum_j [\alpha^{\mathrm{i}}_j(x_j) + \alpha^{\mathrm{d}}_j(x_j)\sum_{k\neq j}\lambda_{jk}\alpha^{\mathrm{a}}_k(x_k)]$$

$$= \pi(x)\sum_j [\tilde{\alpha}^{\mathrm{i}}_j(x_j) + \tilde{\alpha}^{\mathrm{a}}_j(x_j)\sum_{k\neq j}\lambda_{kj}\tilde{\alpha}^{\mathrm{d}}_k(x_k)], \quad x \in \mathbb{E}. \tag{8.14}$$

For a fixed $j \in M$, we will consider the sum of these equations over all $x_\ell \in \mathbb{E}_\ell$, for $\ell \in M\backslash\{j\}$. First, note that

$$\begin{aligned}\sum_{x_\ell:\ell\neq j} [\text{Left side of (8.14)}] &= \sum_{x_\ell:\ell\neq j} \pi(x)\{\alpha^{\mathrm{i}}_j(x_j) + \sum_{j'\neq j}\alpha^{\mathrm{d}}_{j'}(x_{j'})\lambda_{j'j}\alpha^{\mathrm{a}}_j(x_j)\\ &\quad + \alpha^{\mathrm{d}}_j(x_j)\sum_{k\neq j}\lambda_{jk}\alpha^{\mathrm{a}}_k(x_k)\\ &\quad + \sum_{j'\neq j}[\alpha^{\mathrm{i}}_{j'}(x_{j'}) + \alpha^{\mathrm{d}}_{j'}(x_{j'})\sum_{k\neq j,j'}\lambda_{j'k}\alpha^{\mathrm{a}}_k(x_k)]\}\\ &= \pi_j(x_j)[\alpha^{\mathrm{i}}_j(x_j) + \beta^{\mathrm{a}}_j\alpha^{\mathrm{a}}_j(x_j) + \beta^{\mathrm{d}}_j\alpha^{\mathrm{d}}_j(x_j)] + A_j,\end{aligned}$$

where

$$A_j = \sum_{j'\neq j}[\overline{\alpha}^{\mathrm{i}}_{j'} + \overline{\alpha}^{\mathrm{d}}_{j'}\sum_{k\neq j,j'}\lambda_{j'k}\overline{\alpha}^{\mathrm{a}}_k].$$

A similar computation using (8.11) yields

$$\sum_{x_\ell:\ell\neq j} [\text{Right side of (8.14)}] = \pi_j(x_j)[\tilde{\alpha}^{\mathrm{i}}_j(x_j) + \beta^{\mathrm{a}}_j\tilde{\alpha}^{\mathrm{a}}_j(x_j) + \beta^{\mathrm{d}}_j\tilde{\alpha}^{\mathrm{d}}_j(x_j)] + A_j.$$

Since (8.14) is an equality, the preceding sums are equal, and equating them yields

$$\alpha^{\mathrm{i}}_j(x_j)+\beta^{\mathrm{a}}_j\alpha^{\mathrm{a}}_j(x_j)+\beta^{\mathrm{d}}_j\alpha^{\mathrm{d}}_j(x_j) = \tilde{\alpha}^{\mathrm{i}}_j(x_j)+\beta^{\mathrm{a}}_j\tilde{\alpha}^{\mathrm{a}}_j(x_j)+\beta^{\mathrm{d}}_j\tilde{\alpha}^{\mathrm{d}}_j(x_j), \quad x_j \in \mathbb{E}. \tag{8.15}$$

These are the balance equations divided by $\pi_j(x_j)$ for q_j. Hence π_j is the stationary distribution for q_j. □

The following result characterizes a product form distribution for the network process. Here we use the function

$$D_{jk}(x_j, x_k) = (\overline{\alpha}^{\mathrm{d}}_j - \alpha^{\mathrm{d}}_j(x_j))\lambda_{jk}(\overline{\alpha}^{\mathrm{a}}_k - \alpha^{\mathrm{a}}_k(x_k)) - (\overline{\alpha}^{\mathrm{d}}_k - \tilde{\alpha}^{\mathrm{d}}_k(x_k))\lambda_{kj}(\overline{\alpha}^{\mathrm{a}}_j - \tilde{\alpha}^{\mathrm{a}}_j(x_j)).$$

Theorem 8.8. *The stationary distribution of the network process X is $\pi(x) = \prod_{j\in M}\pi_j(x_j)$, $x \in \mathbb{E}$, if and only if each π_j is the stationary distribution of q_j for some coefficients β^{a}_j, β^{d}_j, and the π_j's are such that (8.12) holds and*

$$D_{jk}(x_j, x_k) + D_{kj}(x_k, x_j) = 0, \quad j \neq k \in M,\ x_j \in \mathbb{E}_j,\ x_k \in \mathbb{E}_k. \tag{8.16}$$

The proof of this result uses the following lemma.

Lemma 8.9. *Suppose each π_j is the stationary distribution of q_j for some coefficients β_j^{a}, β_j^{d}. Then $\pi(x) = \prod_{j\in M} \pi_j(x_j)$ is a stationary distribution for X if and only if*

$$\sum_j [\beta_j^{\mathrm{a}}\alpha_j^{\mathrm{a}}(x_j) + \beta_j^{\mathrm{d}}\alpha_j^{\mathrm{d}}(x_j) - \beta_j^{\mathrm{a}}\tilde{\alpha}_j^{\mathrm{a}}(x_j) - \beta_j^{\mathrm{d}}\tilde{\alpha}_j^{\mathrm{d}}(x_j)]$$

$$= \sum_j [\alpha_j^{\mathrm{d}}(x_j)\sum_{k\neq j} \lambda_{jk}\alpha_k^{\mathrm{a}}(x_k) - \tilde{\alpha}_j^{\mathrm{a}}(x_j)\sum_{k\neq j}\lambda_{kj}\tilde{\alpha}_k^{\mathrm{d}}(x_k)], \quad x \in \mathbb{E}. \tag{8.17}$$

If β_j^{a}, β_j^{d} are of the form (8.12), then (8.17) is equivalent to

$$\sum_j \sum_{k\neq j} D_{jk}(x_j, x_k) = 0, \quad x \in \mathbb{E}. \tag{8.18}$$

PROOF. Recall that the balance equations for q and π are (8.14). Then to prove the first assertion, it suffices to show that (8.14) is equivalent to (8.17). To this end, recall that the balance equations for q_j are (8.15) multiplied by $\pi_j(x_j)$. Summing (8.15) on j and subtracting (8.14) divided by $\pi(x)$ from the sum shows that (8.14) is equivalent to (8.17). This proves the first assertion. The second assertion follows since substituting (8.12) into (8.17) and rearranging terms yields (8.18). □

Proof of Theorem 8.8. First, assume the stationary distribution of X is $\pi(x) = \prod_{j\in M}\pi_j(x_j)$. Then by Theorem 8.7, the π_j is the stationary distribution for q_j with coefficients that satisfy (8.12). To prove (8.16), note that (8.18) holds by Lemma 8.9. Also, for any $k \neq \ell$, (8.12) and (8.11) imply

$$\sum_{x_\ell} \pi_\ell(x_\ell)[D_{\ell k}(x_\ell, x_k) + D_{k\ell}(x_k, x_\ell)] = 0. \tag{8.19}$$

Then multiplying (8.18) by $\prod_{\ell\neq j,k}\pi_\ell(x_\ell)$, and summing it on x_ℓ for $\ell \neq j, k$ and using (8.19) yields (8.16).

For the converse, assume each π_j is the stationary distribution of q_j and that (8.12) and (8.16) are satisfied. Since (8.16) implies (8.18), it follows by Lemma 8.9 that $\pi(x) = \prod_{j\in M}\pi_j(x_j)$ is the stationary distribution of X. □

Theorem 8.8 yields the following procedure for establishing the existence of a product form stationary distribution for the network process and obtaining the distribution when it exists.

Procedure for Obtaining a Product Form Distribution

Step 1. For each node j, obtain the stationary distribution π_j of q_j as a function of the coefficients $\boldsymbol{\beta}_j = (\beta_j^{\mathrm{a}}, \beta_j^{\mathrm{d}})$ viewed as a dummy vector. Since π_j is a function of $\boldsymbol{\beta}_j$, so is $\overline{\alpha}_j^{\mathrm{s}}$, and we write it as $\overline{\alpha}_j^{\mathrm{s}}(\boldsymbol{\beta}_j)$, for s = a, d.

Step 2. Find $\boldsymbol{\beta}_j$'s that satisfy the *traffic equations*

$$\beta_j^{\mathrm{a}} = \sum_{k\neq j}\overline{\alpha}_k^{\mathrm{d}}(\boldsymbol{\beta}_k)\lambda_{kj}, \qquad \beta_j^{\mathrm{d}} = \sum_{k\neq j}\lambda_{jk}\overline{\alpha}_k^{\mathrm{a}}(\boldsymbol{\beta}_k), \quad j \in M. \tag{8.20}$$

Step 3. Let π_j be the distribution associated with the $\boldsymbol{\beta}_j$'s obtained from Step 2. Verify (8.16) for these distributions and coefficients.

If these steps are successful, then $\pi(x) = \prod_j \pi_j(x_j)$ is the stationary distribution of the network process.

Equations (8.20) are often called traffic equations, because for queueing networks, β_j^{a} and β_j^{d} are the average number of arrivals and departures, respectively, for node j. Finding $\boldsymbol{\beta}_j$'s that satisfy (8.20) is a fixed point problem, whose solution is usually established by Brouwer's fixed point theorem. For a particular application, one may be able to construct an algorithm to compute a fixed point. There may be more than one solution, but any solution will work.

Such a fixed point exists if the network has a product form stationary distribution. This is due to the following observation.

Restatement of Theorem 8.8. The network process has a product form stationary distribution if and only if there exist $\boldsymbol{\beta}_j$'s that satisfy Steps 1–3 above.

The next result is a variation of Theorem 8.8. It follows immediately from Theorem 8.7 and Lemma 8.9.

Theorem 8.10. *The stationary distribution of X is $\pi(x) = \prod_{j\in M} \pi_j(x_j)$, $x \in \mathbb{E}$, if and only if each π_j is the stationary distribution of q_j for some coefficients β_j^{a}, β_j^{d}, and the π_j's are such that (8.17) holds.*

Remark 8.11. *(Results for Closed Networks).* We have assumed that the state space $\mathbb{E}$ is a product space and π and π_j's are probability distributions. However, from their proofs, it is clear that the sufficient conditions in Theorems 8.8 and 8.10 for a product form distribution apply even when $\mathbb{E}$ is a subset of the product space of the $\mathbb{E}_j$'s and the π and π_j's are invariant measures instead of normalized distributions. In particular, the results with these modifications apply to closed networks. On the other hand, the necessary conditions in these theorems are generally not valid in these situations because, in the proof of Theorem 8.7, the summation of $\prod_{\ell\neq j} \pi_\ell(x_\ell)$ over $x \in \mathbb{E}$ for each fixed $x_j \in \mathbb{E}_j$ may depend on x_j.

8.4 Quasi-Reversibility and Biased Local Balance

In this section, we characterize product form distributions for the network process X when its nodes are quasi-reversible and its transition rates satisfy a biased local balance condition.

We will use the following definition of quasi-reversibility, which is consistent with the classical one in Section 8.1.

Definition 8.12. The transition rate q_j is *quasi-reversible with respect to* π_j if π_j is the stationary distribution of q_j and $\alpha_j^{\mathrm{a}}(x_j)$ and $\tilde{\alpha}_j^{\mathrm{d}}(x_j)$ are independent of x_j. That is,

$$\alpha_j^{\mathrm{a}}(x_j) = \overline{\alpha}_j^{\mathrm{a}}, \quad \text{and} \quad \tilde{\alpha}_j^{\mathrm{d}}(x_j) = \overline{\alpha}_j^{\mathrm{d}}, \quad \text{for each } x_j \in \mathbb{E}_j. \tag{8.21}$$

To see the meaning of this definition, consider a transition rate q_j in which only one of the rate components $q_j^{\mathrm{a}}(x_j, y_j)$, $q_j^{\mathrm{d}}(x_j, y_j)$, and $q_j^{\mathrm{i}}(x_j, y_j)$ is positive for each x_j, y_j. Then $\alpha_j^{\mathrm{a}}(x_j) = \overline{\alpha}_j^{\mathrm{a}}$ implies by Theorem 4.11 that the times of a-transitions for q_j form a Poisson process with rate $\overline{\alpha}_j^{\mathrm{a}}$. Also, $\tilde{\alpha}_j^{\mathrm{d}}(x_j) = \overline{\alpha}_j^{\mathrm{d}}$ implies by Theorem 4.12 that the times of d-transitions in equilibrium for q_j form a Poisson process with rate $\overline{\alpha}_j^{\mathrm{d}}$.

Note that in Theorem 8.8, the key condition (8.16) for the network process X to have a product form stationary distribution is satisfied if each q_j is quasi-reversible.

In addition to the usual balance equations for a process, we will use the following notion.

Definition 8.13. The Markov transition rate q is *biased locally balanced* with respect to a positive probability measure π on $\mathbb{E}$ and real numbers $b = \{b_j : j \in M\}$ satisfying $\sum_j b_j = 0$ if

$$\pi(x)\left(\sum_k \sum_y q_{jk}(x, y) + b_j\right) = \sum_k \sum_y \pi(y) q_{kj}(y, x), \quad x \in \mathbb{E},\ j \in M. \tag{8.22}$$

In this definition, π is necessarily the stationary distribution for X since the global balance equations are the sum of these local balance equations over j. Also, we say q is *locally balanced* with respect to π when the b_j's are 0.

For the next result, we consider the network process X under the added assumption that each $\alpha_j^{\mathrm{a}}(x_j)$ is independent of x_j, or, equivalently,

$$\alpha_j^{\mathrm{a}}(x_j) = \overline{\alpha}_j^{\mathrm{a}}, \quad \text{for each } x_j \text{ and } j \in M. \tag{8.23}$$

This is the first part of the quasi-reversibility condition. Because of Theorem 8.7, we make the natural assumption that the coefficient β_j^{d} of q_j is given by

$$\beta_j^{\mathrm{d}} = \sum_{k \neq j} \lambda_{jk} \overline{\alpha}_k^{\mathrm{a}}, \quad j \in M. \tag{8.24}$$

No restriction is placed on the other coefficient β_j^{a}.

Theorem 8.14. *Under the assumptions (8.23) and (8.24), the following statements are equivalent.*
(i) *The q is biased locally balanced with respect to $\pi(x) = \prod_{j \in M} \pi_j(x_j)$ and b.*
(ii) *Each q_j is quasi-reversible with respect to π_j for some β_j^{a}, and the π_j's are such that*

$$\beta_j^{\mathrm{a}} = \sum_{k \neq j} \overline{\alpha}_k^{\mathrm{d}} \lambda_{kj}, \quad j \in M. \tag{8.25}$$

If these statements hold, then

$$b_j = \overline{\alpha}_j^{\mathrm{a}} \beta_j^{\mathrm{a}} - \overline{\alpha}_j^{\mathrm{d}} \beta_j^{\mathrm{d}}. \tag{8.26}$$

Proof. Suppose (i) holds. By the definitions of q, π and assumptions (8.23), (8.24), it follows like (8.14), that the biased local balance equation (8.22) divided by $\pi(x)$ is

$$\alpha_j^{\mathrm{i}}(x_j) + \beta_j^{\mathrm{d}} \alpha_j^{\mathrm{d}}(x_j) + b_j = \tilde{\alpha}_j^{\mathrm{i}}(x_j) + \tilde{\alpha}_j^{\mathrm{a}}(x_j) \sum_{k \neq j} \tilde{\alpha}_k^{\mathrm{d}}(x_k) \lambda_{kj}. \tag{8.27}$$

Define β_j^{a} by (8.25). Fix $j \in M$. Multiplying (8.27) by $\pi_j(x_j)$, then summing over x_j and using (8.11), we have

$$\overline{\alpha}_j^{\mathrm{d}}\beta_j^{\mathrm{d}} + b_j = \overline{\alpha}_j^{\mathrm{a}} \sum_{k \neq j} \tilde{\alpha}_k^{\mathrm{d}}(x_k)\lambda_{kj}.$$

Fix $\ell \neq j$. Multiplying this equation by $\prod_{k \neq j,\ell} \pi_k(x_k)$, then summing over x_k, for $k \neq j, \ell$ and using (8.11) and (8.25), we obtain

$$\overline{\alpha}_j^{\mathrm{d}}\beta_j^{\mathrm{d}} + b_j = \overline{\alpha}_j^{\mathrm{a}}\beta_j^{\mathrm{a}} + (\tilde{\alpha}_\ell^{\mathrm{d}}(x_\ell) - \overline{\alpha}_\ell^{\mathrm{d}})\lambda_{\ell j} 1(\ell \neq j). \tag{8.28}$$

Summing this on j and using (8.25) yields

$$(\tilde{\alpha}_\ell^{\mathrm{d}}(x_\ell) - \overline{\alpha}_\ell^{\mathrm{d}}) \sum_{j \neq \ell} \lambda_{\ell j} = 0.$$

This proves $\tilde{\alpha}_\ell^{\mathrm{d}}(x_\ell) = \overline{\alpha}_\ell^{\mathrm{d}}$. Thus, (8.21) holds.

Next, note that (8.28) implies (8.26). Furthermore, applying (8.26) to (8.27) yields (8.15), which is the balance equation divided by $\pi_j(x_j)$ for q_j and π_j. Hence π_j is the stationary distribution for q_j. This proves that (i) implies (ii).

Now, assume (ii) holds. Then (8.15) holds, and using $\alpha_j^{\mathrm{a}}(x_j) = \overline{\alpha}_j^{\mathrm{a}}$ and $\tilde{\alpha}_j^{\mathrm{d}}(x_j) = \overline{\alpha}_j^{\mathrm{d}}$ in (8.15), we have

$$\alpha_j^{\mathrm{i}}(x_j) + \beta_j^{\mathrm{a}}\overline{\alpha}_j^{\mathrm{a}} + \beta_j^{\mathrm{d}}\alpha_j^{\mathrm{d}}(x_j) = \tilde{\alpha}_j^{\mathrm{i}}(x_j) + \beta_j^{\mathrm{a}}\tilde{\alpha}_j^{\mathrm{a}}(x_j) + \beta_j^{\mathrm{d}}\overline{\alpha}_j^{\mathrm{d}}, \quad x_j \in \mathbb{E}.$$

Define b_j by (8.26). Applying (8.26) and then (8.24) and $\overline{\alpha}_j^{\mathrm{d}} = \tilde{\alpha}_j^{\mathrm{d}}(x_j)$ to the preceding display yields (8.27). Then substituting (8.23) and (8.24) into (8.27) yields the biased local balance condition (8.22). This completes the proof that (ii) implies (i). □

What is the difference between Theorems 8.8 and 8.14 ? In the former, both of the coefficients β_j^{a}, β_j^{d} of q_j are unspecified dummy variables, while in the latter, β_j^{d} is given by (8.24) and only β_j^{a} is a dummy variable. Consequently, in Theorem 8.8 the conditions (8.12), (8.16) required for a product form distribution for the network are more involved than the conditions (8.21), (8.25) required in Theorem 8.14.

The following is a procedure for applying Theorem 8.14; compare this with the procedure in the preceding section.

Quasi-Reversible Procedure for a Product Form Distribution

Step 1. For each node j, obtain the stationary distribution π_j of q_j as a function of the coefficient β_j^{a} viewed as a dummy variable. Since π_j is a function of β_j^{a}, so is $\overline{\alpha}_j^{\mathrm{d}}$, and we write it as $\overline{\alpha}_j^{\mathrm{d}}(\beta_j^{\mathrm{a}})$.

Step 2. Find β_j^{a}'s that satisfy the traffic equations

$$\beta_j^{\mathrm{a}} = \sum_{k \neq j} \overline{\alpha}_k^{\mathrm{d}}(\beta_k^{\mathrm{a}})\lambda_{kj}, \quad j \in M.$$

Step 3. Let π_j be the distribution associated with the β_j's obtained in Step 2. For these distributions, verify the quasi-reversibility condition

$$\tilde{\alpha}_j^{\mathrm{d}}(x_j) = \overline{\alpha}_j^{\mathrm{d}}(\beta_j^{\mathrm{a}}), \quad \text{for each } x_j \in \mathbb{E}_j \text{ and } j \in M. \tag{8.29}$$

If these steps are successful, then $\pi(x) = \prod_j \pi_j(x_j)$ is the stationary distribution of the network process.

Remark 8.15. There may be solutions to the traffic equations in Step 2 even though (8.29) is not satisfied. In this case, one might be able to obtain a product form stationary distribution by verifying condition (8.16).

Example 8.16. *Network with Random Environments at Nodes.* Suppose the network process X has the following structure. Customers move among the nodes where they are processed, and the state of each node $j \in M$ is a pair $x_j = (n_j, z_j)$, where n_j is the number of customers at the node and z_j is the "environment" of the node. Whenever the network is in state $x = (x_j : j \in M)$, two types of transitions may occur. First, the environment at some node j may change from z_j to z'_j. The time until such a transition is exponentially distributed with rate $\eta_j(z_j, z'_j)$. Second, a single customer may move from some node j to some node k and the environments at nodes j, k change from z_j, z_k to z'_j, z'_k, respectively. The time until such a transition is exponentially distributed with rate $\mu_j(n_j, z_j, z'_j)\lambda_{jk}\lambda_k(z'_k)$. Then the network process has transition rates of the form (8.6), where

$$\begin{aligned} q^{\mathrm{a}}_j(x_j, x'_j) &= \lambda_j(z'_j)1(n'_j = n_j + 1), \\ q^{\mathrm{d}}_j(x_j, x'_j) &= \mu_j(n_j, z_j, z'_j)1(n'_j = n_j - 1 \geq 0), \\ q^{\mathrm{i}}_j(x_j, x'_j) &= \eta_j(z_j, z'_j)1(n'_j = n_j). \end{aligned}$$

Note that $\alpha^{\mathrm{a}}_j(x_j) = \sum_{z'_j} \lambda_j(z'_j)$ is independent of x_j. As above, we define

$$\beta^d_j = \sum_{k \neq j} \lambda_{jk} \overline{\alpha}^{\mathrm{a}}_k,$$

and we consider the coefficient β^a_j as a dummy variable.

Now, suppose each q_j has a stationary distribution π_j. Think of π_j as a function of β^{a}_j. Note that

$$\tilde{\alpha}^{\mathrm{d}}_j(n_j, z_j) = \pi_j(n_j, z_j)^{-1} \sum_{z'_j} \pi_j(n_j + 1, z'_j)\mu_j(n_j + 1, z'_j, z_j).$$

Assume π_j is such that $\tilde{\alpha}^{\mathrm{d}}_j(n_j, z_j)$ is independent of (n_j, z_j). Then it follows that $\tilde{\alpha}^{\mathrm{d}}_j(n_j, z_j) = \overline{\alpha}^{\mathrm{d}}_j$. Denote this quantity by $\overline{\alpha}^{\mathrm{d}}_j(\beta^a_j)$, since it, as well as π_j, is a function of β^a_j. Under these assumptions, q_j is quasi-reversible with respect to π_j.

Finally, assume there exist β^{a}_j's that satisfy the traffic equations

$$\beta^{\mathrm{a}}_j = \sum_{k \neq j} \overline{\alpha}^{\mathrm{d}}_k(\beta^{\mathrm{a}}_k)\lambda_{kj}, \quad j \in M.$$

Let π_j be the distributions associated with these β^{a}_j's. Then it follows by Theorem 8.14 that the stationary distribution of q is $\pi(x) = \prod_{j \in M} \pi_j(x_j)$. □

8.5 Networks with Reversible Routing

In this section, we present corollaries of Theorem 8.8 when the routing rates of the network are reversible.

Here, for simplicity, we assume that λ_{jk} is irreducible and $\lambda_{jj} = 0$. Let w_j, $j \in M$, denote its stationary distribution. Recall that λ_{jk} is reversible if

$$w_j \lambda_{jk} = w_k \lambda_{kj}, \quad j, k \in M.$$

Corollary 8.17. *If λ_{jk} is reversible, then Theorem 8.8 holds with $D_{jk}(x_j, x_k)$ replaced by*

$$D^*_{jk}(x_j, x_k) = w_k(\overline{\alpha}^{\rm d}_j - \alpha^{\rm d}_j(x_j))(\overline{\alpha}^{\rm a}_k - \alpha^{\rm a}_k(x_k)) - w_j(\overline{\alpha}^{\rm d}_k - \tilde{\alpha}^{\rm d}_k(x_k))(\overline{\alpha}^{\rm a}_j - \tilde{\alpha}^{\rm a}_j(x_j)).$$

PROOF. The assertion follows from Theorem 8.8 by substituting $\lambda_{jk} = w_k \lambda_{kj}/w_j$ in $D_{jk}(x_j, x_k)$. □

Corollary 8.18. *Suppose λ_{jk} is reversible. Assume each π_j is the stationary distribution of q_j for some coefficients $\beta^{\rm a}_j$, $\beta^{\rm d}_j$, and the π_j's are such that (8.12) is satisfied and*

$$\tilde{\alpha}^{\rm a}_j(x_j) = w_j^{-1} \alpha^{\rm d}_j(x_j), \quad \tilde{\alpha}^{\rm d}_j(x_j) = w_j \alpha^{\rm a}_j(x_j). \quad x_j \in \mathbb{E}_j, \tag{8.30}$$

Then $\pi(x) = \prod_{j \in M} \pi_j(x_j)$ is the stationary distribution of q. In addition,

$$w_j = \overline{\alpha}^{\rm d}_j / \overline{\alpha}^{\rm a}_j = \beta^{\rm a}_j / \beta^{\rm d}_j, \quad j \in M. \tag{8.31}$$

PROOF. First note that the last equality in (8.30) implies $\overline{\alpha}^{\rm d}_j = w_j \overline{\alpha}^{\rm a}_j$. This and (8.30) imply that each $D^*_{jk}(x_j, x_k) = 0$. Then Corollary 8.17 yields the first assertion. Furthermore, from $\overline{\alpha}^{\rm d}_j = w_j \overline{\alpha}^{\rm a}_j$ and the reversibility of λ_{jk}, we have

$$\beta^{\rm a}_j = \sum_{k \neq j} \overline{\alpha}^{\rm d}_k \lambda_{kj} = \sum_{k \neq j} \overline{\alpha}^{\rm a}_k w_k \lambda_{kj} = \sum_{k \neq j} w_j \lambda_{jk} \overline{\alpha}^{\rm a}_k = w_j \beta^{\rm d}_j.$$

Thus (8.31) holds. □

Recall that Theorem 8.14 is for a network with quasi-reversible nodes, while Corollary 8.18 is for a network whose nodes need not be quasi-reversible, but the routing is restricted to be reversible. The next result is for networks with both types of nodes.

Corollary 8.19. *Suppose there is a subset $J \subset M$ such that the assumptions of Corollary 8.18 hold for the nodes in J, and assumptions (8.23), (8.24) and (ii) of Theorem 8.14 hold for the nodes in*

$$K = \{k : k \notin J, \text{ or } k \in J \text{ and } \lambda_{k\ell} + \lambda_{\ell k} \neq 0, \text{for some } \ell \notin J\}.$$

Then $\pi(x) = \prod_{j \in M} \pi_j(x_j)$ is the stationary distribution of q.

PROOF. The set K contains all the nodes in $M \setminus J$ and those nodes in J that are directly connected to set $M \setminus J$. Hence, the nodes that only satisfy the conditions of Corollary 8.18 are not directly connected to the nodes that only satisfy conditions of Theorem 8.14. This implies that each $D_{jk}(x_j, x_k)$ is 0 for all j, k, so the assertion follows by Theorem 8.8. □

The following is an illustration of Corollary 8.18.

Example 8.20. Suppose the network process represents customers moving in a network in which the state x_j, denoted here by n_j, represents the number of customers at node $j \in \{1, 2, \ldots, m\}$. The network has an outside source, denoted by node 0, which has a single state 0. Assume the transition rates for the network are given by (8.6), where

$$q_0^{\mathrm{d}}(0,0) = \mu_0, \qquad q_0^{\mathrm{a}}(0,0) = 1$$

$$q_j^{\mathrm{d}}(n, n') = \begin{cases} \mu_j(n) q_j(\ell) & (n' = n - \ell \geq 1, \ell \geq 0) \\ \mu_j(n) \sum_{\ell=n}^{\infty} q_j(\ell) & (n' = 0, n \geq 1) \\ 0 & (\text{otherwise}), \end{cases}$$

$$q_j^{\mathrm{a}}(n_j, n') = \lambda_j(n) 1(n' = n + 1),$$

for $j = 1, 2, \ldots, m$. Assume the network does not have internal transitions.

According to these rates, each node $j \neq 0$ in isolation operates as a batch service system. Whenever it contains n customers, arrivals enter at the rate $\lambda_j(n)$; also, batches exit at the rate μ_j, and the size of a batch is $\min\{n, \ell\}$, where ℓ is selected by the batch-size probability distribution $q_j(\ell)$. Assume $q_j(0) > 0$ and that the mean of q_j exceeds 1. Note that in the network process, a batch departure at a node triggers only a single customer arrival at some node. This is because each arrival transition rate q_j^a only allows single-unit increments. Another feature is that a node may have bogus departures when it is not empty. Namely, whenever nodes j and k contain $n_j \geq 1$ and n_k customers, respectively, there is a null departure at node j and an arrival at node k at the rate $\mu_j(n_j) q_j(0) \lambda_{jk} \lambda_k(n_k)$.

We will derive the stationary distribution of the network by appealing to Corollary 8.18. Assume the routing probabilities λ_{jk} are reversible with stationary distribution w_j. Furthermore, for the node j transition rate q_j given by (8.7), we select its beta coefficients such that $\beta_j^{\mathrm{a}} / \beta_j^{\mathrm{d}} = w_j$, which is consistent with (8.31). We conjecture that q_j has a stationary distribution π_j of the form

$$\pi_j(n) = c_j \rho_j^n / \lambda_j(n), \quad n \geq 0, \tag{8.32}$$

for some ρ_j, where c_j is the normalizing constant. In addition, node 0 has the degenerate distribution $\pi_0(0) = 1$.

Before verifying this conjecture, let us see what else is needed to satisfy the assumptions of Corollary 8.18. First consider condition (8.30). Clearly,

$$\tilde{\alpha}_j^a(n) = 0 = w_j^{-1} \alpha_j^d(0),$$

and, for $n \geq 1$, we have $\alpha_j^d(n) = \mu_j(n)$ and

$$\tilde{\alpha}_j^a(n) = \frac{1}{\pi_j(n)} \pi_j(n-1) \lambda_j(n-1) = \lambda_j(n) / \rho_j.$$

Consequently, the first part of (8.30) holds, namely $\tilde{\alpha}_j^a(n) = w_j^{-1}\alpha_j^d(n)$, if and only if

$$\lambda_j(n)/\mu_j(n) = \rho_j w_j^{-1}, \quad n \geq 1. \tag{8.33}$$

Hereafter, we assume this is true. Under this assumption, a similar calculation as above shows that the second part of condition (8.30) is satisfied. Furthermore, another easy check shows that the π_j's satisfy (8.12). Thus, the assumptions of Corollary 8.18 are satisfied.

It remains to show that the stationary distribution π_j is given by (8.32). The balance equations for q_j are

$$\pi_j(0)\beta_j^{\mathrm{a}}\lambda_j(0) = \beta_j^{\mathrm{d}} \sum_{\ell=1}^{\infty} \pi_j(\ell)\mu_j(\ell) \sum_{m=\ell}^{\infty} q_j(m),$$

$$\pi_j(n)[\beta_j^{\mathrm{a}}\lambda_j(n) + \beta_j^{\mathrm{d}}\mu_j(n)] = \beta_j^{\mathrm{a}}\lambda_j(n-1)\pi_j(n-1)$$
$$+ \beta_j^{\mathrm{d}} \sum_{\ell=0}^{\infty} \pi_j(n+\ell)\mu_j(n+\ell)q_j(\ell), \quad n \geq 1.$$

Substituting (8.32) into the first balance equation and using a little algebra and (8.33), we obtain

$$\sum_{\ell=0}^{\infty} \rho_j^{\ell} q_j(\ell) = \rho_j. \tag{8.34}$$

The same equation is obtained by substituting (8.32) into the balance equation for $n \geq 1$ and dividing both sides by ρ_j^n.

Equation (8.34) has a unique solution $\rho_j \in (0, 1)$. Indeed, the left-hand side is a strictly increasing convex function in ρ_j that begins at $q_j(0) > 0$ and ends at 1 with a tangent equal to the mean batch size, which we assumed exceeds 1. Then π_j given by (8.32) will be a valid distribution provided

$$c_j^{-1} = \sum_{n=0}^{\infty} \rho_j^n/\lambda_j(n) < \infty,$$

which we assume is true.

Thus by Corollary 8.18, the stationary distribution of the network is the product of π_j's given by (8.32). This is an example of a network with a product form distribution, but each q_j is neither reversible nor quasi-reversible. □

8.6 Queueing Networks

The results in the preceding sections were stated for a general network in which state changes may be due to mutual interactions at the nodes. This section describes how the results apply to the network under slightly simpler notation traditionally used for queueing networks. As in the other chapters and the first section of this chapter, a queueing network refers to a system in which customers, items or information/commands move from one node to another and trigger the states of the nodes to change. A state change, called a departure event, is initiated at one node, and this event is then "routed" as an arrival event to another node that triggers a state change at the arriving node.

For this section, we will consider the network process X defined above with the following notational changes. Its transition rates q are still given by (8.6), but now the rate component $q_j^{\mathrm{a}}(x_j, x_j')$ will be a probability denoted by $p_j^{\mathrm{a}}(x_j, x_j')$, where $\sum_{x_j'} p_j^{\mathrm{a}}(x_j, x_j') = 1$. Also, the λ_{jk}'s are normalized to be probabilities such that $\sum_{k \in M} \lambda_{jk} = 1$. Feedback loops are natural in routing of units, and so we allow $\lambda_{jj} > 0$. Under these conventions, we call X a *queueing network process.*

The dynamics of the queueing network are as follows:

- When the state of node j is x_j, a departure there changes the state from x_j to y_j with the rate $q_j^{\mathrm{d}}(x_j, y_j)$.
- A departure from node j is transferred to node k as an arrival with probability λ_{jk} (where node 0 represents the outside).
- An arrival at node k changes its state from x_k to y_k with probability $p_k^{\mathrm{a}}(x_k, y_k)$.
- When the state of node j is x_j, there may be an internal change to state y_y with rate $q_j^{\mathrm{i}}(x_j, y_j)$, and this state change does not trigger changes at other nodes. Such a transition could include feedbacks described by the rate

$$q_j^{\mathrm{i}}(x_j, y_j) = q_j^{\star}(x_j, y_j) + \sum_{x_j'} q_j^{\mathrm{d}}(x_j, x_j')\lambda_{jj} p_j^{\mathrm{a}}(x_j', y_j),$$

  where $q_j^{\star}$ is the rate of a pure internal transition at node j.

This queueing network is more general than the conventional ones, because arrivals, departures, and internal changes may occur at the same time. Also, the state changes need not be for actual departures or arrivals in the traditional sense.

Under the notational conventions above, we automatically have

$$\alpha_j^{\mathrm{a}}(x_j) = \bar{\alpha}_j^{\mathrm{a}} = 1, \quad x_j \in \mathbb{E}_j, \ j \in M.$$

Also, each node j in isolation has transition rates

$$q_j(x_j, y_j) = \beta_j^{\mathrm{a}} p_j^{\mathrm{a}}(x_j, y_j) + (1 - \lambda_{jj})q_j^{\mathrm{d}}(x_j, y_j) + q_j^{\mathrm{i}}(x_j, y_j). \tag{8.35}$$

Here β_j^{a} is the average arrival rate and $1 - \lambda_{jj}$ takes the place of β_j^{d}. If $\pi_j(x_j)$ is the stationary distribution of (8.35) with dummy parameter β_j^{a}, then $\overline{\alpha}_j^{\mathrm{d}}$ is the average departure rate from node j, which is a function of β_j^{a}.

Now, let us see how the results above simplify for the queueing network process. First, note that Theorem 8.8 is as follows.

Theorem 8.21. *The stationary distribution of the queueing network process is $\pi(x) = \prod_{j \in M} \pi_j(x_j)$ if and only if each π_j is the stationary distribution of q_j in (8.35) for some coefficient β_j^{a}, and the π_j's are such that*

$$\beta_j^{\mathrm{a}} = \sum_{k \neq j} \bar{\alpha}_k^{\mathrm{d}} \lambda_{kj}, \quad j \in M, \tag{8.36}$$

and, for $j \neq k \in M$, $x_j \in \mathbb{E}_j$, $x_k \in \mathbb{E}_k$,

$$(\tilde{\alpha}_j^{\mathrm{d}}(x_j) - \overline{\alpha}_j^{\mathrm{d}})\lambda_{jk}(\tilde{\alpha}_k^{\mathrm{a}}(x_k) - 1) + (\tilde{\alpha}_k^{\mathrm{d}}(x_k) - \overline{\alpha}_k^{\mathrm{d}})\lambda_{kj}(\tilde{\alpha}_j^{\mathrm{a}}(x_j) - 1) = 0. \tag{8.37}$$

Here only the one equation (8.36) from (8.20) is needed since $\beta_j^d = 1 - \lambda_{jj}$. Equation (8.36) states that β_j^{a} is the total arrival rate at node j from all other nodes. This is why (8.36) and (8.20) are called traffic equations. Keep in mind that (8.36) are nonlinear equations in the β_j^{a}'s since $\overline{\alpha}_j^{\mathrm{d}}$ is a function of β_j^{a}. Note that each one of the following conditions is sufficient for (8.37).

- Both nodes j and k are quasi-reversible.
- Both nodes j and k are noneffective for arrivals. Node j is said to be *noneffective for arrivals* if $\tilde{\alpha}_j^{\mathrm{a}}(x_j) = 1$ for all $x_j \in \mathbb{E}_j$.
- Either node j or node k is quasi-reversible and noneffective for arrivals.

These sufficient conditions can be relaxed further if $\lambda_{jk} = 0$ or $\lambda_{kj} = 0$. Usually, the outside source is noneffective for arrivals. If, in addition, it is quasi-reversible, the outside is a Poisson source, which is the last case above. So, we do not need to check (8.37) for nodes connected only to the Poisson source.

For the queueing network, we have the following result concerning quasi-reversibility.

Theorem 8.22. (Quasi-Reversible Queueing Network) *The assertions in Theorem 8.14 apply to the queueing network process—the only simplification is that equation (8.26) reduces to* $b_j = \beta_j^{\mathrm{a}} - (1 - \lambda_{jj})\overline{\alpha}_j^{\mathrm{d}}$.

Quasi-reversibility is not a necessary condition for a product form distribution of the network, even though this property is part of a sufficient condition for a product form. There are known examples of networks with product form distributions and none of the nodes are quasi-reversible. In certain situations as follows, quasi-reversibility is not far from being necessary.

Corollary 8.23. *Suppose the queueing network process has the stationary distribution* $\pi(x) = \prod_{j \in M} \pi_j(x_j)$. *Assume node* j *satisfies*

$$\lambda_{jk^*} \neq 0 \quad \text{and } \lambda_{k^*j} = 0, \quad \text{for some } k^* \neq j, \text{ and} \tag{8.38}$$

$$\tilde{\alpha}_{k^*}^{\mathrm{a}}(x_{k^*}) \neq 1 \quad \text{for some } x_{k^*} \in \mathbb{E}_{k^*}. \tag{8.39}$$

Then node j *is quasi-reversible with respect to* π_j.

PROOF. Theorem 8.21 ensures that π_j is the stationary distribution for q_j and that (8.37) holds. Under the hypotheses, (8.37) reduces to

$$(\tilde{\alpha}_j^{\mathrm{d}}(x_j) - \overline{\alpha}_j^{\mathrm{d}})\lambda_{jk}(\tilde{\alpha}_k^{\mathrm{a}}(x_k) - 1) = 0.$$

Since this is true for each x_j and each k^* and x_{k^*} that satisfy the hypotheses, it follows that $\tilde{\alpha}_j^{\mathrm{d}}(x_j) = \overline{\alpha}_j^{\mathrm{d}}$, for each x_j. Thus, node j is quasi-reversible with respect to π_j. □

Remark 8.24. Under the first assumption in Corollary 8.23, condition (8.39) is satisfied if, for some $k \neq j$, the Markov transition probabilities $p_k^{\mathrm{a}}(x_k, y_k)$ on $\mathbb{E}_k$

are transient. To see this, first note that $\tilde{\alpha}_k^{\rm a}(x_k) = 1$, for each x_k, if and only if π_k is the positive stationary measure for the transition probabilities $p_k^{\rm a}(x_k, y_k)$. Thus, if these probabilities are transient, then (8.39) is satisfied.

In a conventional queueing network, a customer entering a node always "increases" the number of customers at the node. In such a network, $p_k^{\rm a}$ is clearly transient. Hence if the network has a product form distribution and (8.38) holds, then node j is quasi-reversible.

We conclude this section by showing how quasi-reversibility can be used to obtain a product form distribution for an unconventional queueing network.

Example 8.25. Consider a queueing network with exogenous Poisson arrivals, Markovian routing probabilities λ_{jk}, and constant departure rates μ_j at the nodes. Assume the network operates like a Jackson network with the following exception. Whenever a customer is assigned by the probabilities λ_{jk} to enter node j, it either enters with probability a_j (thereby adding one unit to node j), or it does not enter but it deletes one customer with probability $\bar{a}_j = 1 - a_j$, provided a customer is there. Then the transition rates for the network are given by (8.6), where

$$q_0^{\rm d}(0,0) = \lambda, \qquad p_0^{\rm a}(0,0) = 1,$$
$$q_j^{\rm d}(x_j, y_j) = \mu_j 1(y_j = x_j - 1 \geq 1),$$
$$p_j^{\rm a}(x_j, y_j) = a_j 1(y_j = x_j + 1) + \bar{a}_j 1(y_j = \max\{0, x_j - 1\}).$$

Clearly, q_j defined by (8.35) is the transition rate function for an $M/M/1$ queue with arrival rate $\beta_j^{\rm a} a_j$ and service rate $\mu_j + \beta_j^{\rm a}\bar{a}_j$. Therefore, its stationary distribution is

$$\pi_j(x_j) = (1 - \rho_j)\rho_j^{x_j}, \qquad j \neq 0,$$

provided $\rho_j \equiv \beta_j^{\rm a} a_j/(\mu_j + \beta_j^{\rm a}\bar{a}_j) < 1$, which we assume is true. Each node j is quasi-reversible since, for each x_j,

$$\tilde{\alpha}_j^{\rm d}(x_j) = \pi_j(x_j)^{-1}\pi_j(x_j + 1)q_j^{\rm d}(x_j + 1, x_j) = \rho_j\mu_j$$

In this case, the traffic equation (8.36) is

$$\beta_j^{\rm a} = \lambda_{0,j} + \sum_{k \neq j,0} \frac{\mu_k \beta_k^{\rm a} a_k}{\mu_k + \beta_k^{\rm a}\bar{a}_k}\lambda_{kj}\,.$$

For $\beta_j^{\rm a}$'s that satisfy these equations, let π_j denote the stationary distribution above. Then by Corollary 8.21, we conclude that the stationary distribution of the network is the product of the π_j's. □

8.7 Time-Reversals and Departure–Arrival Reversals

In this section, we show that a product form network process in "reverse time" has the same type of transition rate function as the original process. We also point out that, by reversing the roles of arrival and departure transitions in the network, one obtains a dual network process whose structure is typically different from the original process.

We first consider the network process in reverse time. Suppose the network transition rate q is ergodic and its stationary distribution is π. The *time-reversal of* q is the transition rate

$$\hat{q}(x, y) = \pi(x)^{-1}\pi(y)q(y, x), \qquad x, y \in \mathbb{E}.$$

This $\hat{q}$ has the same stationary distribution as q. Now, assume π is the product of stationary distributions π_j of the node transition rates q_j given by (8.7). The time reversal of q_j is

$$\begin{aligned}\hat{q}_j(x_j, y_j) &= \pi_j(x_j)^{-1}\pi_j(y)q_j(y_j, x_j)\\ &= \beta_j^{\mathrm{a}}\hat{q}_j^{\mathrm{a}}(x_j, y_j) + \beta_j^{\mathrm{d}}\hat{q}_j^{\mathrm{d}}(x_j, y_j) + \hat{q}_j^{\mathrm{i}}(x_j, y_j), \qquad x_j, y_j \in \mathbb{E}_j,\end{aligned}$$

where

$$\hat{q}_j^{\mathrm{s}}(x_j, y_j) = \pi_j(x_j)^{-1}\pi_j(y_j)q_j^{\mathrm{s}}(y_j, x_j), \qquad \mathrm{s} = \mathrm{a, d, i}.$$

Now, an easy check shows that $\hat{q}$ has the same form (8.6) as q, with λ_{jk} and q_j^{s} replaced respectively by λ_{jk} and $\hat{q}_j^{\mathrm{s}}$ ($\mathrm{s} = \mathrm{a, d, i}$). This is consistent with π being the product of the π_j's, which are also the stationary distributions of the $\hat{q}_j$'s.

Next, let us consider the idea of reversing the roles of arrivals and departures in the network. The key part of the transition rate q in (8.6) is the product

$$q_j^{\mathrm{d}}(x_j, y_j)\lambda_{jk}q_k^{\mathrm{a}}(x_k, y_k).$$

Because of the symmetry in this product and the other network assumptions, it is clear that all the results above apply to the process with the roles of a and d reversed. One interpretation of this reversal is that the process is the same, but in the results, a and d are simply interchanged. For instance, in the new Theorem 8.14 the assumption (8.23) would apply to α_j^d and (8.24), (8.25), and (8.26) would apply with a and d interchanged.

A more interesting implication is that the new theorems would apply to any network with routing and transition components $\hat{r}_{jk}$ and $\hat{q}_j^{\mathrm{s}}(x_j, y_j)$, for $\mathrm{s} = \mathrm{a, d, i}$, that satisfy

$$\hat{q}_j^{\mathrm{d}}(x_j, y_j)\hat{r}_{jk}\hat{q}_k^{\mathrm{a}}(x_k, y_k) = q_k^{\mathrm{d}}(x_k, y_k)\lambda_{kj}q_j^{\mathrm{a}}(x_j, y_j). \tag{8.40}$$

Such a network, which has different system dynamics than the original one, could be viewed as a dual of the original network.

8.8 Networks with Multiclass Transitions

In this section, we present extensions of the results of the previous sections to networks with multiclass transitions. The extensions are straightforward, and so the proofs are omitted.

Consider the network we have been discussing with the generalization that each node j has several classes of arrival and departure transitions indexed by the set

T_j. For each $u \in \mathrm{T}_j$, let $q^{\mathrm{s}}_{ju}(x_j, x'_j)$ be the transition rate on $\mathbb{E}_j$ of class u for $\mathrm{s} = \mathrm{a}, \mathrm{d}$. We assume that internal transitions are independent of the class and use the same notation q^{i}_j as in Section 8.2. The routing component λ_{jk} is now extended to $\lambda_{ju,kv}$. One may use different index sets for arrivals and departures, but a single index set T_j can cover these cases by introducing null transitions if necessary.

The transition rates for the Markov network process are defined as

$$q(x, y) = \sum_{j,k} q_{jk}(x, y), \qquad x, y \in \mathbb{E}, \tag{8.41}$$

where

$$q_{jj}(x, y) = q^{\mathrm{i}}_j(x_j, y_j)1(y_\ell = x_\ell, \ell \neq j)$$

$$q_{jk}(x, y) = \sum_{u\in\mathrm{T}_j}\sum_{v\in\mathrm{T}_k} q^{\mathrm{d}}_{ju}(x_j, y_j)\lambda_{ju,kv}q^{\mathrm{a}}_{kv}(x_k, y_k)1(y_\ell = x_\ell, \ell \neq j, k), \quad \text{for } j \neq k.$$

As in Section 8.2, for a distribution π_j on $\mathbb{E}_j$, we define, for $\mathrm{s} = \mathrm{a}, \mathrm{d}$ and for each $u \in \mathrm{T}_j$,

$$\alpha^{\mathrm{s}}_{ju}(x_j) = \sum_{y_j} q^{\mathrm{s}}_{ju}(x_j, y_j),$$

$$\tilde{\alpha}^{\mathrm{s}}_{ju}(x_j) = \pi_j(x_j)^{-1}\sum_{y_j}\pi_j(y_j)q^{\mathrm{s}}_{ju}(y_j, x_j),$$

$$\overline{\alpha}^{\mathrm{s}}_{ju} = \sum_{x_j}\sum_{y_j}\pi_j(x_j)q^{\mathrm{s}}_{ju}(x_j, y_j).$$

These three values for $\mathrm{s} = \mathrm{i}$ are defined (without the subscript u) as in Section 8.2. Assume that $\overline{\alpha}^{\mathrm{s}}_{ju} < \infty$ for $\mathrm{s} = \mathrm{a}, \mathrm{d}, \mathrm{i}$. The transition function q_j of the local process at node j is now changed to

$$q_j(x_j, y_j) = \sum_{u\in\mathrm{T}_j}\left[\beta^{\mathrm{a}}_{ju}q^{\mathrm{a}}_{ju}(x_j, y_j) + \beta^{\mathrm{d}}_{ju}q^{\mathrm{d}}_{ju}(x_j, y_j)\right] + q^{\mathrm{i}}_j(x_j, y_j), \quad x_j, y_j \in \mathbb{E}_j,$$

where coefficients β^{s}_{ju} ($\mathrm{s} = \mathrm{a}, \mathrm{d}$) are determined by the *traffic equations*

$$\beta^{\mathrm{a}}_{ju} = \sum_{k\neq j}\sum_{v\in\mathrm{T}_k}\overline{\alpha}^{\mathrm{d}}_{kv}\lambda_{kv,ju}, \tag{8.42}$$

$$\beta^{\mathrm{d}}_{ju} = \sum_{k\neq j}\sum_{v\in\mathrm{T}_k}\lambda_{ju,kv}\overline{\alpha}^{\mathrm{a}}_{kv}. \tag{8.43}$$

Finally, we redefine D_{jk} as

$$\begin{aligned} D_{ju,kv}(x_j, x_k) = {} & (\overline{\alpha}^{\mathrm{d}}_{ju} - \alpha^{\mathrm{d}}_{ju}(x_j))\lambda_{ju,kv}(\overline{\alpha}^{\mathrm{a}}_{kv} - \alpha^{\mathrm{a}}_{kv}(x_k)) \\ & - (\overline{\alpha}^{\mathrm{d}}_{kv} - \tilde{\alpha}^{\mathrm{d}}_{kv}(x_k))\lambda_{kv,ju}(\overline{\alpha}^{\mathrm{a}}_{ju} - \tilde{\alpha}^{\mathrm{a}}_{ju}(x_j)). \end{aligned}$$

Theorem 8.8 for the multiclass network is as follows.

Theorem 8.26. *The following statements are equivalent.*
(i) *The stationary distribution of q is $\pi(x) = \prod_{j\in M}\pi_j(x_j)$.*

(ii) *Each* π_j *is the stationary distribution for* q_j *with coefficients (8.42) and (8.43), and*

$$\sum_{u \in \mathrm{T}_j} \sum_{v \in \mathrm{T}_k} \big(D_{ju,kv}(x_j, x_k) + D_{kv,ju}(x_k, x_j)\big) = 0, \tag{8.44}$$

for $j \neq k \in M$, $x_j \in \mathbb{E}_j$, $x_k \in \mathbb{E}_k$.

In case the network is the queueing network as in Section 8.6, we assume that, for each $j \in M$

$$\sum_k \sum_{v \in \mathrm{T}_k} \lambda_{ju,kv} = 1, \quad \alpha^{\mathrm{a}}_{ju}(x_j) = 1, \quad u \in \mathrm{T}_j, x_j \in \mathbb{E}_j.$$

Under these conditions, $\beta^{\mathrm{d}}_{ju} = 1 - \sum_{v \in \mathrm{T}_j} \lambda_{ju,jv}$.

Now, the transition rate function q_j defined above is said to be *quasi-reversible with respect to* π_j if π_j is the stationary distribution of q_j and $\alpha^{\mathrm{a}}_{ju}(x_j)$ and $\tilde{\alpha}^{\mathrm{d}}_{ju}(x_j)$ are independent of $x_j \in \mathbb{E}_j$ for each $j \in M$, $u \in \mathrm{T}_j$. The biased local balance condition for this multiclass network has the same form (8.22) with q_{jk} now defined as in (8.41).

Interestingly, in this multiclass setting the biased local balance condition plus product form stationary distribution do not imply quasi-reversibility of the q_j's. For example, it is easy to see that, if

$$\sum_{u \in \mathrm{T}_j} \left[\tilde{\alpha}^{\mathrm{d}}_{ju}(x_j) - \overline{\alpha}^{\mathrm{d}}_{ju}\right] \lambda_{ju,kv} = 0, \tag{8.45}$$

then (8.44) is satisfied, thus the network is a product form. On the other hand, one can show, using the same arguments as in the proof of Theorem 4.2, that (8.45) implies biased local balance. However, (8.45) is clearly weaker than quasi-reversibility. Thus, quasi-reversibility is sufficient but may not be necessary for a product form and biased local balance when there are multiple class of transitions.

Our final result is the multiclass analogue of Corollary 8.18.

Corollary 8.27. *Suppose* $\lambda_{ju,kv}$ *is reversible on* $M' \equiv \{ju : j \in M, u \in \mathrm{T}_j\}$ *with stationary distribution* w_{ju}. *Assume each* q_j *has coefficients (8.42) and (8.43). If* π_j *is the stationary distribution of* q_j *and*

$$\tilde{\alpha}^{\mathrm{a}}_{ju}(x_j) = w^{-1}_{ju} \alpha^{\mathrm{d}}_{ju}(x_j), \quad \tilde{\alpha}^{\mathrm{d}}_{ju}(x_j) = w_{ju} \alpha^{\mathrm{a}}_{ju}(x_j), \quad x_j \in \mathbb{E}_j. \tag{8.46}$$

then $\pi(x) = \prod_{j \in M} \pi_j(x_j)$ *is the stationary distribution of* q. *If this is the case,* $w_{ju} = \overline{\alpha}^{\mathrm{d}}_{ju} / \overline{\alpha}^{\mathrm{a}}_{ju} = \beta^{\mathrm{a}}_{ju} / \beta^{\mathrm{d}}_{ju}$.

8.9 Exercises

1. For the network in Example 8.2, specify conditions under which the input and output processes at a node are Poisson processes. (Quasi-reversibility of a node implies the input and output processes are Poisson for the node in isolation, but not when it is in the network.)

2. Prove Theorem 8.3.

3. Show that the traffic equations in Theorem 8.3 say that the arrival rate into each node j is equal to the sum of the arrival rates into node j from all other nodes.

4. Prove Theorem 8.5.

5. In the context of Theorem 8.5, show that $\overline{\alpha}_j(\beta_j^a)$ is the rate at which customers depart from node j. Also, show that the traffic equations (8.5) say that the arrival rate into each node j is equal to the sum of the arrival rates into node j from all other nodes.

6. Consider the quasi-reversible process defined in Example 8.6, where each customer carries a class label c in a countable set C. Let $N_c(t)$ and $D_c(t)$ denote the numbers of class c customer arrivals and departures, respectively, in the time interval $[0, t]$. Assume the process is stationary. Show that N_c, $c \in C$, are independent Poisson processes and specify their rates. Do the same for the departure processes D_c, $c \in C$.

7. Define a network like Example 8.2 for multiclass customers, using the notation in Example 8.6. State and prove a theorem such as Theorem 8.3, that characterizes an invariant measure for the network.

8. Consider the network process X with transition rates (8.6). Prove that the stationary distribution of X is $\pi(x) = \prod_{j \in M} \pi_j(x_j)$, $x \in E$, if and only if each π_j is the stationary distribution of q_j for some coefficients γ_j^{a}, γ_j^{d}, and π_j is such that (8.17) holds.

9. Consider the network described in Corollary 8.19, where node 0 has a single state (i.e., it is a Poisson source). For this network, justify the assertion in Corollary 8.19 under the weaker supposition that the assumptions of Corollary 8.18 hold for "the nodes in $J \setminus \{0\}$," instead of "the nodes in J." Hint: Consider the $D_{k0}(x_k, 0)$'s.

10. For the queueing network described in Section 8.6, suppose node j is quasi-reversible. Show that the stationary distribution π_j of q_j is also the stationary distribution of the transition rate function q'_j given by

$$q'_j(x_j, y_j) = \beta_j^{\mathrm{a}+} p_j^{\mathrm{a}}(x_j, y_j) + q_j^{\mathrm{d}}(x_j, y_j) + q_j^{\mathrm{n}}(x_j, y_j), \quad x_j, y_j \in \mathbb{E}_j,$$

where $\beta_j^{\mathrm{a}+} = \sum_{k \in M} \overline{\alpha}_k^{\mathrm{d}} \lambda_{kj}$. Here, $\beta_j^{\mathrm{a}+} = \beta_j^{\mathrm{a}} + \lambda_{jj} \overline{\alpha}_j^{\mathrm{d}}$ is the total arrival rate including the feedback rate, and q'_j does not include feedback transitions as internal transitions. The transition rate q'_j is standard in the quasi-reversibility literature, but it is not convenient to use for a non-quasi-reversible node.

8.10 Bibliographical Notes

The classical quasi-reversible network models in the first section were discussed in Kelly (1979), Whittle (1986b), and Walrand (1988). The rest of the chapter is from Chao et al. (1998). Further insights on quasi-reversibility are in Henderson and Taylor (1990). Henderson et al. (1992), Henderson et al. (1995), and Chao and Miyazawa (1996).

9

Space–Time Poisson Models

This chapter covers space–time Poisson models for queueing networks, spatial service or storage systems, and particle systems. Such a model describes the collective movement of units or customers in space and time, where the units enter the system according to a Poisson space–time process and then move about independently of each other. Because of these properties, the evolution of the system can be formulated by certain "random transformations" of Poisson point processes in space and time. We characterize these transformations and then use them in a variety of models. An important example is a network with time-dependent Poisson arrival process and infinite-server nodes with general service times.

We also consider models for systems in which the input process is not Poisson, but the system is sparsely populated. The sparseness leads to Poisson space–time models that are justified by convergence theorems. An example is a network of infinite-server nodes with a non-Poisson arrival process and general service times.

9.1 Introductory Examples

The following are two classic examples of space–time Poisson models that give a glimpse of what lies ahead.

Example 9.1. *Treelike Network of $M/G/\infty$ Service Stations.* Consider an open network of m service stations (or nodes), where the service times at node j are independent and identically distributed with mean μ_j^{-1}. There is no queueing for service, since only a finite number of the servers are busy at any time. For simplicity, assume the network forms a tree with a single root, and each customer enters the root node and moves up the tree on some branch determined by Markov probabilities p_{jk}, where p_{jk} is the probability that a departure from node j enters node k next. Upon reaching the end of the branch, the customer exits the network. Each branch is therefore a route through the network. The probability that a customer visits node j is $p_j \equiv p_{1j_1} \cdots p_{j_n j}$, where $1, j_1, \ldots, j_n, j$ is the unique path from the root node to j. Assume the customers enter the root node over time according to a homogeneous Poisson process with intensity λ. We will consider the system in equilibrium and defined on the entire time axis $\mathbb{R}$.

The following results, based on material in Section 9.6, describe customer flows in this system. These results apply, in particular, to a single $M/G/\infty$ system and to a tandem system of such nodes.

• Let $Q_j(t)$ denote the number of customers at node j at time t. Then $\{Q_j(t) : t \in \mathbb{R}\}$ is a stationary process and $Q_j(t)$ is a Poisson random variable with mean $p_j\lambda/\mu_j$.

• For each time t, the quantities $Q_1(t), \ldots, Q_m(t)$ at the nodes are independent. But at different times they are not independent.

• Let $N_j(I)$ and $D_j(I)$ denote the number of customer arrivals and departures at node j in the time time set I. Then N_j and D_j are homogeneous Poisson processes with intensity λp_j.

• For each fixed time t, the departure process at each node j up to time t is independent of its future traffic after time t. That is, $\{D_j(I) : I \in (-\infty, t]\}$ is independent of $\{Q_j(u) : u > t \text{ and } N_j(I) : I \in (t, \infty)\}$.

• Suppose $J_1, \ldots, J_n$ are disjoint subsets of nodes such that a customer who visits one subset cannot visit any of the others. Then the families of processes $\{(Q_j, N_j, D_j) : j \in J_i\}$, $1 \le i \le n$ are independent.

These properties imply that each station in isolation behaves like a single $M/G/\infty$ system, and that sectors of the network are independent if customers cannot move among them, even though they come from one Poisson source.

What can we say about the quantities above when the input process is not Poisson? Generally, they do not have tractable distributions. However, the results above are good approximations when the input is not Poisson but the flows in the network are sparse as described in Section 9.10. □

Example 9.2. *Markovian Particle System.* Consider a particle system on a finite set $\mathbb{E}$, where $N_t(i)$ denotes the number of particles at the location $i \in \mathbb{E}$ at time $t \geq 0$. Suppose the particles move independently in the space $\mathbb{E}$, in continuous time, according to an ergodic Markov transition rate function that has a stationary distribution $\pi(i)$, $i \in \mathbb{E}$. Assume the system begins at time 0 under the special condition that each quantity $N_0(i)$ is a Poisson random variable with mean $\pi(i)$, and the quantities $N_0(i)$, $i \in \mathbb{E}$, are independent. That is, the particles form a "spatial" Poisson process N_0 on the finite set $\mathbb{E}$, and its mean measure is π. Then by Theorem 9.14 below, it follows that, at each time t, the locations of particles in the space represented by N_t also form a spatial Poisson process with the same mean measure π. Related particle systems are the subject of Section 9.7. □

The main theme of this chapter is that the results in the preceding examples and in the sections to follow are properties of random transformations of Poisson processes, including thinnings, partitionings, and translations. For instance, in Example 9.1 above, the departure process D_1 at node 1 consists of times $T_n + W_n$, where T_n and W_n are the arrival and waiting times of unit n. In other words, D_1 is a "random translation" of the arrival process (each arrival time T_n is translated by the time W_n). Since the arrival process is Poisson, it follows that D_1 is also Poisson. This result is a consequence of Theorem 9.12 below on Poisson invariance under random transformations.

9.2 Laplace Functionals of Point Processes

A Laplace transform is a tool for characterizing the distribution and moments of a nonnegative random variable. These transforms are also useful for establishing convergence in distribution of random variables. The analogous tool for point processes is a Laplace functional. This section reviews a few properties of the Laplace functionals we need for identifying Poisson processes and studying convergence of point processes to Poisson processes.

Using the terminology of Section 4.1, suppose that N is a point process on a space $\mathbb{E}$, and denote the locations of its points by the sequence $\{X_n\}$. The *Laplace functional* of N is defined, for $f : \mathbb{E} \to \mathbb{R}_+$, by

$$L_N(f) = E\left(\exp[-\int_{\mathbb{E}} f(x)N(dx)]\right).$$

Here $\int_{\mathbb{E}} f(x)N(dx) = \sum_n f(X_n)$. Laplace functionals play a similar role for point processes that Laplace transforms (or moment generating functions) play for nonnegative random variables.

The basic principle is that the Laplace functional of N uniquely determines its distribution. Recall that the probability distribution of N, namely $P\{N \in \cdot\}$, is determined by its finite-dimensional distributions. For what follows, we let $\mathcal{C}$ denote the set of all continuous functions $f : \mathbb{E} \to \mathbb{R}_+$ with compact support (i.e., $\{x : f(x) > 0\}$ is contained in a compact set). Recall that $\stackrel{\mathcal{D}}{=}$ denotes equality in distribution.

Proposition 9.3. *For point processes N and N' on $\mathbb{E}$, each one of the following statements is equivalent to $N \stackrel{\mathcal{D}}{=} N'$.*
(a) $\int_{\mathbb{E}} f(x)N(dx) \stackrel{\mathcal{D}}{=} \int_{\mathbb{E}} f(x)N'(dx)$, for $f \in \mathcal{C}$.
(b) $L_N(f) = L_{N'}(f)$, for $f \in \mathcal{C}$.

Laplace functionals are often more convenient to use than finite-dimensional distributions in deriving the distribution of a point process constructed as a function of random variables or point processes. A standard approach for establishing a point process is Poisson is to verify that its Laplace functional has the following form; this also yields its mean measure.

Example 9.4. If N is a Poisson process with mean measure μ, then its Laplace functional is

$$L_N(f) = \exp[-\int_{\mathbb{E}} (1 - e^{-f(x)})\mu(dx)]. \tag{9.1}$$

This follows by proving it first for indicator functions f, then for linear combinations of indicator functions, and finally for general nonnegative functions by monotone convergence. □

A few places in our analysis involve the notions of weak and vague convergence, which are defined as follows. Suppose that $\mu, \mu_1, \mu_2, \ldots$ are measures on $\mathbb{E}$ that are finite on compact sets. The measures μ_n *converge vaguely* to μ as $n \to \infty$, denoted by $\lim_{n\to\infty} \mu_n = \mu$, if

$$\lim_{n\to\infty} \mu_n(A) = \mu(A), \qquad \text{for each } A \in \mathcal{E} \text{ such that } \mu(\partial A) = 0,$$

where ∂A is the boundary of A. If these measures are all probability measures, this vague convergence is *weak convergence*. A sequence of point processes N_n on $\mathbb{E}$ *converges in distribution to* N *as* $n \to \infty$, denoted by $N_n \Rightarrow N$, if the distribution $P\{N_n \in \cdot\}$ of N_n converges weakly to the distribution $P\{N \in \cdot\}$ of N.

Laplace functionals are also useful tools for proving convergence of point processes based on the following result.

Theorem 9.5. *If* N_n *are point processes on* $\mathbb{E}$ *such that* $\lim_{n\to\infty} L_{N_n}(f) = L_N(f)$, *for each* f, *then* $N_n \Rightarrow N$ *as* $n \to \infty$.

9.3 Transformations of Poisson Processes

This section discusses random transformations of Poisson processes that are the basis of space–time Poisson models.

We begin by considering the following question for nonrandom transformations. If the points of a Poisson process are mapped to some space by a nonrandom transformation, then do these points also form a Poisson process? The answer is yes, provided only that the mean measure for the new process is finite on compact sets.

To see this, suppose N is a Poisson process on $\mathbb{E}$ with mean measure $\mu(B) = EN(B)$, $B \in \mathcal{E}$. Consider a map g from $\mathbb{E}$ to a space $\mathbb{E}'$ (possibly $\mathbb{E}$). Denote its inverse by

$$g^{-1}(B) \equiv \{x \in \mathbb{E} : g(x) \in B\}, \qquad B \in \mathcal{E}'.$$

Now, assume that each point X_n of N is mapped to the location $g(X_n) \in \mathbb{E}'$. We represent this transformation of N by the point process M on $\mathbb{E} \times \mathbb{E}'$ defined by

$$\begin{aligned} M(A \times B) &\equiv \sum_n 1((X_n, g(X_n)) \in A \times B) \qquad (9.2) \\ &= N(A \cap g^{-1}(B)), \quad A \in \mathcal{E},\ B \in \mathcal{E}'. \end{aligned}$$

Keep in mind that $\sum_n = \sum_{n=1}^{N(\mathbb{E})}$. The quantity $M(A \times B)$ denotes the number of points of N in $A \in \mathcal{E}$ that are mapped into $B \in \mathcal{E}'$. Then the transformed points in the space $\mathbb{E}'$ are represented by the point process N' defined by

$$\begin{aligned} N'(B) \equiv M(\mathbb{E} \times B) &= \sum_n 1(g(X_n) \in B) \\ &= N(g^{-1}(B)), \quad B \in \mathcal{E}'. \end{aligned} \tag{9.3}$$

The N' is a point process if it is finite on compact sets. To study N', it is convenient to use the larger process M rather than only N'. Note that because we allow multiple points at a single location, we need not assume g is a one-to-one mapping.

Theorem 9.6. *Under the preceding assumptions, the transformation process M defined by (9.2) is a Poisson process with mean measure*

$$E[M(A \times B)] = \mu(A \cap g^{-1}(B)), \quad A \in \mathcal{E}, \ B \in \mathcal{E}'.$$

Hence, the process N' defined by (9.3) is a Poisson process with mean measure $EN'(B) = \mu(g^{-1}(B))$, $B \in \mathcal{E}'$, provided this measure is finite for each compact B.

PROOF. We will show that M satisfies the two conditions in the definition of a Poisson process. Since N is a Poisson process, $M(A \times B) = N(A \cap g^{-1}(B))$ has a Poisson distribution with mean $\mu(A \cap g^{-1}(B))$. This mean is finite for any B when A is compact. It remains to verify that M has independent increments. It suffices to show that $M(A_i \times B_i) = N(A_i \cap g^{-1}(B_i))$, $i = 1, \ldots, k$, are independent for disjoint $A_1, \ldots, A_k$ in $\mathcal{E}$ and disjoint $B_1, \ldots, B_k$ in $\mathcal{E}'$. This independence follows since $A_i \cap g^{-1}(B_i)$, $i = 1, \ldots, k$, are disjoint and N has independent increments. Thus, M has independent increments and hence is a Poisson process.

Next, note that the process $N'(B) = M(\mathbb{E} \times B)$ has independent increments since M does, and $N'(B)$ has a Poisson distribution with $EN'(B) = \mu(g^{-1}(B))$. Thus, N' is a Poisson process when $\mu(g^{-1}(B))$ is finite for each compact B. □

Example 9.7. Suppose that N is a Poisson process of points $X_n = (X_n^1, X_n^2)$ in the nonnegative quadrant $\mathbb{R}_+^2$ of the plane. The *projection* of N on the x_1-axis is defined by $N'(A) = \sum_n 1(X_n^1 \in A)$, for $A \subset \mathbb{R}_+$. In other words, the points of N are mapped from $\mathbb{R}_+^2$ to $\mathbb{R}_+$ by the projection map $g(x_1, x_2) = x_1$. By Theorem 9.6, the process $M(A \times B) = \sum_n 1(X_n \in A, X_n^1 \in B)$ is Poisson. Furthermore, N' is a Poisson process with mean measure $EN'(B) = \mu(g^{-1}(B)) = \mu(\mathbb{E} \times B)$, provided this is finite for bounded sets B. Unfortunately, this mean is infinite when N is a homogeneous process with $\mu(A \times B) = \lambda |A||B|$. In this case, one can still consider $N'(B) \equiv M([0, b] \times B)$ as the projection of the points of N that lie in the region $[0, b] \times \mathbb{R}_+$, and then N' will be a Poisson process.

Next, consider the map $g(x_1, x_2) = \sqrt{x_1^2 + x_2^2}$, which records the distance from the origin to the point (x_1, x_2). Let N' denote the point process of these distances associated with the points of N. Theorem 9.6 ensures that N' is a Poisson process since

$$EN'([0,r]) = \mu(f^{-1}([0,r])) = \mu(\{(x_1,x_2) : \sqrt{x_1^2+x_2^2} \le r\})$$

is finite for each r. □

We are now ready to consider random transformations of point processes. Suppose that N is a point process on $\mathbb{E}$. Assume that each of its points is mapped into a space $\mathbb{E}'$ according to a probability kernel $p(x, B)$ from $\mathbb{E}$ to $\mathbb{E}'$, where $p(x, B)$ is the probability that a point at $x \in \mathbb{E}$ is mapped into the subset $B \in \mathcal{E}'$, independently of the other points. We represent this random transformation by the point process M on $\mathbb{E} \times \mathbb{E}'$, where $M(A \times B)$ denotes the number of points of N in $A \in \mathcal{E}$ that are mapped into $B \in \mathcal{E}'$. That is,

$$M(A \times B) = \sum_n 1((X_n, Z_n) \in A \times B), \quad A \in \mathcal{E},\ B \in \mathcal{E}', \tag{9.4}$$

where X_n's are the point locations of N, and the point at X_n is mapped to Z_n. The assumption we made on the mapping means that the Z_n's are conditionally independent given N, and

$$P\{Z_n \in B \mid N\} = p(X_n, B), \quad B \in \mathcal{E}',\ n \le N(\mathbb{E}).$$

Another way of writing this probability is $P\{Z_n \in B \mid N, n \le N(\mathbb{E})\}$, where $n \le N(\mathbb{E})$ is included only when $N(\mathbb{E})$ can be finite. Note that M contains the initial process $N(\cdot) = M(\cdot \times \mathbb{E}')$ as well as the process of transformed points $N'(\cdot) = M(\mathbb{E} \times \cdot)$.

Definition 9.8. The point process M on $\mathbb{E} \times \mathbb{E}'$ defined by (9.4) is a *marked point process associated with* N. The $\{Z_n\}$ are *location-dependent marks* of N, and N' is the point process of the marks. We sometimes call M a *p-transformation of* N, where $\{Z_n\}$ are the *transformed points* of N.

In some settings, such as the following one, the transformed points represent auxiliary marks or information related to the original points.

Example 9.9. Suppose the point process N on the time axis $\mathbb{R}$ represents the times T_n at which customers enter a network. Assume the nth particle entering at time T_n has an associated mark $Z_n \equiv ((S_n^i, W_n^i) : i = 1, \dots, L)$, where $S_n^1, \dots, S_n^L$ are the stations or nodes the unit visits in that order, $W_n^1, \dots, W_n^L$ are the respective waiting times at the stations, and L is the length of the route, which may be random. Assume the Z_n's are location-dependent marks of N. Then according to the notation above, $\mathbb{E}'$ would denote the space of all possible vectors Z_n, and $p(t, B)$ would denote the probability that a particle arriving at time t selects a vector in the set of vectors $B \in \mathcal{E}'$. □

The following remark points out some technicalities about marks.

Remark 9.10. *(Construction of Marks).* One can construct marks Z_n for the points X_n as follows. Define (measurable) random functions $\{\gamma_n(\cdot) : n \ge 1\}$ from $\mathbb{E}$ to $\mathbb{E}'$ on the same underlying probability space as N (or an enlargement of that space). Define these random functions such that they are independent, identically distributed, independent of N, and

$$P\{\gamma_n(x) \in B\} = p(x, B), \quad x \in \mathbb{E},\ B \in \mathcal{E}'.$$

Then $Z_n \equiv \gamma_n(X_n)$ are clearly location-dependent marks of N.

Another method for constructing marks is to define random elements Y_n with values in some space $\tilde{\mathbb{E}}$ on the same probability space as N, and define a nonrandom function $g : \mathbb{E} \times \tilde{\mathbb{E}} \to \mathbb{E}'$ such that the Y_n's are independent, identically distributed, independent of N, and

$$P\{g(x, Y_n) \in B\} = p(x, B), \quad B \in \mathcal{E}'.$$

Then $Z_n \equiv g(X_n, Y_n)$ are location-dependent marks of N.

The Laplace functional of the marked point process M is related to that of N as follows. This relation is useful for deriving properties of M, when N has a tractable Laplace functional.

Proposition 9.11. *The Laplace functional of the marked point process M associated with N is*

$$L_M(f) = E\left\{\exp\left[\int_{\mathbb{E}} \log[\int_{\mathbb{E}'} e^{-f(x,z)} p(x, dz)] N(dx)\right]\right\}. \tag{9.5}$$

That is, $L_M(f) = L_N(h)$, where $h(x) = -\log[\int_{\mathbb{E}'} e^{-f(x,z)} p(x, dz)]$.

Proof. Conditioning on N and using the property that the Z_n's are conditionally independent given N, we have

$$\begin{aligned} L_M(f) &= E\left\{E\left[e^{-\sum_n f(X_n, Z_n)} \mid N\right]\right\} \\ &= E\left\{\prod_n \int_{\mathbb{E}'} e^{-f(X_n,z)} p(X_n, dz)\right\} \\ &= E\left\{\exp\left[\sum_n \log \int_{\mathbb{E}'} e^{-f(X_n,z)} p(X_n, dz)\right]\right\}. \end{aligned}$$

Using the property $\sum_n g(X_n) = \int_{\mathbb{E}} g(x) N(dx)$, for $g : \mathbb{E} \to \mathbb{R}$, the last expectation equals the right side of (9.5), and hence (9.5) is true. □

The following is the major result that random transformations of Poisson processes are also Poisson.

Theorem 9.12. *If N is a Poisson process and M is a p-transformation of N, then M is a Poisson process with mean measure*

$$E[M(A \times B)] = \int_A p(x, B)\mu(dx), \quad A \in \mathcal{E},\ B \in \mathcal{E}'. \tag{9.6}$$

Hence, the point process N' of mark values is a Poisson process on $\mathbb{E}'$ with mean measure $EN'(\cdot) = \int_{\mathbb{E}} p(x, \cdot)\mu(dx)$, provided this measure is finite on compact sets.

PROOF. From Proposition 9.11, we know that $L_M(f) = L_N(h)$, where $L_N(h)$ is a Poisson Laplace functional of the form (9.1). Then

$$\begin{aligned} L_M(f) &= \exp\left[-\int_{\mathbb{E}} (1 - e^{-h(x)})\mu(dx)\right] \\ &= \exp\left[-\int_{\mathbb{E}\times\mathbb{E}'} (1 - e^{-f(x,z)})p(x, dz)\mu(dx)\right]. \end{aligned}$$

But this is the Laplace functional of a Poisson process with mean given by (9.6). This proves the first assertion of the theorem. The second assertion that $N'(\cdot) = M(\mathbb{E} \times \cdot)$ is a Poisson process follows since it is the Poisson process M on a subset of its space $\mathbb{E} \times \mathbb{E}'$. □

The following example is a generalization of Example 9.2 above.

Example 9.13. *Markovian Particle Movements.* Consider a system in which at time 0 particles are located in the space $\mathbb{E}$ such that they form a Poisson process N on $\mathbb{E}$ with mean measure μ. The number of particles in the entire space is infinite when $\mu(\mathbb{E}) = \infty$. Suppose the particles move independently in the space $\mathbb{E}$ such that a particle initially located at x moves according to a time-homogeneous Markov process with transition probability $P(t, x, B)$ of being in the set B at time t. Let $N_t(A \times B)$ denote the number of particles that initially began in the set $A \in \mathcal{E}$ and are located in the set $B \in \mathcal{E}$ at time t, and let $N_t'(B) = N_t(\mathbb{E} \times B)$, which is the number of particles in B at time t regardless of where they initially began. The Markovian movements of the individual particles lead to the following Markovian behavior of the entire system.

Theorem 9.14. (Markov/Poisson Location Processes) *The family of point processes $\{N_t : t \in \mathbb{R}_+\}$ is a time-homogeneous, measure-valued Markov process, and each N_t is a Poisson process on $\mathbb{E}^2$ with mean measure*

$$E[N_t(A \times B)] = \int_A P(t, x, B)\mu(dx), \qquad A, B \in \mathcal{E}. \tag{9.7}$$

If the mean measure μ of initial particles is an invariant measure of the probabilities $P(t, x, B)$, then the family of location point processes $\{N_t' : t \in \mathbb{R}_+\}$ is a stationary Markov process. Furthermore, each N_t' is a Poisson process on $\mathbb{E}$ with mean measure μ.

PROOF. For each $t \geq 0$, the process N_t is a transformation of the Poisson process N based on the probabilities $p(x, B) \equiv P(t, x, B)$. Then by Theorem 9.12, N_t is a Poisson process on $\mathbb{E}^2$ with mean given by (9.7), and N_t' is also a Poisson process on $\mathbb{E}$.

Next, note that, for each $0 \leq t < u$, the point process N_u represents location-dependent marks of N_t based on the probabilities

$$p((x, y), A \times B) \equiv P(u - t, y, B)1(x \in A).$$

Then the conditional distribution of N_u given $\{N_s : s \le t, N_t = \nu\}$ is equal to the distribution of N_{u-t} given $N_0 = \nu$. Thus, $\{N_t : t \in \mathbb{R}_+\}$ is a time-homogeneous Markov process.

A similar argument shows that $\{N_t' : t \in \mathbb{R}_+\}$ is a time-homogeneous Markov process. Moreover, under the assumption that μ is an invariant measure of $P(t, x, B)$, it follows that, for each t and B,

$$EN_t'(B) = E[N_t(\mathbb{E} \times B)] = \int_{\mathbb{E}} P(t, x, B)\mu(dx) = \mu(B) = EN_0'(B).$$

Then N_t' is equal in distribution to N_0', since they are Poisson processes with the same mean measure μ. Because $\{N_t' : t \in \mathbb{R}_+\}$ is a Markov process and each N_t' has the same distribution, it follows that the Markov process $\{N_t' : t \in \mathbb{R}_+\}$ is stationary. □

We end this section with another insight into transformations. Theorem 9.12 says that a random transformation of a Poisson process is a Poisson process on a product space. We now prove the converse that essentially any Poisson process on a product space is a transformation of a Poisson process.

Theorem 9.15. *If M is a Poisson process on $\mathbb{E} \times \mathbb{E}'$ such that the mean measure $\mu(\cdot) \equiv E[M(\cdot \times \mathbb{E}')]$ on $\mathbb{E}$ is finite on compact sets, then M is a marked point process associated with its marginal process $N(\cdot) \equiv M(\cdot \times \mathbb{E}')$. The conditional distribution $p(x, B)$ of the marks is defined by*

$$E[M(A \times B)] = \int_A p(x, B)\mu(dx), \quad A \in \mathcal{E},\ B \in \mathcal{E}'. \tag{9.8}$$

Proof. The mean measure of M can always be factored as in (9.8), where, for each fixed B, the $p(x, B)$ as a function of x is the Radon–Nikodym derivative of $E[M(\cdot \times B)]$ with respect to μ. Consequently, the mean measure of M has the same form as that of the marked Poisson process in Theorem 9.12, and so M is equal in distribution to that marked Poisson process. This proves the assertion. □

9.4 Translations, Partitions, and Clusters

This section describes several fundamental transformations of Poisson processes. For this discussion, we assume that N is a Poisson process on $\mathbb{E}$ with mean measure μ and point locations X_n.

We first consider translations of the points $\{X_n\}$ of N by location-dependent marks $\{Z_n\}$ of N. Define point processes M and N' by

$$M(A \times B) = \sum_n 1(X_n \in A, X_n + Z_n \in B), \quad A, B \in \mathcal{E},$$

$$N'(B) = \sum_n 1(X_n + Z_n \in B), \quad B \in \mathcal{E}.$$

The process N' is the *translation of* N by the Z_n's, and M represents the before and after of the translations. For these processes to be well defined, we require that the addition operation is defined on $\mathbb{E}$ (which it is, if $\mathbb{E}$ is a Euclidean space) and that $\mathbb{E}$ is large enough to include all the translated points. We also let

$$G(B|x) \equiv P\{Z_n \in B \mid X_n = x\}$$

and $B - x \equiv \{b - x : b \in B\}$. Note that this conditional distribution does not depend on n.

Corollary 9.16. **(Translation of a Poisson Process)** *Under the preceding assumptions, the processes M and N' are Poisson processes with respective mean measures*

$$E[M(A \times B)] = \int_A G(B - x|x)\mu(dx), \quad A, B \in \mathcal{E}, \tag{9.9}$$

and $EN'(B) = EM(\mathbb{E} \times B)$, $B \in \mathcal{E}$, provided these measures are finite on compact sets.

Proof. Clearly $X_n + Z_n$ are location-dependent marks of N with conditional distribution $p(x, B) = G(B-x|x)$. Also, N' is the point process of these marks and M is the p-transformation of N. Thus, the assertion follows by Theorem 9.12. □

A basic property of Poisson processes is that a sum of independent Poisson processes is also a Poisson process; see Exercise 1. We now describe a reverse operation under which a process is partitioned into several subprocesses. Specifically, consider a transformation of the Poisson process N that represents a partitioning of it into m subprocesses $N_1, \ldots, N_m$ by the following rule.

Partitioning Rule: If N has a point at the location $x \in \mathbb{E}$, then it is assigned to subprocess i with probability $p(x, i)$, where $\sum_{i=1}^{m} p(x, i) = 1$.

In other words, the point at location X_n is assigned to the subprocess Z_n, where the Z_n's are location-dependent marks of the X_n's with conditional distribution

$$p(x, i) = P\{Z_n = i \mid X_n = x\}, \quad x \in \mathbb{E},\ i = 1, \ldots, m,\ N(\mathbb{E}) \geq n.$$

The resulting subprocesses are

$$N_i(A) = \sum_n 1((X_n, Z_n) \in A \times \{i\}), \quad A \in \mathcal{E},\ i = 1, \ldots, m.$$

Clearly $N_1, \ldots, N_m$ form a partition of N in that $N = N_1 + \ldots + N_m$.

Corollary 9.17. **(Partition of a Poisson Process)** *Under the preceding assumptions, the partition $N_1, \ldots, N_m$ of the Poisson process N consists of independent Poisson processes with mean measures*

$$EN_i(A) = \int_A p(x, i)\mu(dx), \quad A \in \mathcal{E},\ i = 1, \ldots, m.$$

Proof. By Theorem 9.12, we know that $M(\cdot) = \sum_n 1((X_n, Z_n) \in \cdot)$ is a Poisson process. Then the assertion follows since the processes $N_i(\cdot) = M(\cdot \times \{i\})$ represent M on the disjoint subsets $\mathbb{E} \times \{i\}$, for $1 \leq i \leq m$. □

A special case of the preceding result is as follows.

Example 9.18. *Thinning of a Poisson Process.* Suppose the points of the Poisson process N are deleted according to the rule that a point of N at x is retained with probability $p(x)$, and the point is deleted with probability $1 - p(x)$. Let N_1 and N_2 denote the resulting processes of retained and deleted points, respectively. Then N_1 and N_2 are independent Poisson processes with respective mean measures

$$EN_1(A) = \int_A p(x)\mu(dx), \quad EN_2(A) = \int_A (1 - p(x))\mu(dx), \quad A \in \mathcal{E}. \quad \square$$

Splitting and merging of flows in a network, as we now describe, are examples of partitioning and summing of point processes.

Example 9.19. *Routing in Acyclic Graphs.* Consider the directed graph shown in Figure 9.1 in which units are routed in the direction of the arrows. Let $N_{jk}(t)$ denote the number of units that are routed on the arc from node j to node k in the time interval $(0, t]$. Assume that the input processes N_{0j}, $j = 1, 2, 3$, from outside are independent Poisson processes on $\mathbb{R}_+$ with respective intensities λ_{0j}, $j = 1, 2, 3$. Upon entering the graph, each arrival is routed independently through the graph according to the probabilities on the arcs, and there are no delays at the nodes (the travel through the graph is instantaneous). For instance, a unit entering node 3 is routed to node 5 or node 6 with respective probabilities p_{35} and p_{36}, where $p_{35} + p_{36} = 1$.

The results above on partitioning and merging of Poisson processes yield the following properties. Each flow N_{jk} from j to k is a Poisson process with an intensity λ_{jk}, which is obtainable in the obvious manner. For instance,

$$\lambda_{13} = p_{13}\lambda_{01}, \qquad \lambda_{36} = p_{36}[\lambda_{03} + p_{13}\lambda_{01}],$$
$$\lambda_{35} = p_{35}[\lambda_{03} + p_{13}\lambda_{01}], \quad \lambda_{60} = \lambda_{36} + p_{56}\lambda_{35}.$$

Also, some of the flows are independent (denoted by $\perp$). Examples are $N_{12} \perp N_{13}$, $N_{36} \perp N_{56}$, $N_{36} \perp N_{24}$, and $N_{13} \perp N_{24}$. On the other hand, many flows are not independent (denoted by $\not\perp$). Examples are $N_{12} \not\perp N_{24}$, $N_{35} \not\perp N_{40}$, $N_{13} \not\perp N_{60}$, and $N_{23} \not\perp N_{40}$.

In addition, the flow $N_k = \sum_i N_{jk}$ through each node k is a Poisson process with intensity $\sum_j \lambda_{jk}$. Clearly all the N_k's are dependent. If the arc between 5 and 4 did not exist, however, then N_4 would be independent of N_3, N_5, and N_6. $\square$

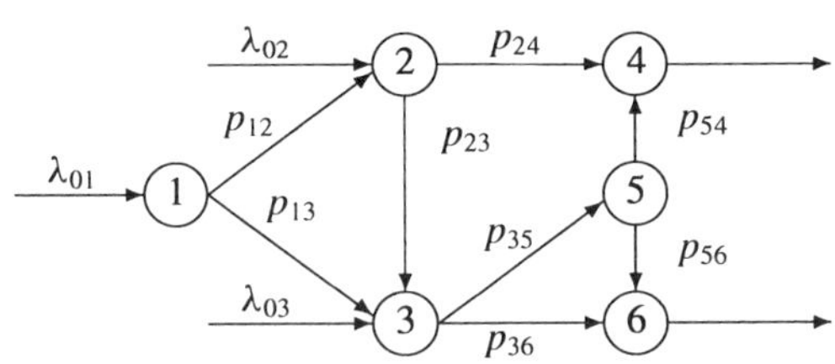

FIGURE 9.1. Partitioning and Merging of Flows

There are several relatives of Poisson processes that are defined via marks. Here are some examples.

Example 9.20. *Compound Poisson Process.* Suppose that Z_n are real-valued, location-dependent marks of N. Then

$$M'(B) = \sum_n Z_n 1(X_n \in B), \quad B \in \mathcal{E},$$

is a *compound Poisson process*. This is also called a signed Poisson random measure on $\mathbb{E}$. The mass Z_n (possibly negative) is located at the point X_n, and $EM'(B) = \int_B \int_{\mathbb{R}} zp(x, dz)\mu(dx)$. The distribution of $M'(B)$ is the standard compound Poisson distribution when the marks Z_n are independent of N. Such compound Poisson processes can also be defined in contexts where the marks Z_n are random vectors or random elements of a group with the addition operation. □

Example 9.21. *Poisson Cluster Processes.* Suppose the marks Z_n of the Poisson process N are point processes on a space $\mathbb{E}'$. Then

$$N'(A \times B) = \sum_n 1(X_n \in A) Z_n(B),$$

is a *Poisson cluster process* on $\mathbb{E} \times \mathbb{E}'$, and $N'(\mathbb{E} \times \cdot) = \sum_n Z_n(\cdot)$ is the cluster process on $\mathbb{E}'$. The Laplace functional of this cluster process is given in Exercise 4. One can use this functional to obtain moments, but the distribution of N' may be intractable for complicated Z_n's. Clearly, N' is not a Poisson process, even if each Z_n is one. □

Recall that under a partitioning of N, each of its points is assigned to a "single" category or is labeled by a "single" attribute. We now consider the situation in which each point is split into several parts, or each point carries multiple attributes, leading to multivariate phenomena.

Example 9.22. *Multivariate Processes.* Suppose that each point X_n of the Poisson process N is associated with one or more attributes from a countable set $\mathcal{I}$ and let $Z_{ni} = 1$ or 0 if the point X_n does or does not have attribute i. Assume that $Z_n = (Z_{ni} : i \in \mathcal{I})$ are location-dependent marks of N. Define the point process N_i' on $\mathbb{E}$ by

$$N_i'(A) = \sum_n 1(X_n \in A, Z_{ni} = 1), \quad A \in \mathcal{E},\ i \in \mathcal{I}.$$

The collection $\{N_i' : i \in \mathcal{I}\}$ is a *multivariate Poisson process*. Each N_i' is a Poisson process with $EN_i'(\cdot) = \int_{(\cdot)} p(i, x)\mu(dx)$, where

$$p(i, x) = P\{Z_{ni} = 1 \mid X_n = x\}$$

is the probability that a point at x has attribute i. These processes are "dependent" Poisson processes. Several of them may have a point at the same location.

A natural generalization would be to assume the points X_n are associated with location-dependent marks $Z_n = (Z_{ni} : i \in \mathcal{I})$, where Z_{ni} is a point process on a space $\mathbb{E}_i$. Then the family of point processes

$$N_i'(A \times B) = \sum_n 1(X_n \in A) Z_{ni}(B), \quad A \in \mathcal{E}_i', \ i \in \mathcal{I},$$

is a *multivariate Poisson cluster process* on $\mathbb{E} \times \mathbb{E}_i'$, $i \in \mathcal{I}$. Multivariate compound Poisson processes are defined similarly. All of these multivariate processes can be viewed via the cluster process in the preceding example, where Z_n is defined on $\mathcal{I} \times \cup_{i \in \mathcal{I}} \mathbb{E}_i'$. □

9.5 Service Systems with No Queueing

An important use of translations is for modeling the number of customers in a service system with no queueing. This section contains several examples.

Example 9.23. *$M/G/\infty$ System: Energy and Communication Model.* Companies that provide natural gas, electricity, or computer/telephone service are interested in predicting their service loads or demands over time. Consider such a system in which customers request services at times T_n that form a Poisson process N on $\mathbb{R}_+$ with mean measure μ. Assume that a request at any time t requires use of the service for a duration that has a distribution $G(\cdot|t)$. Let W_n denote the service duration (sojourn or waiting time in the system) for an arrival at time T_n. Then the W_n's are location-dependent marks of N. Consider the number of customers Q_t that are receiving services at time t. The process $\{Q_t : t \in \mathbb{R}_+\}$ is a $M/G/\infty$ system with time-dependent arrivals and services. Clearly,

$$Q_t = \sum_n 1(T_n \le t, T_n + W_n > t) = M((0,t] \times (t, \infty)), \tag{9.10}$$

where $M(\cdot) \equiv \sum_n 1((T_n, T_n + W_n) \in \cdot)$. Then by Corollary 9.16, M is a Poisson process with mean given by (9.9). Therefore, Q_t has a Poisson distribution with mean

$$EQ_t = \int_{(0,t]} [1 - G(t-s|s)]\mu(ds).$$

Does the random variable Q_t have a limiting distribution as t tends to infinity? It does if its mean has a limit. Here is an explanation. Using the change of variable $s = t - u$ in the preceding integral, we have

$$\alpha_t \equiv EQ_t = \int_{(0,t]} [1 - G(u|t-u)]\mu(du).$$

Now, assume the limit $G(u) \equiv \lim_{t\to\infty} G(u|t-u)$ exists at each u that is a continuity point of G. Also, assume that $\alpha \equiv \int_{(0,\infty)} [1 - G(u)]\mu(du)$ is finite. Then $\alpha_t \to \alpha$ as $t \to \infty$. Consequently,

$$P\{Q_t = n\} = \alpha_t^n e^{-\alpha_t}/n! \to \alpha^n e^{-\alpha}/n! \quad \text{as } t \to \infty.$$

Next, consider the number of service terminations (or departures) in the time interval $(0, t]$, which is

$$D(0, t] = \sum_n 1(T_n + W_n \le t) = M((0, t] \times (0, t]). \tag{9.11}$$

According to Corollary 9.16, D is a Poisson process with

$$ED(0, t] = EN(t) - EQ_t = \int_{(0,t]} G(t - s|s)\mu(ds).$$

Another consequence of M being a Poisson process is that the future of Q is independent of the past of D, denoted by $Q_+ \perp D_-$. This means, as in Definition 4.9, that $\{Q_u : u > t\}$ is independent of $\{D(A) : A \subset (0, t]\}$, for each $t \in \mathbb{R}_+$. To see this, note that, for each t, the process $\{D(A) : A \subset (0, t]\}$ given by (9.11) is a function of M on $B_t \equiv (0, t]^2$. Also, $\{Q_u : u > t\}$ given by (9.10) is a function of M on the set $\cup_{u>t}(0, u] \times (u, \infty) \subset B_t^c$. Since M is independent on the disjoint sets B_t and B_t^c, it follows that $\{Q_u : u > t\}$ is independent of $\{D(A) : A \subset (0, t]\}$. Hence $Q_+ \perp D_-$. □

Example 9.24. *Spatial $M/G/\infty$ System.* The preceding $M/G/\infty$ energy and communication model for time-dependent service loads has the following space–time analogue. As above, suppose customers request service according to a Poisson process N with mean measure μ. Assume that a customer request at time t comes from a subregion B of the service region $\mathbb{E}$ with probability $r(t, B)$, where $r(t, \mathbb{E}) = 1$. Also, assume that a request at time t from a location $x \in \mathbb{E}$ requires service for a time that has a distribution $G(\cdot|t, x)$. Consider the number of customers $N_t(B)$ that are receiving services in the subregion B at time t. We call $\{N_t : t \in \mathbb{R}_+\}$ a *spatial $M/G/\infty$ system.* A tacit assumption is that customers are at fixed locations while they receive service (e.g., receiving natural gas at houses). Possible models of mobile customers are particle systems discussed in Sections 9.7 and 9.6.

Now, we can write

$$N_t(B) = \sum_n 1(T_n \le t, X_n \in B, T_n + W_n > t) = M((0, t] \times B \times (t, \infty)),$$

where (T_n, X_n, W_n) denotes the nth customer's respective arrival time, location, and service duration; and $M(\cdot) = \sum_n 1((T_n, X_n, T_n + W_n) \in \cdot)$. Clearly, the $(X_n, T_n + W_n)$'s are location-dependent marks of N with conditional distribution

$$P\{X_n \in B, T_n + W_n \in C|T_n = s\} = \int_B G(C - s|s, x)r(s, dx).$$

Then by Theorem 9.12, M is a Poisson process. Consequently, the point process N_t representing locations of customers in service at time t is a spatial Poisson process on $\mathbb{E}$ with mean measure

$$EN_t(B) = \int_{(0,t]} \int_B [1 - G(t - s|s, x)]r(s, dx)\mu(ds).$$

Properties of the departure process for this spatial system are the subject of Exercise 6. □

We now describe an optimization problem for a production system that is formulated as a $M/G/\infty$ model.

Example 9.25. *Production Scheduling*. Consider a production system in which units (parts, material, orders, etc.) of type i enter the system at times $0 \le T_{i1} \le T_{i2} \le \dots$ that form a Poisson process N_i on a finite horizon $[0, T]$ with mean measure μ_i, where $i \in \mathcal{I}$ (a finite set). The N_i's are independent. Each unit spends some time in the system and then exits as a certain type of output ℓ in a finite set $\mathcal{L}$. Each input unit yields one output unit, but several types of inputs may yield the same type of output. For the nth type i unit that arrives at time T_{in}, let W_{in} denote its sojourn or waiting time in the system and let L_{in} denote the type of output that it produces. The system may consist of one or more stations where the units are processed, possibly several times before they are finished and exit the system. The output type L_{in} may be random. For instance, the type i might represent a node in which the unit enters, and L_{in} might be its exit or last node.

We will not invoke further microlevel assumptions of the processing, but simply view the production system as an $M/G/\infty$ input–output model. Accordingly, we will assume that, for each $i \in \mathcal{I}$, the (L_{in}, W_{in}) are marks of N_i such that

$$\begin{aligned} P\{L_{ni} = \ell,\, W_{in} \le w \mid T_{in} = t,\ T_{i'k};\ k \ne n,\ i' \in \mathcal{I}\} \\ = P\{L_{ni} = \ell,\, W_{in} \le w \mid T_{in} = t\}. \end{aligned}$$

We will write this probability as the product of the conditional probabilities

$$p_i(\ell|t) \equiv P\{L_{in} = \ell \mid T_{in} = t\}, \qquad G_{i\ell}(w|t) \equiv P\{W_{in} \le w | L_{in} = \ell, T_{in} = t\}.$$

These probabilities are the input data. For instance, $p_i(\ell|t)$ may be determined by a unit's route in a network, and $G_{i\ell}(w|t)$ may be determined by its total service times on the route.

For this production system, the cumulative output of type ℓ up to time t is

$$D_\ell(t) = \sum_{i\in\mathcal{I}} \sum_n 1(L_{in} = \ell, T_{in} \le t, T_{in} + W_{in} \le t), \qquad 0 \le t \le T.$$

As in the preceding example, D_ℓ is a Poisson process with

$$E D_\ell(t) = \sum_{i\in\mathcal{I}} \int_0^t p_i(\ell|s) G_{i\ell}(t-s|s) \mu_i(ds).$$

Also, the quantity of i-units in the system at time t is

$$Q_i(t) = \sum_n 1(T_{in} \le t, T_{in} + W_{in} > t), \qquad 0 \le t \le T, \qquad i \in \mathcal{I}.$$

As in the preceding example, $Q_i(t)$ is a Poisson random variable with

$$E Q_i(t) = \sum_{\ell\in\mathcal{L}} \int_0^t p_i(\ell|s)[1 - G_{i\ell}(t-s|s)]\mu_i(ds).$$

We will consider a problem of optimizing the inputs to meet certain output requirements at minimal cost. This is sometimes called a *Material Requirements*

Planning problem. A standard approach is to consider the input times T_{in} as being deterministic variables that are to be optimized. Ideally, this would be a Markov decision process in which the system is dynamically observed and controlled over time. This leads, however, to an intractable mathematical programming problem. To get around this difficulty, we will consider the input times $0 \leq T_{i1} \leq T_{i2} \leq \ldots$ as a Poisson process just as above, and consider the mean value functions $\mu_i(t) \equiv EN_i(0, t]$ as variables to be optimized. Furthermore, we will consider a "static" optimization over a fixed time horizon rather than a "dynamic" optimization. We will comment further on this approach after formulating the problem.

We will address the production scheduling problem of selecting the functions $\mu_i(t)$ that minimize the expected work in progress (WIP), which is

$$E[\sum_{i\in\mathcal{I}} \int_0^T Q_i(t)\,dt] = \int_0^T [\sum_{i\in\mathcal{I}} \mu_i(t) - \sum_{\ell\in\mathcal{L}} ED_\ell(t)]\,dt.$$

In addition, we want to ensure that the cumulative mean output of product ℓ at time t attains the level $d_\ell(t)$. That is,

$$ED_\ell(t) \geq d_\ell(t), \quad 0 \leq t \leq T, \quad \ell \in \mathcal{L}.$$

In other words, the preceding problem is the mathematical programming problem

$$\min_{\mu_i, i\in\mathcal{I}} \sum_{i\in\mathcal{I}} \int_0^T [\mu_i(t) - \sum_{\ell\in\mathcal{L}} \int_0^t p_i(\ell|s)G_{i\ell}(t-s|s)\mu_i(ds)]\,dt$$

subject to

$$\sum_{i\in\mathcal{I}} \int_0^t p_i(\ell|s)G_{i\ell}(t-s|s)\mu_i(ds) \geq d_\ell(t), \quad 0 \leq t \leq T, \quad \ell \in \mathcal{L}.$$

This is an infinite-parameter linear programming problem. For practical purposes, however, it can be formulated as a standard linear programming problem as follows. Assume the functions $\mu_\ell(t)$ are step functions that only change at integer points $t = 1, 2, \ldots, T$, and make similar assumptions for the functions $r_i(t)$, $p_i(\ell|t)$, $G_{i\ell}(u-t|t)$. Let $\mu_{it} = \mu_i(t) - \mu_i(t-1)$. Then the preceding problem reduces to the linear program

$$\min_{\mu_{it}} \sum_{i\in\mathcal{I}} \sum_{t=1}^{T} [\sum_{s=1}^{t} \mu_{is} - \sum_{\ell\in\mathcal{L}} \sum_{s=1}^{t} p_i(\ell|s)G_{i\ell}(t-s|s)\mu_{is}]$$

subject to $\mu_{it} \geq 0$ and

$$\sum_{i\in\mathcal{I}} \sum_{s=1}^{t} p_i(\ell|s)G_{i\ell}(t-s|s)\mu_{is} \geq d_\ell(t), \quad t = 1, \ldots, T, \quad \ell \in \mathcal{L}.$$

One would use this model as follows. Upon optimizing the means μ_{it}, the actual input times can be generated by simulating the associated Poisson process. The production schedule would be implemented for an initial part of the time horizon, say the first week of a ten week horizon. At the end of the first week, the problem would be resolved for the next ten week horizon with updated requirements etc. Again, the schedule would be implemented for one week. This *rolling horizon model* is a natural approximation for a Markov decision process. The beauty of this approach is that it is tractable for actual computations.

In addition to determining production quantities, this model also provides Poisson distributions of the quantities in the system and the outputs. For instance, upon determining an optimal schedule, one might be interested in the probabilities $P\{D_\ell(t) \geq d_\ell(t)\}$ that the cumulative outputs up to certain times t for the schedule attain the required levels. If these probabilities were too low, one might consider adjusting the input rates.

9.6 Network of $M/G/\infty$ Service Stations

In this section, we discuss the movements of customers in a network of $M/G/\infty$ service stations. The properties of this network we present here follow from the results in the next section on particle systems. We view the movements of customers in a network of $M/G/\infty$ stations as particles moving in a space (the set of stations), and a particle's "location process" is determined by a stochastic routing mechanism and the service times at the stations. Special cases of this network are the treelike network discussed in Example 9.1 and the single $M/G/\infty$ system discussed in Example 9.23.

Consider a network of m service stations (called nodes) that operate as follows. Customers enter the network according to a Poisson process N on the time axis $\mathbb{R}$ with mean measure μ and arrival times

$$\cdots \leq T_{-1} \leq T_0 \leq 0 < T_1 \leq T_2 \leq \cdots .$$

Let $\mathbb{E} = \{0, 1, \ldots, m\}$ denote the space of nodes, where 0 denotes outside of the network. The nth customer entering the network at time T_n selects, or is assigned, a random route $\mathbf{S}_n = (S_{n1}, \ldots, S_{nL_n})$ through the network, where $S_{ni} \in \mathbb{E}$ denotes the ith service station or node the customer visits, and the length $1 \leq L_n \leq \infty$ may be random and depend on $\mathbf{S}_n$. Multiple visits to node 0 are allowed and after visiting node S_{nL_n}, the customer enters 0 and stays there forever. In addition, the customer selects, or is assigned, a vector of nonnegative "waiting" times (service, delay, or sojourn times) $\mathbf{W}_n = (W_{n1}, \ldots, W_{nL_n})$, where W_{ni} is the customer's waiting time at node S_{ni}. We assume that the route and waiting time vectors $\{(\mathbf{S}_n, \mathbf{W}_n)\}$ associated with the arrival times $\{T_n\}$ are marks of N. This implies that there are no interactions among the customers (e.g., queueing for service) that determine their waiting times.

This network is an example of a particle system on the discrete space $\mathbb{E}$, where the location processes are determined by the routes and waiting times. Under the preceding assumptions, the probability that a customer entering at time s is in the sector $B \subset \mathbb{E}$ at time t is

$$P_t(s, B) = \sum_i P\{i \le L_n,\ S_{ni} \in B,\ \sum_{k=1}^{i-1} W_{nk} \le t - s < \sum_{k=1}^{i} W_{nk} \mid T_n = s\}$$

$$+ 1(0 \in B)P\{\sum_{k=1}^{L_n} W_{nk} \le t - s \mid T_n = s\}. \tag{9.12}$$

Note that these probabilities do not depend on n because $\{(\mathbf{S}_n, \mathbf{W}_n)\}$ are marks. These probabilities and the arrival-intensity measure μ are the basic input data for the network.

For a typical application, the routing of units in the network would depend on the structure of the network and the nature of the customers and services. A standard assumption is that the customer routes are independent and Markovian, where p_{jk} denotes the probability of a customer moving to node k upon departing from node j. Then the probability of a route $(s_1, \ldots, s_\ell)$ of nonrandom length ℓ is $p_{0s_1} \cdots p_{s_\ell 0}$. Another convention is that there are several types of customers and all customers of the same type take the same route. In this case, the probability of a route is the probability that the customer entering the network is the type that takes that route. The simplest service times at a node are those that are independent and identically distributed, depending on the node and independent of everything else. Then the sums of service times are characterized by convolutions of the distributions. The next level of generality is that the service times are independent at the nodes, but their distributions may depend on the route as well as the node. An example of dependent service times is that a customer entering a certain subset of routes is initially assigned a service time according to some distribution and then that time is its service time at "each" node on its route.

We will now describe the customer movements in the network using the results of the last section.

Where are the Customers at Time t? Let $N_t(I \times B)$ denote the number of customers who enter during the time set $t - I$ and are located in the sector $B \in \mathcal{E}$ at time $t \in \mathbb{R}$. By Theorem 9.27 below, this N_t is a Poisson process on $\mathbb{R}_+ \times \mathbb{E}$, with

$$E[N_t(I \times B)] = \int_I P_t(s, B)\mu(ds),$$

where $P_t(s, B)$ is the probability given by (9.12) that a customer entering at time s is in B at time t. In particular, the numbers of customers $N_t(I \times \{j\})$, $j \in \mathbb{E}$, who enter during $t - I$ and are at the respective nodes at time t are independent Poisson random variables.

Arrivals, Departures, and Node Changes Let $\bar{N}(I \times I' \times B \times C)$ denote the number of customers that initially enter the network during the time set I and afterward during the time set I' move from sector B to sector C. Note that the process $\bar{N}$ does not depend on time t because it does not record information dynamically with respect to time. Depending on the routing, a customer may move from B to C several times. This $\bar{N}$ is a Poisson cluster process such as those in

Example 9.32. Special cases of this process represent the flow of departures from B, and the flow of arrivals to C. For one-time occurrences, these cluster processes may simply be Poisson processes. For instance, suppose that each customer enters the network only once and eventually leaves it. Let $\bar{N}(I \times I' \times B)$ denote the number of customers that enter the network during the time set I and depart during I' from the sector B. Then as in Example 9.33, it follows that $\bar{N}$ is a Poisson process with

$$E[\bar{N}(I \times I' \times B)] = \int_I P\{S_{nL_n} \in B, s + \sum_{k=1}^{L_n} W_{nk} \in I' \mid T_n = s\}\mu(ds).$$

In particular, the departure processes $D_j(\cdot) \equiv N_t(\mathbb{R} \times (\cdot) \times \{j\})$, $1 \leq j \leq m$, from all the nodes are independent Poisson processes.

As another example, suppose customers move from sector B to sector C at most once, and $\tilde{N}(I \times I')$ denotes the number of customers that enter the network during I and move from B to C during I'. Then $\tilde{N}$ is a Poisson process on $\mathbb{R}^2$ with

$$E[\tilde{N}(I \times I')] = \int_I \sum_i P\{i \leq L_n,\ S_{ni} \in B,\ S_{n(i+1)} \in C,$$
$$s + \sum_{k=1}^{i} W_{nk} \in I' \mid T_n = s\}\mu(ds).$$

There are several independence properties such as those in Corollary 9.34 for this network. For instance, in the last example, for each fixed t, the movements of customers from B to C during the past $(-\infty, t]$ is independent of the populations in B during the future (t, ∞).

Flows on the Routes The preceding results for population sizes, arrivals, departures, node changes, etc. for "nodes" have obvious analogues for "routes." For instance, let $N_t(I \times B)$ denote the number of customers who enter during the time set $t - I$ and are located in the set of routes B at time $t \in \mathbb{R}$. Then N_t is a Poisson process on $\mathbb{R}_+ \times$ (routes), with

$$E[N_t(I \times B)] = \int_{t-I} P\{\mathbf{S}_n \in B,\ s + \sum_{k=1}^{L_n} W_{nk} > t \mid T_n = s)\mu(ds).$$

As another example, let $\bar{N}(I \times I' \times B)$ denote the number of customers who enter during I and during I' depart from a route in B. Then $\bar{N}$ is a Poisson process with

$$E[\bar{N}(I \times I' \times B)] = \int_I P\{\mathbf{S}_n \in B, s + \sum_{k=1}^{L_n} W_{nk} \in I' \mid T_n = s\}\mu(ds).$$

The preceding are just a few properties of the network that follow by space–time Poisson reasoning. There are many other properties that can be obtained by similar arguments.

9.7 Particle Systems

This section discusses particle systems in which particles enter a space according to a space–time Poisson process, and their subsequent movements in the space may depend only on their initial entry times and entry locations. Because the particles do not interact as they move, various events concerning their evolution are described by events associated with marked Poisson processes. The network in the preceding section is a classic example.

We will consider a system in which particles enter a space $\mathbb{E}$ over the doubly-infinite time axis $\mathbb{R}$ such that the pairs of entry times and locations, denoted by (T_n, X_n), form a space–time Poisson process N on $\mathbb{R} \times \mathbb{E}$. The pair (T_n, X_n) is the *space–time entry point* of the nth particle. The total number of arrivals $N(I \times \mathbb{E})$ in a finite time interval I may be infinite; this is equivalent to its mean being infinite, which is allowed. We assume that a particle entering at time $t \in \mathbb{R}$ at the location $x \in \mathbb{E}$ moves in $\mathbb{E}$ according to a stochastic location process that may depend on (t, x) but nothing else. This model also covers systems in which particles may leave the space $\mathbb{E}$ of interest and possibly reenter and exit many times; one simply enlarges $\mathbb{E}$ to contain a location 0 that represents the "outside" state. Because of this convention, we need not define exit times of the particles explicitly, other than through their location processes.

To describe the locations of the particles or some of their attributes over time, we will use system attribute processes defined as follows.

Definition 9.26. Suppose $Z_n \equiv \{Z_n(t) : t \in \mathbb{R}\}$ are location-dependent marks of N, where Z_n is a stochastic process with state space $\mathbb{E}'$ such that $Z_n(t)$ is an "attribute" or set of attributes associated with the nth particle at time t. For each $t \in \mathbb{R}$, define a point process N_t on $\mathbb{R}_+ \times \mathbb{E} \times \mathbb{E}'$ by

$$N_t(I \times A \times B) = \sum_n 1(T_n \in t - I,\, X_n \in A,\, Z_n(t) \in B).$$

This quantity denotes the number of particles that enter somewhere in the set $A \in \mathcal{E}$ during the time interval $t - I$ and, at time t, their attributes are in the set $B \in \mathcal{E}'$. The point process N_t is the *system attribute process at time t* associated with the *particle attribute processes* Z_n.

An attribute of a particle may be an extra label, or classification designating its "type" as it moves in the space $\mathbb{E}$. The type may change over time independently of the other particles, but possibly dependent on its location process in $\mathbb{E}$. An attribute process may also be a multivariate process that keeps track of several attributes, including a particle's location. The preceding definition refers to particle entries during time sets of the form $t - I$, since they are more appropriate for bookkeeping than time sets I. The quantity $N_t(I \times A \times B)$ is finite for compact sets I, A, B. However, the "total number" of particles $N_t(\mathbb{R}_+ \times \mathrm{E} \times B)$ with attributes in B at time t may be infinite when the number of arrivals up to time t is infinite. For simplicity, $Z_n(t)$ is assumed to be defined for all $t \in \mathbb{R}$; it can be set to any value for $t < T_n$ if it is not relevant before the particle entry time T_n. The process N_t does not count attributes of particles that enter after time t.

An attribute process Z_n is called the *location process* for the nth particle in $\mathbb{E}$ when $Z_n(t)$ is its location at any time t, where, in particular, $Z_n(T_n) = X_n$. In this case, $N_t(I \times A \times B)$ records the number of particles that enter in $(t - I) \times A$ and are located in $B \in \mathcal{E}$ at time t. Batch entries and batch movements of particles are also allowed. Suppose that at each space–time entry point (T_n, X_n) a batch of particles enters and they move thereafter in $\mathbb{E}$ possibly depending on others in the batch, but independently of the other particles. Then the location process would be $Z_n \equiv \{Z_n^i(t) : t \in \mathbb{R}, i = 1, \ldots \beta_n\}$, where $Z_n^i(T_n) = X_n$ and β_n is the (possibly random) batch size.

Many attributes are functionals of location processes. Examples of such an attribute $Z_n(t)$ for the nth particle are as follows:

- Amount of time it spends in a set B prior to time t.
- Longest time it has stayed in a state without moving prior to time t.
- Length of time it will ever spend in a set B.
- Time to its next movement after time t.
- Subspace it visits in the time interval $[t - s, t + u]$ for positive, fixed s, u.
- The last state it visits in B before exiting this set for the final time.
- Number of distinct states it visits after time t.

For the rest of this section, we consider the particle system described above with space–time Poisson input process N. We will characterize the system attribute processes N_t, $t \in \mathbb{R}$, associated with particle attribute processes Z_n. The attribute $Z_n(t)$ may represent a single parameter or a complicated family of random elements.

Without any further assumptions on the particle system, we have the following basic result for system attributes. Here

$$p_t(s, x, B) \equiv P\{Z_n(t) \in B \mid T_n = s, X_n = x\}, \quad N(\mathbb{R} \times \mathbb{E}) \geq n. \tag{9.13}$$

Theorem 9.27. (General Attributes) *For each $t \in \mathbb{R}$, the system attribute process N_t is a Poisson process on $\mathbb{R}_+ \times \mathbb{E} \times \mathbb{E}'$ with*

$$E[N_t(I \times A \times B)] = \int_{(t-I)\times A} p_t(s, x, B)\mu(ds\, dx). \tag{9.14}$$

If $p_t(t - s, x, \cdot)$ converges vaguely to some kernel $p(s, x, \cdot)$ as $t \to \infty$, for each s, x, then

$$\lim_{n\to\infty} E[N_t(I \times A \times B)] = \int_{I\times A} p(s, x, B)\mu(ds\, dx). \tag{9.15}$$

In this case, N_t converges in distribution as $t \to \infty$ to a Poisson process, whose mean measure is given by the preceding limit.

Proof. For fixed $t \in \mathbb{R}$, the points $(T_n, X_n, Z_n(t))$ form a marked point process M_t on $\mathbb{R} \times \mathbb{E} \times \mathbb{E}'$, and M_t is a Poisson process with mean measure $p_t(s, x, dz)\mu(ds\, dx)$ by Theorem 9.12. Clearly, N_t is the transformation of M_t

under the mapping $g(s, x, z) = (t - s, x, z)$. Then it follows by Theorem 9.6 that N_t is a Poisson process with mean measure (9.14).

The second assertion of the theorem follows by applying the dominated convergence theorem to (9.14), which equals $\int_{I\times A} p_t(t-u, x, B)\mu(du\,dx)$ under the change of variable $u = t - s$. The third assertion follows by the result in Exercise 2 that says a sequence of Poisson processes converge to a Poisson process if their associated mean measures converge. □

An immediate consequence of the preceding result is that the "total-system" attribute process $N'_t(B) \equiv N_t(\mathbb{R}_+ \times \mathbb{E} \times B)$ at time t is a Poisson process on $\mathbb{E}'$, provided its mean measure (expression (9.14) with $A = \mathbb{E}$) is finite for each compact B. For real-valued attributes such as costs, one may be interested in the cumulative attribute process $Z_t \equiv \sum_n Z_n(t)1(T_n \in (a, t])$, for $t \geq a$, where a is fixed. This is a compound Poisson process provided the sum exists.

We now consider Markovian attributes for the particle system. The Z_n's are *Markov attribute processes* if, for any particle arrival point $(T_n, X_n) = (s, x) \in \mathbb{R}\times\mathbb{E}$, its attribute process $\{Z_n(t) : t \geq s\}$ is a Markov process, possibly depending on (s, x), but not necessarily time homogeneous. The gist of the next result is that the system attribute processes N_t inherit the Markovian property of the particle attribute processes.

Theorem 9.28. (Markovian Attributes) *If the particle attribute processes Z_n are Markovian, then the family of system attributes $\{N_t : t \in \mathbb{R}\}$ is a measure-valued Markov process, which is not necessarily time homogeneous. Each N_t is a Poisson process on $\mathbb{R}_+ \times \mathbb{E} \times \mathbb{E}'$ as described in Theorem 9.27.*

PROOF. We will use the fact that the distribution of a point process is determined by its Laplace functional. Consider fixed $t < u$ and $f : \mathbb{R}_+ \times \mathbb{E} \times \mathbb{E}' \to \mathbb{R}_+$. Note that N_u is the number of attributes for arrivals up to time t plus the number of attributes for arrivals in the interval $(t, u]$. Then

$$
\begin{aligned}
&E\left(\exp[-\int_{\mathbb{R}_+\times\mathbb{E}\times\mathbb{E}'} f(s, x, z)N_u(ds\,dx\,dz)]\right.\\
&\qquad\qquad \left.\mid N_s : s < t,\ N_t(\cdot) = \sum_n 1((t_n, x_n, z_n) \in \cdot)\right)\\
&= \prod_n \int_{\mathbb{E}'} e^{-f(u-t_n, x_n, z)} P\{Z_n(u) \in dz \mid T_n = t_n, X_n = x_n, Z_n(t) = z_n\}\\
&\qquad \times E\left(\exp[-\sum_n f(u - T_n, X_n, Z_n(u))1(T_n \in (t, u])\right).
\end{aligned}
$$

Now, the right side of this equality, and hence the distribution of N_u, is a function of t, u, N_t and is independent of N_s, $s < t$. This proves that $\{N_t : t \in \mathbb{R}\}$ is a Markov process. □

Next, we consider Markovian location processes. We will assume the entry process N for the particle system is stationary in time. That is, the distribution of

$N((I_1+t)\times B_1),\ldots,N((I_k+t)\times B_k)$ is independent of t for any time and space sets $I_1, B_1, \ldots, I_k, B_k$. A necessary and sufficient condition for this stationarity is that $EN(I \times A) = \lambda|I|\eta(A)$, for some $\lambda > 0$ and measure η on $\mathbb{E}$ that is finite on compact sets; see Exercise 8. Consider attributes Z_n that are the location processes for the particles, and let N_t denote the associated system location process at time t. Assume that Z_n is a time-homogeneous Markov process with transition probabilities

$$P(t, x, B) \equiv P\{Z_n(s+t) \in B \mid T_n = s, X_n = x\},$$

that are independent of s. Keep in mind that $Z_n(T_n) = X_n$. Also, consider the point process

$$N'_t(I \times B) \equiv N_t(I \times \mathbb{E} \times B), \quad I \subset \mathbb{R}_+,\ B \in \mathcal{E}.$$

This records the entry times and locations of particles at time t, regardless of where they entered.

Theorem 9.29. (Markovian/Poisson Location Processes) *Consider the particle system under the assumptions above that the Poisson entry process N is stationary in time with $EN(I \times A) = \lambda|I|\eta(A)$, and the particle location processes are Markovian. Then the family $\{N_t : t \in \mathbb{R}\}$ of system location processes is a stationary, measure-valued Markov process. Also, each N_t is a Poisson process with mean measure*

$$E[N_t(I \times A \times B)] = \int_{I\times A} P(s, x, B)\lambda\, ds\, \eta(dx), \quad I \subset \mathbb{R}_+,\ A,\ B \in \mathcal{E}.$$

If in addition, η is an invariant measure for the Markovian location processes, then the family $\{N'_t : t \in \mathbb{R}\}$ of entry-location processes is a stationary, measure-valued Markov process, where each N'_t is a Poisson process with $E[N'_t(I \times A)] = \lambda|I|\eta(A)$.

Proof. The first assertion follows by Theorem 9.28. The second assertion follows from the first assertion and the fact that $\int_{\mathbb{E}} P(u, x, B)\eta(dx) = \eta(B)$, since η is a stationary distribution for $P(u, x, B)$. □

Example 9.30. *Brownian/Poisson Particle System.* Consider the particle system described above, where the particles move in the space $\mathbb{E} = \mathbb{R}^d$. Suppose the Poisson entry process N is homogeneous with intensity λ: It is stationary in space as well as time, and $EN(I \times A) = \lambda|I|\,|A|$. Assume each particle that enters at a location $x \in \mathbb{R}^d$ moves according to a d-dimensional Brownian motion process $\{W_t : t \in \mathbb{R}_+\}$ with mean vector x. This process has continuous sample paths such that $W_0 = x$ w.p.1, and, for each $t < u$, the increment $W_u - W_t$ is independent of $\{W_s : s \le t\}$ and its distribution (denoted by $P(t, x, B)$) is a d-dimensional normal distribution with mean vector x and covariance matrix t times the identity matrix. It is well known (and easy to check) that an invariant measure for this Brownian motion is the Lebesgue measure on $\mathbb{R}^d$. Therefore, by Theorem 9.29, the family $\{N'_t : t \in \mathbb{R}\}$ is a stationary, measure-valued Markov process, where each location process N'_t is a homogeneous Poisson process on $\mathbb{R}_+ \times \mathbb{R}^d$ with intensity λ. □

We now consider stationary attributes of the particle system. For the next result, assume the entry and attribute processes N and Z_n are jointly stationary in time. That is, the process

$$Y_t \equiv \{T_n + t, X_n, \{Z_n(t+u) : u \in \mathbb{R}\}, n \in \mathbb{Z}\}, \quad t \in \mathbb{R}, \tag{9.16}$$

is stationary. Let $\lambda > 0$ and η be a measure on $\mathbb{E}$ that is finite on compact sets such that $N(I \times A) = \lambda |I| \eta(A)$. In addition, assume that

$$p(x, B) \equiv P\{Z_n(t) \in B \mid T_n = s, X_n = x\}, \quad x \in \mathbb{E},\ B \in \mathcal{E}',$$

is independent of s and t. Here $N(\mathbb{R} \times \mathbb{E}) = \infty$ since its mean is infinite. The following is a stationary/Poisson analogue of the preceding Markovian/Poisson results.

Theorem 9.31. (Stationary Systems) *Under the preceding assumptions, the system attribute family $\{N_t : t \in \mathbb{R}\}$ is a measure-valued stationary process, and each N_t is a Poisson process with*

$$E[N_t(I \times A \times B)] = \lambda |I| \int_A p(x, B)\eta(dx).$$

PROOF. It suffices to show, for each $f : \mathbb{R}_+ \times \mathbb{E} \times \mathbb{E}' \to \mathbb{R}_+$, that the distribution of $U_t \equiv \sum_n f(t - T_n, X_n, Z_n(t))$ is independent of t. But this follows from the fact that U_t is a function of the stationary process Y defined by (9.16). In the terminology of Chapter 5, U is a stationary functional of Y. The mean measure of N_t is a special case of (9.14). □

Example 9.32. *Point Process Attributes.* For the particle system we are discussing, an item of interest is the number of times that a particle enters a certain set B. This type of point process attribute can be expressed as follows. Suppose the attribute Z_n for the nth particle is a space–time point process on $\mathbb{R} \times \mathbb{E}'$ such that $Z_n(I \times B)$ is the number of occurrences in the time interval I of a certain event or "B-attribute" for the particle. It is natural to define the system attribute point process at time t by

$$N_t(I \times A \times I' \times B) = \sum_n 1(T_n \in t - I, X_n \in A) Z_n((I' + t) \times B), \tag{9.17}$$

where $I,\ I' \subset \mathbb{R}_+,\ A \in \mathcal{E},\ B \in \mathcal{E}'$. This records the number of particles that enter A in the time set $t - I$ and have a certain B-attribute in the time set $I' + t$.

The following are examples of point process attributes that are functionals of the particle location processes:

- Times of departures from a certain set B.
- Times of transitions (instantaneous movements) from a set B to a set C.
- Distinct states the particle visits.
- Times at which sojourns in a set are longer than w time units.
- The number of traverses of a certain "route" in $\mathbb{E}$ (e.g., from B to C).

Since the entry process N is Poisson and the Z_n's are marks of it, the system attribute processes N_t have several nice properties. First, each N_t is a Poisson cluster process. The proof of this and the specification of its mean are the aim of Exercise 9.

Another property is that $\{N_t : t \in \mathbb{R}\}$ is a stationary process if the entry and attribute processes N and Z_n are jointly stationary in time, and the conditional distribution of Z_n given $T_n = s$, $X_n = x$ is independent of s. The proof of this is similar to that of Theorem 9.31.

The system attribute processes N_t use the time t as a reference point. Point process attributes without such a time reference can be modeled by the point process $\bar{N}$ defined by

$$\bar{N}(I \times A \times I' \times B) = \sum_n 1(T_n \in I, X_n \in A) Z_n(I' \times B),$$

where $I,\ I' \subset \mathbb{R},\ A \in \mathcal{E},\ B \in \mathcal{E}'$. This records the number of particles that enter A in the time set I and have a certain B-attribute in the time set I'. □

Example 9.33. *One-Time-Occurrence Attributes.* Generally, the Poisson cluster process N_t of point process attributes given by (9.17) is not a Poisson process. It is, however, for attributes that occur only once as follows. Suppose the attribute for the nth particle is a point (T_n', Y_n') in $\mathbb{R} \times \mathbb{E}'$ signifying the time at which some special event occurs and the value of something at the event. For instance, if B is a set that is visited exactly once by each particle, then T_n' could denote the nth particle's exit time and $Y_n' \in B$ denote its exit location from B. The one-time-occurrence attribute is represented by the point process $Z_n(\cdot) = 1((T_n', Y_n') \in \cdot)$ that contains only one point. Then the attribute point process $N_t(I \times A \times I' \times B)$ records the number of special events in the time interval $t + I'$ that take values in B for the system. In this case, the Poisson cluster process N_t reduces to being a Poisson process. □

Let us return to the particle system we have been studying with space–time Poisson entry process N and particle location processes $\{Z_n\}$ that are location-dependent marks of N. There are a variety of independent events and processes associated with the system due to the independent increment property of a Poisson process. For instance, let $N_t'(I)$ denote the number of particles that entered in the time interval I and have visited all points of a fixed set B at least once prior to time t. Then by Theorem 9.12, N_t' is a Poisson process on $\mathbb{R}$, provided its mean measure is nonzero and finite. Consequently, the numbers of such particles $N_t'(I_1), \ldots, N_t'(I_k)$ associated with disjoint entry-time intervals $I_1, \ldots, I_k$ are independent Poisson random variables. There are more subtle examples of independence such as the following one. A special case of this is in Example 9.23.

Corollary 9.34. (Populations Independent of Past Departures) *Suppose the set $B \in \mathcal{E}$ is such that each particle enters it at most once. Let Q_t denote the number of particles in B at time t, and let $D(I)$ denote the number of departures from B during the time interval I. Suppose these processes are finite valued. Then for any fixed time t, the departures from B during $(-\infty, t]$ are independent of the population in B after time t (i.e., $Q_+ \perp D_-$).*

PROOF. Let M denote the point process with points at (T_n, X_n, Z_n), where the Z_n's are location processes. This M is a Poisson process by Theorem 9.12. Now, we can write

$$Q_t = M(\mathbb{R} \times \mathbb{E} \times V_t(B)),$$

where $V_t(B) = \{z : z(t) \in B\}$ is the set of location sample paths that are in B at time t. Similarly, the number of departures from B at some time $s \in I$ that are not in B thereafter, is

$$D(I) = M(\mathbb{R} \times \mathbb{E} \times \cup_{s \in I} V_s(B) \cap_{u \geq s} V_u(\mathbb{E} \backslash B)).$$

By the preceding expressions, it follows that $\{Q_u : u > t\}$ is determined by the process M on the set $\mathbb{R} \times \mathbb{E} \times \cup_{u > et} V_u(B)$, and $\{D(I) : I \subset (-\infty, t]\}$ is determined by M on a "subset" of $\mathbb{R} \times \mathbb{E} \times \cap_{u \geq t} V_u(\mathbb{E} \backslash B)$. Since these two sets are disjoint and the Poisson process M has independent increments, it follows that the departures from B during $(-\infty, t]$ are independent of the population in B after time t. □

9.8 Poisson Convergence of Space–Time Processes

In the rest of this chapter, we address the following issue. Suppose N is a point process on $\mathbb{E}$ that need not be Poisson, and let M be a p-transformation or marked point process associated with N. Suppose the space of marks $\mathbb{E}'$ is a "large space" in the sense that the probability $p(x, C)$ of a point assignment to any compact set C is very small regardless of the structure of the initial points in the space $\mathbb{E}$. Then the points in the large space $\mathbb{E}'$ would be sparse, and hence M might form a Poisson process on $\mathbb{E} \times \mathbb{E}'$. Our aim is to express this idea more precisely in terms of limit theorems for sequences of transformed point processes.

This section begins with an example of the limit theorems that lie ahead. Following this example is a limit theorem describing a sequence of space–time processes that converge to a space–time Poisson process. The material in this section is a precursor of the convergence theorems for general random transformations in the next two sections.

Example 9.35. *Routing in a Large Network.* Consider a network whose point process of arrivals N on the time axis $\mathbb{R}_+$ is stationary and ergodic with intensity $\lambda = EN(1)$. Each arrival to the network is independently assigned to a route $j \in \{1, \ldots, n\}$ with probability $p_n(j)$, where $\sum_{j=1}^n p_n(j) = 1$. We do not make any assumptions on the nature of the route or waiting times on the routes. For instance, this network could simply be a flow of customers arrivals that are partitioned into n substreams that are routed to n service stations.

We will consider the flows arriving to the routes when the network is large in the sense that the number of routes n is large and the probabilities $p_n(j)$ are very small. To make the idea of a large network more precise, we consider a sequence of networks that is growing such that the number of routes n tends to infinity. Assume there are constants $a_n \to \infty$ such that, for each j, the limit

$$r_j \equiv \lim_{n\to\infty} a_n p_n(j)$$

exists (it may be 0 for some j's). Let $N_{nj}(t)$ denote the number of arrivals to route j up to time $a_n t$ in the n-route network. Now, consider the p_n-transformation of $N(a_n\cdot)$ defined by $M_n(A \times B) \equiv \sum_{j\in B} N_{nj}(A)$. Then it follows by Theorem 9.36 below that M_n converges in distribution to a Poisson process. This is equivalent to the convergence

$$(N_{n1}, \ldots, N_{nn}) \Rightarrow (N_1, N_2 \ldots), \quad \text{as } n \to \infty, \tag{9.18}$$

where $N_1, N_2 \ldots$ are independent Poisson processes with respective intensities $\lambda r_1, \lambda r_2, \ldots$.

The convergence statement (9.18) justifies that, for a network with a large number of routes n and uniformly small routing probabilities $p_n(j)$, the point processes $N_{n1}, \ldots, N_{nn}$ can be approximated by independent Poisson processes $N^*_{n1}, \ldots, N^*_{nn}$ with intensities $\lambda a_n p_{n1}, \ldots, \lambda a_n p_{nn}$, respectively. A natural measure for the quality of this approximation is the total variation distance. This distance for random elements X and Y is defined by

$$d(X, Y) = \sup_B |P\{X \in B\} - P\{Y \in B\}|.$$

In this case, it follows by a property of Poisson random variables that, for each $k \leq n$ and t,

$$d((N_{n1}(t), \ldots, N_{nk}(t)), (N^*_{n1}(t), \ldots, N^*_{nk}(t))) \leq t\lambda a_n (\sum_{j=1}^{k} p_{nj})^2. \qquad \square$$

We now present a limit theorem underlying the preceding example and other space–time systems as well. Consider a system in which items (or customers) arrive over time according to a point process N on $\mathbb{R}_+$. An arrival at time t generates a mark (or point) in a Borel subset B in a space $\mathbb{E}$ with probability $p(t, B)$. Let $\tilde{M}$ denote the p-transformation of N that represents the arrival times and marks. That is, $\tilde{M}(A \times B)$ is the number of arrivals during the time set A that generate marks in the set B. For instance, this system might represent truck deliveries, fires, emergency telephone calls, computer packet destinations, defects in a pipeline, asteroid landings, sunk ships, bugs in computer code, etc. in appropriate spaces over time.

Our interest is in this system when the space $\mathbb{E}$ is large or, equivalently, that the marks are sparse in $\mathbb{E}$. Accordingly, we will consider a sequence of such systems $\{\tilde{M}_n : n \geq 1\}$, where each $\tilde{M}_n$ on $\mathbb{R}_+ \times \mathbb{E}_n$ is a p_n-transformation of N. Typically, $\mathbb{E}_n$ are increasing subsets of some space. Without loss of generality, we assume $\mathbb{E}_n \equiv \mathbb{E}$. To express that $\mathbb{E}$ is large, we make the sparseness assumption that the mark probabilities $p_n(t, C)$ converge to 0 uniformly in t as $n \to \infty$, for any compact set C. This property is implicit in assumption (9.19) below. It implies, however, that the point processes $\tilde{M}_n$ converge in distribution to 0 as $n \to \infty$. We therefore model the sequence of systems by the normalized point processes

$$M_n(A \times B) \equiv \tilde{M}_n(a_n A \times B), \quad A \subset \mathbb{R}_+, \; B \in \mathcal{E},$$

where a_n are normalizing constants that tend to infinity and $a_n A \equiv \{a_n t : t \in A\}$. The M_n is the process $\tilde{M}_n$ with a_n as its new time unit. This rescaling of time is comparable to a traditional normalization like $a_n^{-1} \sum_{k=1}^n X_k$, which is a rescaling of the space of values. Note that M_n is a p_n-transformation of $N(a_n \cdot)$.

Under this formulation, a limit of the sequence M_n would be an appropriate model for the large space $\mathbb{E}$. The following result describes when such a limit is a Poisson process.

Theorem 9.36. *Suppose $t^{-1}N(t) \Rightarrow \lambda$ as $t \to \infty$, for some positive constant λ, and there are constants $a_n \to \infty$ and a kernel $r(x, B)$ such that the vague-convergence limit*

$$r(t, \cdot) \equiv \lim_{n \to \infty} a_n p_n(a_n t, \cdot) \tag{9.19}$$

exists uniformly in t. Then $M_n \Rightarrow M$ as $n \to \infty$, where M is a Poisson process with $E[M(A \times B)] = \int_A r(s, B)\lambda\, ds$.

This result is a special case of Theorem 9.37 in the next section. Note that the limiting process M depends on the random structure of N only through its intensity λ. This is due to the fact that M_n is asymptotically independent of N, which means $P\{M_n \in \cdot \mid N\} \Rightarrow P\{M \in \cdot\}$. This convergence is implicit in the proof of Theorem 9.37.

9.9 Transformations into Large Spaces

This section describes transformations of point processes that converge to Poisson processes as the space becomes large.

Consider a sequence of point processes $\{M_n : n \geq 1\}$ on $\mathbb{E} \times \mathbb{E}'$ such that each M_n is a p_n-transformation of N_n. We assume the space $\mathbb{E}'$ is large in the sense that, for any compact sets $C \in \mathcal{E}$ and $C' \in \mathcal{E}'$,

$$\sup_{x \in C} p_n(x, C') \to 0, \quad \text{as } n \to \infty. \tag{9.20}$$

This says that the probability of a point of N_n in C being mapped into C' tends to 0 as $n \to \infty$. In the preceding section, the analogous assumption is $p_n(a_n t, C') \to 0$, which is implied by (9.19). That model normalized the space $\mathbb{E}$ by the constants a_n, but this is not done here. Instead, such a normalization is implicit in N_n, which now carries the subscript n.

We will refer to the mean measure of M_n conditional on N_n, which is the "random" measure

$$E[M_n(A \times B) \mid N_n] = \sum_{i=1}^{N_n(A)} p_n(X_{ni}, B) = \int_A p_n(x, B) N_n(dx), \tag{9.21}$$

where $\{X_{ni} : i \geq 1\}$ are the point locations of N_n. The following result says that M_n converges in distribution to a Poisson process if its random mean measure $N_n(dx)p_n(x, dx')$ converges in distribution.

Theorem 9.37. *Let M_n be a p_n-transformation of N_n. Suppose (9.20) holds, and the random measure $N_n(dx)p_n(x, dx')$ on $\mathbb{E} \times \mathbb{E}'$ converges in distribution to some nonrandom measure μ on $\mathbb{E} \times \mathbb{E}'$. Then $M_n \Rightarrow M$ as $n \to \infty$, where M is a Poisson process on $\mathbb{E} \times \mathbb{E}'$ with mean measure μ.*

Remark 9.38. The assumptions (9.20) and $N_n(dx)p_n(x, dx') \Rightarrow \mu(dx\, dx')$ imply that $N_n(A) \Rightarrow \infty$ when $\mu(A \times \mathbb{E}') > 0$. This was apparent in Theorem 9.36, where $N(a_n t) \Rightarrow \infty$.

PROOF. It suffices, by Theorem 9.5, to show that the Laplace functional of M_n converges to that of M. By Proposition 9.11, the Laplace functional of M_n is

$$\begin{aligned} L_{M_n}(f) &= E[\exp\{-\int_{\mathbb{E}\times\mathbb{E}'} f(x, x')M_n(dx\, dx')\}] \\ &= E[\exp\{\int_{\mathbb{E}} \log[1 - g_n(x)]N_n(dx)\}], \end{aligned} \tag{9.22}$$

where f is a bounded nonnegative function on $\mathbb{E} \times \mathbb{E}'$ with compact support and

$$g_n(x) \equiv \int_{\mathbb{E}'} [1 - e^{-f(x,x')}]p_n(x, dx'), \quad x \in \mathbb{E}.$$

Using the series expansion for $\log[1 - g_n(x)]$ in (9.22), we have

$$L_{M_n}(f) = E[\exp\{-\int_{\mathbb{E}} g_n(x)N_n(dx) - Z_n\}], \tag{9.23}$$

where

$$Z_n = \int_{\mathbb{E}} \sum_{k=2}^{\infty} \frac{g_n(x)^{k-1}}{k} g_n(x)N_n(dx).$$

Now, under the hypotheses of the theorem, it follows that

$$\int_{\mathbb{E}} g_n(x)N_n(dx) \Rightarrow \int_{\mathbb{E}\times\mathbb{E}'} (1 - e^{-f(x,x')})\mu(dx\, dx'), \quad \text{as } n \to \infty. \tag{9.24}$$

Next, let $C \in \mathcal{E}$ and $C' \in \mathcal{E}'$ be compact sets such that $f(x, x') = 0$ for (x, x') not in $C \times C'$. Then it follows by assumption (9.20) that

$$g_n(x) \leq \varepsilon_n \equiv \sup_{x\in C} p_n(x, C') \to 0, \quad \text{as } n \to \infty.$$

Using this and (9.24), we have

$$0 \leq Z_n \leq \frac{\varepsilon_n}{1 - \varepsilon_n} \int_{\mathbb{E}} g_n(x)N_n(dx) \Rightarrow 0, \quad \text{as } n \to \infty. \tag{9.25}$$

Applying (9.24) and (9.25) to (9.23) and using dominated convergence, it follows that

$$L_{M_n}(f) \to \exp[-\int_{\mathbb{E}\times\mathbb{E}'}(1 - e^{-f(x,x')})\mu(dx\,dx')], \quad \text{as } n \to \infty.$$

But this limit is the Laplace functional of a Poisson process on $\mathbb{E} \times \mathbb{E}'$ with mean measure μ; recall (9.1). Thus $L_{M_n}(f) \to L_M(f)$, which proves $M_n \Rightarrow M$. □

9.10 Particle Flows in Large Spaces

This section describes two models of particle flows in large spaces. The first one has a spatial normalization, and the second one has a time normalization. Using Theorem 9.37, we show how the flows are described by Poisson processes.

The following example is an extension of the Markovian particle system in Theorem 9.29; here the initial particle locations need not be Poisson and the movements need not be Markovian, but the space is large.

Example 9.39. *Particle System with Spatial Normalization.* Consider a system that contains a finite or infinite number of particles that move about in the Euclidean space $\mathbb{R}^d$. At time 0, the particles are located in $\mathbb{R}^d$ according to a point process N. The particles move by some random mechanism such that, conditioned on N, the particles move independently, and $p_t(x, B)$ is the probability that a particle located at x at time 0 is located in a set B at time t. The system may be closed, or it may be open, meaning that particles may eventually leave the system, and $1 - p_t(x, \mathbb{R}^d)$ is the probability that a particle starting at x exits the system by time t.

We assume the particles become widely dispersed over time in the sense that the probabilities $p_t(x, C)$ converge to 0 as $t \to \infty$ for any compact C. For instance, the particles may all eventually exit the system. More specifically, we assume there are constants $a_t \to \infty$ such that, for each x and B, the limit

$$r(x, B) \equiv \lim_{t\to\infty} a_t p_t(x, a_t B), \tag{9.26}$$

exists uniformly in x for each B with $r(x, \delta B) = 0$, and the limit is finite when B is compact. The scaling $a_t B$ is possible because the set B is in $\mathbb{R}^d$. In addition, assume

$$a_t^{-1} N(a_t \cdot) \Rightarrow \mu(\cdot), \quad \text{as } t \to \infty, \tag{9.27}$$

where μ is a nonrandom measure on $\mathbb{R}^d$. This holds, for instance, if N is a stationary ergodic point process.

Let $M_t(A \times B)$ denote the number of particles initially in A that are in the rescaled set $a_t B$ at time t. The point process M_t on $\mathbb{R}^d \times \mathbb{R}^d$ is clearly a p_t-transformation of $N(a_t \cdot)$. Note that (9.26) and (9.27) imply

$$N(a_t\,dx)p_t(x, a_t\,dx') \Rightarrow \mu(dx)r(x, dx'), \quad \text{as } t \to \infty.$$

Then it follows by Theorem 9.37 that

$$M_t \Rightarrow M, \quad \text{as } t \to \infty,$$

where M is a Poisson process on $\mathbb{R}^d \times \mathbb{R}^d$ with

$$E[M(A \times B)] = \int_A \mu(dx) r(x, B). \qquad \square$$

The next example is an extension of the network of $M/G/\infty$ service systems discussed in Section 9.6.

Example 9.40. *Particle System with Time Normalization.* Consider a system in which particles enter the space $\mathbb{E}$ over time according to a point process N on $\mathbb{R}_+ \times \mathbb{E}$, where a point (s, x) of N is an "entry point" of a particle denoting its entry time s and entry location x. The number of particle arrivals $N([0, t] \times \mathbb{E})$ in the finite time interval $[0, t]$ may be infinite.

Upon entering the space $\mathbb{E}$, each particle moves in it according to some stochastic mechanism and eventually exits the system. For convenience, assume that $\mathbb{E}$ contains the state 0 that denotes the outside of the system. Conditioned that a particle enters at (s, x), the probability that it is in a set $B \in \mathcal{E}$ at time t is $p_t(s, x, B)$. Our interest is in the locations of the particles over time when the space is large.

To this end, consider a sequence of systems we just described, where the probabilities for the nth system are $p_{n,t}(s, x, B)$. Assume there are constants $a_n \to \infty$ such that, for each t, s, x, and B, the vague-convergence limit

$$r_t(s, x, \cdot) \equiv \lim_{n\to\infty} a_n p_{n,a_n t}(a_n s, x, \cdot), \tag{9.28}$$

exists uniformly in s, x. In addition, assume that

$$a_n^{-1} N_n \Rightarrow \mu, \quad \text{as } n \to \infty, \tag{9.29}$$

where $N_n(I \times A) \equiv N(a_n I \times A)$ and μ is a nonrandom measure on $\mathbb{R}_+ \times \mathbb{E}$. For each n and t, let $M_{n,t}$ be a point process on $\mathbb{R}_+ \times \mathbb{E}^2$ such that $M_{n,t}(I \times A \times B)$ denotes the number of particles that enter in the time and space set $a_n I \times A$ and are in B at time $a_n t$. In other words, $M_{n,t}$ is a $p_{n,a_n t}$-transformation of N_n. The assumptions (9.28) and (9.29) ensure that

$$N_n(ds\, dx) p_{n,a_n t}(a_n s, x, dx') \Rightarrow \mu(ds\, dx) r_t(s, x, dx'), \quad \text{as } t \to \infty$$

Then by Theorem 9.37 it follows that, for each t,

$$M_{n,t} \Rightarrow M_t, \quad \text{as } n \to \infty,$$

where M_t is a Poisson process with

$$E[M_t(I \times A \times B)] = \int_{I\times A} \mu(ds\, dx) r_t(s, x, B).$$

Note that this convergence statement is for each fixed time t. To describe the joint convergence of the processes $M_{n,t_1}, \ldots M_{n,t_k}$ for times $t_1, \ldots, t_k$, one would have to know more about the dependencies in the movements of the particles.

However, here is one elementary observation we can make about M_t. Suppose the vague convergence limit

$$r(s, x, \cdot) \equiv \lim_{t\to\infty} r_t(s, x, \cdot)$$

exists. Then since each M_t is a Poisson process, it follows by Exercise 2 that $M_t \Rightarrow M$, where M is Poisson with mean measure $\mu(ds\,dx)r(s, x, B)$.

Next, note that point process M_n' describing the particles that exit the system (enter the outside state 0) is defined by

$$M'(I \times A \times [0, t]) \equiv M_{n,t}(I \times A \times \{0\}).$$

This is the number of particles that enter in $a_n I \times A$ and exit by time $a_n t$. Define

$$p_n'(s, x, [0, t]) \equiv p_{n,a_n t}(s, x, \{0\}).$$

Clearly M_n' is a p_n'-transformation of N_n. Now, the assumptions above ensure that M_n' satisfies the conditions of Theorem 9.37. Consequently,

$$M_n' \Rightarrow M', \quad \text{as } n \to \infty,$$

where M' is a Poisson process with

$$E[M'(I \times A \times [0, t])] = \int_{I\times A} r_t(s, x, \{0\})\mu(ds\,dx).$$

The results in this example extend to a normalization of the space, as well as the time, provided that $\mathbb{E} = \mathbb{R}^d$. The key assumption would be that there exist constants a_n, b_n, and c_n that tend to infinity such that, for each t, s, x, and B, the vague-convergence limit

$$r_t(s, x, \cdot) \equiv \lim_{n\to\infty} c_n p_{n,a_n t}(a_n s, b_n x, b_n(\cdot)),$$

exists uniformly in s, x. Then one would consider $M_{n,t}(I \times A \times B)$ as the number of particles that enter in $a_n I \times b_n A$ and are in $b_n B$ at time $a_n t$. □

9.11 Exercises

1. Let $N_1, \ldots, N_m$ denote a finite collection of independent Poisson processes on a space $\mathbb{E}$ with mean measures $\mu_1, \ldots, \mu_m$. Consider their sum or superposition $N = N_1 + \cdots + N_m$. Show that N is a Poisson process with mean measure $\mu = \mu_1 + \cdots + \mu_m$. Give one proof using only the definition of a Poisson process, and give a second proof by using Laplace functionals. Is this result meaningful for $m = \infty$?
2. For each $n \geq 1$, let N_n denote a Poisson process on a space $\mathbb{E}$ with mean measure μ_n. Show that if μ_n converges vaguely to a measure μ as $n \to \infty$, then N_n converges in distribution to a Poisson process with mean measure μ.
3. In the context of Proposition 9.11, find the Laplace functional of the process N' of marks in M.

4. Consider the cluster process $N'(A \times B) = \sum_n Z_n(B)1(X_n \in A)$ in Example 9.21. Show that its Laplace functional is

$$L_{N'}(f) = \exp[-\int_{\mathbb{E}}(1 - H(x))\mu(dx)],$$

where

$$H(x) \equiv E\left(\exp[-\int_{\mathbb{E}} f(x, x')Z_n(dx')] \mid X_n = x, N(\mathbb{E}) \geq n\right).$$

5. The moments of a point process N on $\mathbb{E}$ are given by

$$\begin{aligned} &E[N(A_1)^{n_1} \cdots N(A_k)^{n_k}] \\ &= (-1)^{n_1+\cdots+n_k} \frac{\partial^{n_1+\cdots+n_k}}{\partial t_1^{n_1} \cdots \partial t_k^{n_k}} L_N(f)|_{t_1=\ldots=t_k=0}, \end{aligned}$$

where $f(x) = \sum_{i=1}^k t_i 1(x \in A_i)$. Use this fact to find expressions for the mean and variance of the cluster process quantity $N'(A)$ in Exercise 4.

6. Consider the spatial $M/G/\infty$ system in Example 9.24. Let $D(B \times (0, t])$ denote the number of service terminations from B in the time interval $(0, t]$. Show that D is a Poisson process on $\mathbb{E} \times \mathbb{R}_+$ and determine its mean. Show that, for each t, the process D on $\mathbb{E} \times (0, t]$ is independent of the point processes N_u of customers in service for $u > t$.

7. Consider a production system in which parts enter according to a Poisson process on the time axis $\mathbb{R}_+$ with intensity λ. A part that enters at time t undergoes a service operation whose duration is exponentially distributed with rate $\mu(t)$. Following this operation, there is a nonrandom delay of d time units before the part can exit the system. Let Q_t denote the number of parts in the system that are undergoing a service operation at time t, and let Q'_t denote the number of parts in the system that are undergoing a delay at time t. Find the distributions of these quantities. Let D denote the point process of times at which parts complete the service operation and begin their delay period. Is D a Poisson process? Assume the service rate $\mu(t)$ is a constant independent of time t. Determine the limits

$$\lim_{t\to\infty} P\{Q_t = n\}, \quad \lim_{t\to\infty} P\{D((a + t, b + t]) = n\}.$$

8. Consider a space–time Poisson process N on $\mathbb{R} \times \mathbb{E}$ with point locations (T_n, X_n). Show that the following statements are equivalent.
(a) N is stationary in time (its distribution is invariant under shifts in the time axis).
(b) The times T_n form a homogeneous Poisson process on $\mathbb{R}$ independent of the X_n's.
(c) $EN(I \times A) = \lambda|I|\eta(A)$, for some $\lambda > 0$ and measure η on $\mathbb{E}$ that is finite on compact sets.
Use the fact that a measure μ on $\mathbb{R}$ that satisfies $\mu(I + t) = \mu(I)$, for any interval I and $t \in \mathbb{R}$, is a multiple of the Lebesgue measure.

9. Show that the attribute process N_t in Example 9.32 is a Poisson cluster process. Find its mean measure.

9.12 Bibliographical Notes

General references on point processes are given in Chapter 4. Results on $M/G/\infty$ queues as manifestations of translations of Poisson processes are reviewed for instance in Foley (1982), Daley and Vere-Jones (1988), Cinlar (1995), and Serfozo (1999). Extensions to networks are reviewed in Massey and Whitt (1993) and Keilson and Servi (1994); similar results for spatial systems are in Massey and Whitt (1993) and Leung et al. (1994). Closely related are the results on particle systems beginning with Derman (1955) and later in Brown (1970). More intricate particle systems such as those studied in Zirbel and Çinlar (1996) and Liggett (1997) require different analysis. Most of the material on Poisson processes is a distillation and extension of well-known results. The results on Poisson convergence of space–time processes is an extension of partitions of point processes discussed in Serfozo (1985).

10

Spatial Queueing Systems

This chapter describes a spatial queueing model for stochastic service systems in which customers or units move about and receive services in a region or a general space. The state of such a system is a point process on a space that evolves over time as a "measure-valued" Markov jump process. Each unit moves in the space according to a Markovian routing mechanism and it receives services at the locations it visits. The service times are exponentially distributed and the rates, as in a queueing system, depend on the congestion or configuration of the points in the system. The types of dependencies are extensions of those in Jackson and Whittle queueing networks.

Spatial systems are beginning to receive attention from researchers in wireless communications for analyzing the performance of cellular or mobile phone systems. Other areas of possible applications are logistics (trucks moving in a country), computer systems (messages moving in a virtual network considered as a region), warehouses (movements of pallets or containers), biology (movements of animals, fish, or diseases), and economics (movements of labor, capital, or businesses). Furthermore, spatial problems arise in certain stochastic networks. For instance, when a network is very large and the portion of customers at each node is small, it may be convenient to consider the network as a spatial system in the plane. Another example is a network in which customers are associated with nondiscrete quantities (e.g., oil, travel times, resources for services, chemicals, gas, temperature, stress) that change over time and affect the customers' routings and services. When these quantities or marks are countable, the network can often be modeled by an extended network, but when the marks are uncountable, a spatial system model is more appropriate.

The word "spatial" is sometimes associated with polling and related service systems in which servers travel to the customers. Our focus, however, is on the reverse situation in which customers travel to server locations, or are served as they move by fixed regional servers.

10.1 Preliminaries

The spatial queueing process we will discuss in this chapter is defined as follows.

We will consider a system in which discrete units move in a space $\mathbb{E}$ where they are processed. Typically, $\mathbb{E}$ would be a subset of the plane or $\mathbb{R}^n$. We will assume that $\mathbb{E}$ is a complete, separable metric space (a Polish space), and let $\mathcal{E}$ denote its family of Borel sets. In case $\mathbb{E}$ is a finite set of points, the system is a network. We represent the state of the system by a finite counting measure μ on $\mathbb{E}$, where $\mu(A)$ denotes the number of units in a set $A \in \mathcal{E}$. We also write this state as

$$\mu(A) = \sum_{k=1}^{n} \delta_{x_k}(A), \qquad A \in \mathcal{E},$$

where $x_1, \dots, x_n$ are the locations of the units in $\mathbb{E}$, and $\delta_x(A) = 1(x \in A)$ is the Dirac measure with unit mass at x. The order of the subscripts on the locations x_k is invariant under permutations. Keep in mind that there may be more than one point at a location.

The system is said to be *closed* if the number of units in it is always the same. Otherwise, the system is *open*, and the outside is represented by a point 0 in $\mathbb{E}$ such that $\{0\} \in \mathcal{E}$. The state measure μ does not record points in 0. In this case, units enter $\mathbb{E}\backslash\{0\}$ from 0 and move around for services and eventually exit by reentering 0. For each $n \geq 0$, we let $\mathbb{M}_n$ denote the set of all counting measures μ on $(\mathbb{E}, \mathcal{E})$ such that $\mu(\mathbb{E}\backslash\{0\}) = n$ and $\mu(\{0\}) = 0$. The set $\mathbb{M}_0$ consists only of the zero measure, which we denote by 0.

The theory of closed and open systems is closely related, and so we discuss them together. Accordingly, we will assume the system may be any one of the following types with state space $\mathbb{M}$.

- Closed with ν units and $\mathbb{M} = \mathbb{M}_\nu$.
- Open with capacity ν and $\mathbb{M} = \cup_{n=0}^{\nu} \mathbb{M}_n$.
- Open with unlimited capacity and $\mathbb{M} = \cup_{n=0}^{\infty} \mathbb{M}_n$.

We will represent the evolution of the system by a continuous-time stochastic process $X = \{X_t : t \geq 0\}$ with state space $(\mathbb{M}, \mathcal{M})$. Here $\mathcal{M}$ denotes the smallest σ-field on $\mathbb{M}$ under which the map $\mu \to \mu(A)$ is measurable for each $A \in \mathcal{E}$. In other words, X is a measurable map from a probability space $(\Omega, \mathcal{F}, P)$ to $(\mathbb{M}, \mathcal{M})$, and $X_t(A)$ is the number of units in $A \in \mathcal{E}$ at time t. The $\mathbb{M}$ is endowed with the vague topology and $\mathcal{M}$ denotes the associated Borel sets. Since $\mathbb{E}$ is a Polish space, $\mathbb{M}$ is also a Polish space.

The measure-valued stochastic process X is a Markov jump process, and hence it will be defined by its transition kernel

$$q(\mu, \mathcal{C}) = \lim_{t \downarrow 0} t^{-1} P\{X_t \in \mathcal{C} | X_0 = \mu\}, \quad \mathcal{C} \in \mathcal{M}.$$

Think of the set $\mathcal{C}$ as a collection of measures in $\mathcal{M}$. The form of the kernel q is determined by the dynamics of the system, which we now describe.

We assume that X represents a system whose evolution is such that whenever it is in state μ, the time to the next movement of one unit from a location $x \in \mathbb{E}$ into a set $A \in \mathcal{E}$ is exponentially distributed with rate $\phi_x(\mu)\lambda(x, A)$. This key assumption is analogous to the standard one used in defining queueing networks. Think of $\phi_x(\mu)$ as the *service rate* or departure rate of a unit located at x. Assume it is positive except that $\phi_x(\mu) = 0$ if $\mu(\{x\}) = 0$ and $x \neq 0$. Also, think of $\lambda(x, A)$ as a *routing kernel*, which is the rate at which a unit departing from x enters a set A (or the probability of such a movement). Assume that $\lambda(x, \mathbb{E})$ is finite for each x, and that $\lambda(0, \mathbb{E}) = 0$ when the system is closed. With a slight abuse of terminology, we refer to ϕ_x and λ as rates even though they are only parts of the compound rate $\phi_x(\mu)\lambda(x, A)$.

Whenever X is in state μ and a unit moves from x to y, we say that X jumps from μ to the state

$$T_{xy}\mu \equiv \mu - \delta_x + \delta_y.$$

This transition is feasible provided $T_{xy}\mu \in \mathbb{M}$. We also define

$$T_{AB}\mu \equiv \{T_{xy}\mu \in \mathbb{M} : x \in A, y \in B\}, \quad \mu \in \mathbb{M}, \ A, B \in \mathcal{E}.$$

This is the collection of states in $\mathbb{M}$ that X may enter from μ when a unit moves from A to B.

Under the assumption above on exponential times to movements of units, it follows that X is indeed a Markov process and its transition rates are

$$q(\mu, \mathcal{C}) = \sum_{x \in \mathbb{E}} \phi_x(\mu) \int_{\mathbb{E}} \lambda(x, dy) 1(T_{xy}\mu \in \mathcal{C}), \quad \mu \in \mathbb{M}, \ \mathcal{C} \in \mathcal{M}. \tag{10.1}$$

This sum is finite, since no more than $\mu(\mathbb{E})$ terms in the sum are not 0 (recall that $\phi_x(\mu) = 0$ when $\mu(\{x\}) = 0$ and $x \neq 0$). A tacit assumption in these rates is that if there are $\mu(\{x\}) > 1$ units at a location x, then each one is equally likely to be the departing unit. Consequently, (10.1) implies that, for $\mu \in \mathbb{M}$ and $A, B \in \mathcal{E}$,

$$q(\mu, T_{AB}\mu) = \sum_{k=1}^{\mu(\mathbb{E})} \phi_{x_k}(\mu)\mu(\{x_k\})^{-1}\lambda(x_k, B)1(x_k \in A) + \phi_0(\mu)\lambda(0, B)1(0 \in A),$$

where $\mu \equiv \sum_{k=1}^{\mu(\mathbb{E})} \delta_{x_k}$ and $x_0 = 0$. The last term in the preceding display is automatically 0 when the system is closed. This form of the transition rates is sometimes more convenient than (10.1).

Since X is a Markov jump process, it evolves as follows. Whenever it is in state μ, it remains there for a time that is exponentially distributed with rate

$$q(\mu, \mathbb{M}) = \sum_{k=1}^{\mu(\mathbb{E})} \phi_{x_k}(\mu)\mu(\{x_k\})^{-1}\lambda(x_k, \mathbb{E}) + \phi_0(\mu)\lambda(0, \mathbb{E}). \tag{10.2}$$

Then it jumps into some set $\mathcal{C} \in \mathcal{M}$ with probability $q(\mu, \mathcal{C})/q(\mu, \mathbb{M})$. In particular, the probability that a unit of μ at location x jumps into the set A is

$$p(x, A) \equiv q(\mu, T_{xA}\mu)/q(\mu, T_{x\mathbb{E}}\mu) = \lambda(x, A)/\lambda(x, \mathbb{E}).$$

This is the conditional probability that a unit moves from x into A given that it does move. Since $p(x, A)$ is independent of the state μ, it follows that the routes of the units in $\mathbb{E}$ are independent, and the successive locations of a unit on its route is a Markov chain determined by the probability kernel $p(x, A)$.

Consider a Markov process ξ_t on $\mathbb{E}$ with transition kernel $\lambda(x, A)$. We call ξ_t the *routing process of* X_t. Without loss of generality, we assume this routing process has a finite invariant measure w, whose support is $\mathbb{E}$, that satisfies the balance equations (or *traffic equations*)

$$\int_A w(dx)\lambda(x, \mathbb{E}) = \int_{\mathbb{E}} w(dy)\lambda(y, A), \quad A \in \mathcal{E}. \tag{10.3}$$

When the system is open, we assume 0 is an atom of w and, for simplicity, we set $w(\{0\}) = 1$. Note that the sequence of states visited by ξ_t is a discrete-time Markov chain with transition probabilities $p(x, A)$, and this chain has an invariant measure $w(dx)\lambda(x, \mathbb{E})$.

To obtain a tractable stationary distribution for X, we will assume the service rates satisfy the following balance property, which is similar to that for Whittle networks.

Definition 10.1. The service rates ϕ_x are Φ-*balanced* if Φ is a positive function on $\mathbb{M}$ such that, for each $\mu \in \mathbb{M}$ and $x, y \in \mathbb{E}$ with $T_{xy}\mu \in \mathbb{M}$,

$$\Phi(\mu)\phi_x(\mu) = \Phi(T_{xy}\mu)\phi_y(T_{xy}\mu). \tag{10.4}$$

Also, $\Phi(0) = 1$ when the system is open.

To see what this condition means, consider the transition kernel

$$\tilde{q}(\mu, \mathcal{C}) = \sum_{x \in \mathbb{E}} \phi_x(\mu) 1(T_{x\mathbb{E}}\mu \in \mathcal{C}), \quad \mu \in \mathbb{M},\ \mathcal{C} \in \mathcal{M}.$$

This is (10.1) with $\lambda(x, \cdot) \equiv 1$. Let $\tilde{\pi}(d\mu) = \Phi(\mu)H(d\mu)$, where H is a Haar measure on $\mathbb{M}$ (i.e., $H(\mathcal{C} + \mu) = H(\mathcal{C})$ for each $\mathcal{C} \in \mathcal{M}$ and μ). Then (10.4) is equivalent to

$$\tilde{\pi}(d\mu)\tilde{q}(\mu, d\eta) = \tilde{\pi}(d\eta)\tilde{q}(\eta, d\mu).$$

But this is the definition of $\tilde{q}$ being reversible with respect to $\tilde{\pi}$. Thus, ϕ_x are Φ-balanced if and only if $\tilde{q}$ is reversible with respect to $\Phi(\mu)H(d\mu)$.

An easy check shows that ϕ_x are Φ-balanced if and only if each ϕ_x is of the form

$$\phi_x(\mu) = \Psi(\mu - \delta_x)/\Phi(\mu),$$

for some function Ψ on $\{\mu - \delta_x : \mu \in \mathbb{M}, x \in \mathbb{E}\}$. This expression is a special case of the canonical form of a reversible transition function in Theorem 1.5. Examples of Φ-balanced service rates are in Section 10.3.

We will use the following terminology.

Definition 10.2. The measure-valued Markov process X is a *spatial queueing process* if its transition rates are of the form (10.1), where the routing kernel has an invariant measure and the service rates are Φ-balanced.

Queueing network models are special spatial queueing models as the following example shows.

Example 10.3. *Jackson and Whittle Networks.* Suppose the spatial queueing process X has a finite state space $\mathbb{E} = \{0, 1, \ldots, m\}$. Then X is a Markov network process. Using notation close to that in Chapter 1, we call $n_j = \mu(\{j\})$ the number of units at node j and consider X_t as a process with vector-valued states $\mathbf{n} = (n_1, \ldots, n_m)$. Its transition rates (10.1) are

$$q(\mathbf{n}, \mathbf{n}') = \begin{cases} \lambda(j, j')\phi_i(\mathbf{n}) & \text{if } \mathbf{n}' = T_{jj'}\mathbf{n} \\ 0 & \text{otherwise.} \end{cases}$$

Here $T_{jj'}\mathbf{n}$ is the vector $\mathbf{n}$ with $n_j, n_{j'}$ replaced by $n_j - 1, n_{j'} + 1$. This covers closed networks ($|\mathbf{n}| \equiv \sum_j n_j = \nu$) and open networks ($|\mathbf{n}| \leq \nu$ or $|\mathbf{n}| < \infty$).

The routing rates $\lambda(j, j')$ are assumed to be such that there are positive w_j that satisfy the traffic equations

$$w_j \sum_{j'} \lambda(j, j') = \sum_{j'} w_k \lambda(j', j), \quad j \in \mathbb{E},$$

where $w_0 = 1$ if the network is open. As in Chapter 1, we assume that ϕ_j is Φ-balanced by a positive function $\Phi(\mathbf{n})$ in the sense that, for each $\mathbf{n}$ and $j, j' \in \mathbb{E}$,

$$\Phi(\mathbf{n})\phi_j(\mathbf{n}) = \Phi(T_{jj'}\mathbf{n})\phi_{j'}(T_{jj'}\mathbf{n}).$$

The assumption is consistent with the Φ-balance assumption above on the $\phi_x(\mu)$'s.

Under these assumptions, X is a Whittle network process. It is a Jackson network process if, in addition, each $\phi_j(\mathbf{n})$ depends only on n_j, and $\phi_0(\cdot) = 1$ when the network is open. In either case, we saw in Chapter 1 that an invariant measure for X is

$$\pi(\mathbf{n}) = \Phi(\mathbf{n}) \prod_{j \in \mathbb{E}} w_j^{n_j}.$$

We will soon see that spatial queueing systems have analogous invariant measures. □

10.2 Stationary Distributions and Ergodicity

In this section, we present a closed form expression for an invariant measure of a spatial queueing process. We also give sufficient conditions for the process to be ergodic.

Consider the spatial queueing process X defined above. Recall that w is a finite measure on $\mathbb{E}$ that satisfies the traffic equations (10.3). The following result describes an invariant measure for X in terms of measures π_n on $\mathbb{M}$ defined by

$$\pi_n(C) = \int_{\mathbb{E}^n} w(dx_1)\cdots w(dx_n)1(\mu_{\mathbf{x}} \in C)\Phi(\mu_{\mathbf{x}})\prod_{z\in\mathbb{E}} \mu_{\mathbf{x}}(\{z\})!, \quad C \in \mathcal{M},$$

for $n \geq 1$, where $\mu_{\mathbf{x}} \equiv \sum_{k=1}^{n} \delta_{x_k}$, and $\pi_0(\cdot) \equiv 1(0 \in \cdot)$ (the last 0 is the zero measure). We assume that w and Φ are such that each measure π_n is finite, and $\sum_{n=0}^{\infty} \pi_n$ is a finite measure in case the system is open with unlimited capacity. These conditions are needed to ensure that the integrals below with respect to these measures are finite.

Theorem 10.4. *An invariant measure for the spatial queueing process X is*

$$\pi = \begin{cases} \pi_\nu & \text{if the system is closed with } \nu \text{ units} \\ \displaystyle\sum_{n=0}^{\nu} \pi_n & \text{if the system is open with capacity } \nu \leq \infty\,. \end{cases} \tag{10.5}$$

Furthermore, π satisfies the property that, for each $A \in \mathcal{E}$ and $C \in \mathcal{M}$,

$$\int_C \pi(d\mu)q(\mu, T_{A\mathbb{E}}\mu) = \int_{\mathbb{M}} \pi(d\eta) \int_C q(\eta, d\mu)1(\eta \in T_{A\mathbb{E}}\mu). \tag{10.6}$$

Equation (10.6) is a *partial balance property* that says, for an ergodic and stationary process X, the average number of movements of units out of the set A that takes X out of C is equal to the average number of movements into A that take X into C.

PROOF. First note that the total balance equation that π must satisfy to be an invariant measure is equation (10.6) with $A = \mathbb{E}$. Thus, to prove the theorem, it suffices to show that π given by (10.5) satisfies (10.6). Note that (10.6) is equivalent to the two statements

$$\pi(\{0\})q(0, T_{A\mathbb{E}}0) = \int_{\mathbb{M}} \pi(d\eta)q(\eta, \{0\})1(\eta \in T_{A\mathbb{E}}0), \quad A \in \mathcal{E}, \tag{10.7}$$

$$\int_{\mathbb{M}\setminus\{0\}} \pi(d\mu)f(\mu)q(\mu, T_{A\mathbb{E}}\mu) = \int_{\mathbb{M}} \pi(d\eta) \int_{\mathbb{M}\setminus\{0\}} q(\eta, d\mu)f(\mu)1(\eta \in T_{A\mathbb{E}}\mu), \tag{10.8}$$

for $A \in \mathcal{E}$ and $f(\cdot) \geq 0$. Statement (10.7) is relevant only for an open system.

Clearly (10.7) holds since by $\pi(\{0\}) = w(\{0\}) = \Phi(0) = 1$, the traffic equations (10.3), and Φ-balance with $x = 0$, we have

$$\pi(\{0\})q(0, T_{A\mathbb{E}}0) = w(\{0\})\Phi(0)\lambda(0, \mathbb{E})\phi_0(0)1(0 \in A)$$

$$= \int_{\mathbb{E}} w(dy)\lambda(y, \{0\})\Phi(\delta_y)\phi_y(\delta_y)1(0 \in A)$$

$$= \int_{\mathbb{M}} \pi(d\eta)q(\eta, \{0\})1(\eta \in T_{A\mathbb{E}}0).$$

The last equality follows because

$$\pi(d\delta_y) = w(dy)\Phi(\delta_y), \qquad q(\delta_y, \{0\}) = \phi_y(\delta_y)\lambda(y, \{0\}),$$

and $\eta \in T_{A\mathbb{E}}0$ if and only if $0 \in A$ and η has the form δ_y.

To prove (10.8), observe that we can write $\pi = \sum_{n=0}^{\infty} \pi_n$, which covers both the closed and open systems. Then by the definitions of π and q,

$$\text{Left side of (10.8)} = \sum_{n=1}^{\infty} \int_{\mathbb{E}^n} w(dx_1) \cdots w(dx_n)\Phi(\mu_{\mathbf{x}}) \prod_{z \in \mathbb{E}} \mu_{\mathbf{x}}(\{z\})!$$

$$\times f(\mu_{\mathbf{x}}) \sum_{k=1}^{n} 1(x_k \in A)\phi_{x_k}(\mu_{\mathbf{x}})\mu_{\mathbf{x}}(\{x_k\})^{-1}\lambda(x_k, \mathbb{E})$$

$$= \sum_{n=1}^{\infty} \sum_{k=1}^{n} \int_{\mathbb{E}^{n-1}} w(dx_1) \cdots w(dx_{k-1})w(dx_{k+1}) \cdots w(dx_n)$$

$$\times \int_{A} w(dx_k)\lambda(x_k, \mathbb{E})h(x_k, \overline{\mu}), \tag{10.9}$$

where $\overline{\mu} = \sum_{i=1}^{n} \delta_{x_i} 1(i \neq k)$ and

$$h(x_k, \overline{\mu}) = \Phi(\overline{\mu} + \delta_{x_k})\phi_{x_k}(\overline{\mu} + \delta_{x_k}) f(\overline{\mu} + \delta_{x_k}) \prod_{z \in \mathbb{E}} \overline{\mu}(\{z\})!.$$

Note that the preceding sums on n do not include $n = 0$ since the zero measure does not appear in (10.8).

Now, by Φ-balance, we can write, for any y_k,

$$h(x_k, \overline{\mu}) = \Phi(\overline{\mu} + \delta_{y_k})\phi_{y_k}(\overline{\mu} + \delta_{y_k}) f(\overline{\mu} + \delta_{x_k}) \prod_{z \in \mathbb{E}} \overline{\mu}(\{z\})!. \tag{10.10}$$

Also, by the traffic equations (10.3) in the form

$$\int_{\mathbb{E}} w(dx)g(x)\lambda(x, \mathbb{E}) = \int_{\mathbb{E}} w(dy) \int_{\mathbb{E}} \lambda(y, dx)g(x),$$

for nonnegative g, it follows that the last integral in (10.9) equals

$$\int_{\mathbb{E}} w(dy_k) \int_{A} \lambda(y_k, dx_k)h(x_k, \overline{\mu}).$$

Using this and (10.10) in (10.9), changing the dummy variables in the integrals to y's, and letting $\eta_{\mathbf{y}} = \sum_k \delta_{y_k}$, we have

$$\text{Left side of (10.8)} = \sum_{n=1}^{\infty} \int_{\mathbb{E}^n} w(dy_1) \cdots w(dy_n)\Phi(\eta_{\mathbf{y}}) \prod_{z \in \mathbb{E}} \eta_{\mathbf{y}}(\{z\})!$$

$$\times [\sum_{k=1}^{n} \phi_{y_k}(\eta_{\mathbf{y}})\eta_{\mathbf{y}}(\{y_k\})^{-1} \int_A \lambda(y_k, dx_k) f(T_{y_k x_k}\eta_{\mathbf{y}})]$$

$$= \int_{\mathbb{M}} \pi(d\eta) \int_{\mathbb{M}\setminus\{0\}} q(\eta, d\mu) f(\mu) 1(\eta \in T_{A\mathbb{E}}\mu).$$

The last equality follows by the definitions of π and q. This proves (10.8). □

Knowing that the process X has an invariant measure, the next question is: When is the process ergodic? By definition, it is ergodic if its invariant measure in Theorem 10.4 is finite and the associated normalized probability measure π is its limiting distribution in the sense that

$$\lim_{t\to\infty} \sup_{C\in\mathcal{M}} |P\{X_t \in C\} - \pi(C)| = 0.$$

This is convergence in total variation distance between the probability measures.

Sufficient conditions for ergodicity of X are as follows.

Theorem 10.5. *Consider the transition kernel*

$$\hat{\lambda}(x, A) \equiv b_x \lambda(x, A), \quad x \in \mathbb{E},\ A \in \mathcal{E},$$

where

$$b_x \equiv \inf_{\mu\neq 0} \phi_x(\mu) > 0, \quad x \in \mathbb{E}.$$

In case the system is open with unlimited capacity, define b_0 differently as $b_0 \equiv \sup_\mu \phi_0(\mu)$, and assume that b_0 is finite and that

$$b_0 \int_{\mathbb{E}\setminus\{0\}} b_x^{-1} w(dx) < 1. \tag{10.11}$$

If the kernel $\hat{\lambda}(x, A)$ defines a Markov process on $\mathbb{E}$ that is ergodic, then the spatial queueing process X is ergodic.

PROOF. We will establish the ergodicity of X by comparing it with a spatial queueing process $\{\hat{X}_t : t \geq 0\}$ that has routing rates $\hat{\lambda}(x, A)$ and service rates $\hat{\phi}_x(\cdot) \equiv 1$. The stationary distribution of the routing rates $\hat{\lambda}(x, A)$ is clearly

$$\hat{w}(dx) \equiv \frac{b_x^{-1} w(dx)}{\int_{\mathbb{E}} b_y^{-1} w(dy)},$$

where w satisfies the traffic equations for λ. In case the system is open, we have $w(\{0\}) = 1$ and

$$\hat{w}(\{0\}) = [1 + b_0 \int_{\mathbb{E}\setminus\{0\}} b_x^{-1} w(dx)]^{-1}. \tag{10.12}$$

Then by Theorem 10.4, we know that the process $\hat{X}$ has an invariant measure (10.5), where $\Phi(\cdot) = 1$, and $w = \hat{w}$ in case the system is closed, or $w = \hat{w}(\{0\})^{-1}\hat{w}$ in case the system is open.

First, consider the case in which the system is closed with ν units. The invariant measure of $\hat{X}$ is clearly finite since $\hat{w}$ is. Note that the processes $\hat{X}$ and X have the same communication structure, since the sequence of states visited by $\hat{X}$ is equal in distribution to the sequence of states visited by X. Also, note that $\hat{X}$ moves "slower" than X. This follows since the exponential sojourn time of $\hat{X}$ in each state μ has the rate

$$\hat{q}(\mu, \mathbb{M}) = \sum_{k=0}^{\mu(\mathbb{E})} b_{x_k} \mu(\{x_k\})^{-1} \lambda(x_k, \mathbb{E}),$$

which is smaller than the corresponding rate (10.2) for X. Consequently, to prove X is ergodic, it suffices to show $\hat{X}$ is ergodic.

For this case in which the system is closed, we can write $\hat{X}_t = \sum_{k=1}^{\nu} \delta_{Z_t^k}$, where $Z_t^1, \ldots, Z_t^\nu$ denote the locations of the units at time t. Since $\hat{\phi}_x(\cdot) = 1$, it follows that $Z_t^1, \ldots, Z_t^\nu$ are independent ergodic Markov processes and each one has the transition kernel $\hat{\lambda}(x, A)$. Therefore, the vector-valued process $(Z_t^1, \ldots, Z_t^\nu)$ is ergodic. Clearly, $\hat{X}$ is a time-shift-invariant function, or a stationary functional, of this vector-valued process (recall Proposition 6.1). This proves the desired property that $\hat{X}$ is ergodic, and hence so is X.

Next, consider the case in which the system is open with capacity ν. The ergodicity for this case follows by the preceding argument, since one can view this open system as a closed system that contains node 0, and $X_t(\{0\}) = \nu - X_t(\mathbb{E}\backslash\{0\})$.

Finally, consider the case in which the system is open with unlimited capacity. Note that the rate $\hat{\lambda}(0, \mathbb{E})$ at which units enter the $\hat{X}$ system is faster than the corresponding rate $\lambda(0, \mathbb{E})$ for X, and units in the $\hat{X}$ system move slower than those in the X system. Then to prove X is ergodic, it suffices to show that $\hat{X}$ is ergodic.

By Theorem 10.4, $\hat{X}$ has an invariant measure π that satisfies $\pi(\{0\}) > 0$, and so the measure 0 is an atom of $\hat{X}$ (or π). Let $\hat{\tau}$ denote the time between successive visits of the process $\hat{X}$ to the state 0. Then to prove $\hat{X}$ is ergodic, it suffices by Kac's theorem to show that $\hat{\tau}$ has a finite expectation.

To this end, first note that in the $\hat{X}$ system, units enter $\mathbb{E}$ according to a Poisson process with rate $\hat{\lambda}(0, \mathbb{E})$. Each arrival moves independently in $\mathbb{E}\backslash\{0\}$ according to the Markov routing process with kernel $\hat{\lambda}(x, A)$ for a time S, and then it exits the system. The stationary distribution of this routing process is $\hat{w}$. It follows by a standard property of stationary distributions that the expected time between two successive entries to state 0 in the routing process is

$$\frac{1}{\hat{w}(\{0\})\hat{\lambda}(0, \mathbb{E})} = ES + \frac{1}{\hat{\lambda}(0, \mathbb{E})}. \tag{10.13}$$

The last term is the expected time the system is empty.

Now, a little thought shows that the time $\hat{\tau}$ between successive visits of $\hat{X}$ to 0 is also the duration of a busy period in an $M/G/\infty$ service system with Poisson arrivals at the rate $\hat{\lambda}(0, \mathbb{E})$ and independent service times distributed as S. For such a system, it is known that $\hat{\tau}$ has a finite expectation if $\hat{\lambda}(0, \mathbb{E})ES < 1$. But this inequality is equal to assumption (10.11), because of (10.13) and (10.12). Thus, $\hat{\tau}$ has a finite expectation, which is what we needed to prove that $\hat{X}$ and hence X is ergodic. □

The sufficient conditions above for ergodicity are quite natural and, in some cases, they are also necessary conditions. For instance, consider a closed system with $\nu = 1$ unit and $\phi_x(\cdot) = a$, for each x. If the system is ergodic, then the routing process defined by the kernel $\hat{\lambda}(x, A)$ is ergodic because it is the same as the spatial queueing system. The condition (10.11) is also sometimes a necessary condition for ergodicity. Indeed, consider the infinite-capacity open spatial system $\hat{X}$ as in the preceding proof. Then condition (10.11), which is equivalent to $\hat{\lambda}(0, \mathbb{E})ES < 1$, is necessary as well as sufficient for $\hat{X}$ to be ergodic.

10.3 Properties of Stationary Distributions and Examples

This section describes some elementary properties of stationary distributions for spatial queueing systems and gives a few examples.

For this discussion, assume that X is a spatial queueing system that is stationary and ergodic. Its stationary distribution is given by Theorem 10.4. Now, the distribution of X_t can be expressed via probabilities of the form

$$\begin{aligned} P\{X_t(A_i) = n_i\ ;\ 1 \le i \le m\} &= c\pi_n\{\mu : \mu(A_i) = n_i\ ;\ 1 \le i \le m\} \\ &= \frac{c}{n_1! \cdots n_m!} \int_{A_1^{n_1} \times \cdots \times A_m^{n_m}} g(\mathbf{x}) w(dx_1) \cdots w(dx_n), \end{aligned} \tag{10.14}$$

where $g(\mathbf{x}) \equiv \Phi(\mu_{\mathbf{x}}) \prod_{z \in \mathbb{E}} \mu_{\mathbf{x}}(\{z\})!$, $n = \sum_{i=1}^m n_i$, the $A_1, \ldots, A_m$ is a partition of $\mathbb{E}$, and c is the normalization constant. For instance,

$$P\{X_t(\mathbb{E}) = n\} = \frac{c}{n!} \int_{\mathbb{E}^n} g(\mathbf{x}) w(dx_1) \cdots w(dx_n).$$

Also, when there are n units in the system, the probability distribution of their locations is given by

$$\begin{aligned} &P\{X_t(A_1) = 1, \ldots, X_t(A_n) = 1 | X_t(\mathbb{E}) = n\} \\ &\quad = P\{X_t(\mathbb{E}) = n\}^{-1} c \int_{A_1 \times \cdots \times A_n} \Phi(\mu_{\mathbf{x}}) w(dx_1) \cdots w(dx_n), \end{aligned}$$

for disjoint sets $A_1, \ldots, A_n$.

Next, we observe that there is a zero probability that the system has a unit at a specific location x if this location is not an atom of w. In addition, there may be more than one unit at a location that is an atom of w.

Proposition 10.6. *For each* $x \in \mathbb{E}$, $P\{X_t(\{x\}) > 0\} = 0$ *if and only if* $w(\{x\}) = 0$.

PROOF. It follows from (10.14) that

$$P\{X_t(\{x\}) = 0\} = \sum_{n=0}^{\infty} P\{X_t(\{x\}) = 0,\, X_t(\mathbb{E}) = n\}$$
$$= c \sum_{n=0}^{\infty} \int_{(\mathbb{E}\setminus\{x\})^n} g(\mathbf{x}) w(dx_1) \cdots w(dx_n).$$

By the definition of c, the preceding summation term is c^{-1} if and only if $w(\{x\}) = 0$. Thus, $P\{X_t(\{x\}) = 0\} = 1$ if and only if $w(\{x\}) = 0$, and this proves the assertion. □

The stationary probabilities sometimes simplify for routing and service rates that have closed-form expressions for w and Φ. Here is a class of service rates that are analogous to those in Jackson networks.

Example 10.7. *Locally-dependent Service Rates and Product Forms.* The service rates are said to be *locally dependent* if they are of the form

$$\phi_x(\mu) = \gamma_x(\mu(\{x\})),$$

where $\gamma_x(n)$ is the service rate at location x whenever there are n units at that location, and $\gamma_0(\cdot) = 1$ in case the system is open. An easy check shows that these service rates are balanced by

$$\Phi(\mu) = \prod_{x\in\mathbb{E}} \prod_{n=1}^{\mu(\{x\})} \gamma_x(n)^{-1}.$$

Then the probability (10.14) for the partition $A_1, \ldots, A_m$ of $\mathbb{E}$ has the product form

$$P\{X_t(A_1) = n_1, \ldots, X_t(A_m) = n_m\} = c \prod_{i=1}^{m} \pi_{A_i}(n_i), \tag{10.15}$$

where

$$\pi_A(n) = \frac{1}{n!} \int_{A^n} w(dx_1) \cdots w(dx_n) \prod_{z\in A} \prod_{m=1}^{\mu_{\mathbf{x}}(\{z\})} m\gamma_x(m)^{-1}. \tag{10.16}$$

For the infinite-capacity open system, this product form implies that X has *independent increments*: $X_t(A_1), \ldots, X_t(A_m)$ are independent for any disjoint sets $A_1, \ldots, A_m$ and "fixed" t. This is not true, of course, for a closed or finite-capacity open system since these random variables must sum to certain values.

In case the system is closed with ν units, it follows by (10.15) and the definition of the convolution operator (denoted by $\star$) that the normalization constant c is given by

$$c^{-1} = \sum_{n_1+\cdots+n_m=\nu} \prod_{i=1}^{m} \pi_{A_i}(n_i) = \pi_{A_1} \star \cdots \star \pi_{A_m}(\nu).$$

To evaluate this, one would choose the partition A_i to consist of sets for which the π_{A_i}'s and their convolutions would be easy to compute. Similarly, the normalization constant for an open system with capacity ν is obtained by

$$c^{-1} = \sum_{n=0}^{\nu} \pi_{A_1} \star \cdots \star \pi_{A_m}(n).$$

For the infinite-capacity open system, the normalization constant is $c = c_1 \cdots c_m$, where c_i is the normalization constant of π_{A_i}.

By Proposition 10.6, we know that X may have clusters of units at locations in $\mathbb{E}$ that are atoms of the measure w. To see this effect, let $\hat{\mathbb{E}} \subset \mathbb{E}$ denote the set of atoms of w. In particular, from (10.16), we know that, for any $x \in \hat{\mathbb{E}}$,

$$\pi_{\{x\}}(n) = w(\{x\})^n \prod_{m=1}^{n} \gamma_x(m)^{-1}.$$

From this and (10.15), it follows that the probability that there are n units in $\hat{\mathbb{E}}^c$ and n_x units at each atom x is

$$\begin{aligned} &P\{X_t(\hat{\mathbb{E}}^c) = n,\ X_t(\{x\}) = n_x;\ x \in \hat{\mathbb{E}}\} \\ &= c \int_{\hat{\mathbb{E}}^c} \prod_{k=1}^{n} \gamma_{x_k}(1)^{-1} w(dx_1) \cdots w(dx_n) \prod_{x \in \hat{\mathbb{E}}} w(\{x\})^{n_x} \prod_{k=1}^{n_x} \gamma_x(k)^{-1}. \end{aligned}$$

Note that this model with a finite state space is the Whittle network model in Example 10.3. □

Here is an example of an open system whose arrival rate depends dynamically on the number of units in the system.

Example 10.8. *System-dependent Arrivals and Locally-dependent Service Rates.* For systems with locally-dependent service rates as above, a realistic variation is that the arrival rate into the system is dependent on the current system load in that

$$\phi_0(\mu) = \gamma_0(\mu(\mathbb{E})).$$

This means that $\gamma_0(n)\lambda(0, \mathbb{E})$ is the arrival rate of units into the system whenever it contains n units. In this case, the service rates are balanced by

$$\Phi(\mu) = \prod_{m=1}^{\mu(\mathbb{E})} \gamma_0(m-1) \prod_{0 \neq x \in \mathbb{E}} \prod_{n=1}^{\mu(\{x\})} \gamma_x(n)^{-1}. \qquad \square$$

In some systems the service rates at a location may depend only on the numbers of units in certain subsets of the state space instead of depending on the specific locations of the units. The next example describes a large class of such service rates that are useful for modeling congestion-dependent services.

Example 10.9. *Sector-dependent Service Rates.* Suppose $\mathcal{S} \subset \mathcal{E}$ is a collection of subsets (or sectors) of $\mathbb{E}\backslash\{0\}$ whose population sizes (or loads) at any instant determine service rates as follows. Associated with each $S \in \mathcal{S}$ is a "service-intensity" $\gamma_S(n)$ that is a function of the number of units n in S. Whenever the system is in state μ, the service rate at a location x is influenced by each set S that contains x in the sense that the intensities of these sets are compounded (or multiplied) to yield the compound service rate

$$\phi_x(\mu) = \prod_{S \in \mathcal{S}: x \in S} \gamma_S(\mu(S)).$$

The sets S not containing x have no influence.

In addition, assume as in the preceding example that $\phi_0(\mu) = \gamma_0(\mu(\mathbb{E}))$. Then an easy check shows that these service rates are balanced by

$$\Phi(\mu) = \prod_{m=1}^{\mu(\mathbb{E})} \gamma_0(m-1) \prod_{S \in \mathcal{S}} \prod_{n=1}^{\mu(S)} \gamma_S(n)^{-1}.$$

In using these sector-dependent service rates, one would typically choose the sector family $\mathcal{S}$ that adequately models the dependencies at hand. One could also let $\mathcal{S}$ denote all possible subsets of $E\backslash\{0\}$ and define $\gamma_S(\cdot) \equiv 1$ for all sets S that do not influence the services. □

Example 10.10. *Local–Regional Service Rates.* Suppose the space $\mathbb{E}$ is partitioned into service regions, and the service rate at x is influenced by the number of units in the region R containing x as well as the number of units exactly at x. Specifically, using the notation in the preceding example, assume the service rate at location x in a region R is the local–regional compound rate

$$\phi_x(\mu) = \gamma_x(\mu(\{x\}))\gamma_R(\mu(R)).$$

Also, assume $\phi_0(\cdot) \equiv 1$. These rates are a special case of the preceding example in which $\mathcal{S}$ consists of all the singleton sets $\{x\}$ and all the service regions R. In this case, the rates are balanced by

$$\Phi(\mu) = \prod_{x \in \mathbb{E}} \prod_{n=1}^{\mu(\{x\})} \gamma_x(n)^{-1} \prod_{R} \prod_{n'=1}^{\mu(R)} \gamma_R(n')^{-1}.$$ □

A little thought should convince one that any example of a Jackson or Whittle network has a corresponding analogue as a spatial queueing system. One just translates the discrete structure of the routing and service rates into nondiscrete rates. Here are two illustrations.

Example 10.11. *Closed Circular System.* Consider a closed spatial queueing process X in which ν units move indefinitely on a circle $\mathbb{E}$ of circumference 1. Think of the circle as the interval [0, 1] with the Borel σ-field. The circle is partitioned into regions $R_1, \ldots, R_m$; a region need not be connected. The service time of a unit at a location $x \in R_i$ is exponential with rate $\phi_x(\mu) = \gamma_{R_i}(\mu(R_i))$. Upon finishing its service at a location x, a unit moves into a set A with probability $\lambda(x, A) = F(A - x)$, where F is a distribution on $\mathbb{E}$ whose support is $\mathbb{E}$. The transition kernel of the spatial process X is

$$q(\mu, C) = \sum_{i=1}^{m} \sum_{x \in R_i} \int_{\mathbb{E}} F(dy - x)\gamma_{R_i}(\mu(R_i))1(T_{xy}\mu \in C).$$

Now the Lebesgue measure on [0, 1], denoted by dx, is the stationary probability measure of the routing kernel since

$$\int_A \lambda(x, \mathbb{E})dx = \int_A dx = \int_{\mathbb{E}} F(dz) \int_{A-z} dx$$
$$= \int_{\mathbb{E}} dx \int_{A-x} F(dz) = \int_{\mathbb{E}} dx \lambda(x, A).$$

Also, the service rates are clearly balanced by

$$\Phi(\mu) = \prod_{i=1}^{m} \prod_{n=1}^{\mu(R_i)} \gamma_{R_i}(n)^{-1}.$$

Therefore, by Theorems 10.4 and 10.5, it follows that X is ergodic with stationary distribution

$$\pi(\mathcal{C}) = c \int_{E^\nu} dx_1 \cdots dx_\nu 1(\mu_{\mathbf{x}} \in \mathcal{C}) \prod_{i=1}^{m} \prod_{n=1}^{\mu_{\mathbf{x}}(R_i)} \gamma_{R_i}(n)^{-1}. \qquad \square$$

Example 10.12. *Treelike System.* Consider an infinite-capacity open system shown in Figure 10.1 in which the space $\mathbb{E}$ is shaped like a tree. The routing of units is such that each unit enters the system somewhere in the root set $\mathbb{E}_1$ and then proceeds up the tree to one of the leaf sets $\mathbb{E}_9, \ldots, \mathbb{E}_{14}$ and then exits the system. Whenever the system is in state μ, the time to the next arrival into $\mathbb{E}_1$ is exponentially distributed with rate $\gamma_0(\mu(\mathbb{E}))$ and the probability that it enters some $A \subset \mathbb{E}_1$ is $\lambda(0, A)$. Therefore, $\phi_0(\mu) = \gamma_0(\mu(\mathbb{E}))\lambda(0, \mathbb{E})$. Upon receiving its service at a location $x \in \mathbb{E}_1$, a unit moves up one level in the tree into $\mathbb{E}_2 \cup \mathbb{E}_3$ at a location selected by the probability kernel $\lambda(x, \cdot)$. Thereafter, the unit is routed up the tree one level at a time, receiving a single service in each set, until it exits the system from a leaf set.

The service time of a unit at any $x \in \mathbb{E}_i$ is influenced by the number of units in the subtree S_i that contains x, where S_i is the union of all branches that go through $\mathbb{E}_i$. For instance,

$$S_1 = \mathbb{E}, \qquad S_6 = \mathbb{E}_1 \cup \mathbb{E}_2 \cup \mathbb{E}_6 \cup \mathbb{E}_{11} \cup \mathbb{E}_{12}.$$

We assume that, whenever the system is in state μ, the service time of a unit at $x \in \mathbb{E}_i$ is exponentially distributed with rate $\phi_x(\mu) = \gamma_x(\{x\})\gamma_{S_i}(\mu(S_i))$. This is a compounding of service intensities of the location x and its subtree S_i. Then the service rates are balanced by the Φ in Example 10.9, where $\mathcal{S}$ consists of all singleton sets $\{x\}$ and all subtree sets S_i.

The routing kernel is assumed to be irreducible on $\mathbb{E}$ and its associated routing process has a finite number of jumps between visits to state 0. Therefore, it is ergodic. Because the communication graph of the routing process is a tree, an invariant measure for it is obtainable by considering the traffic equations on each branch separately just as one does for a discrete-time Markov chain with a treelike communication graph. That is, if $B_1, \ldots, B_n$ is a branch (e.g., $\mathbb{E}_1, \mathbb{E}_2, \mathbb{E}_4, \mathbb{E}_9$),

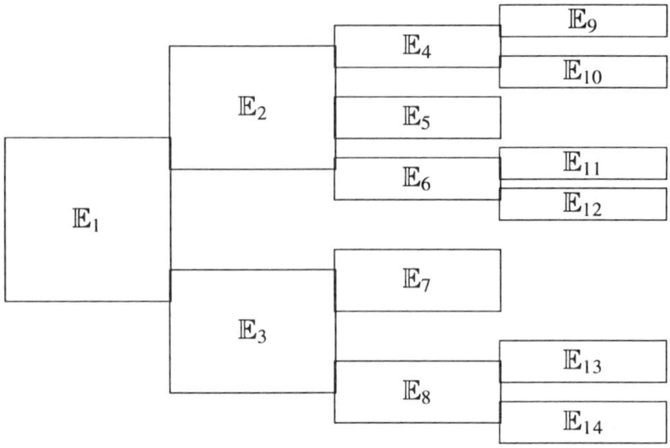

FIGURE 10.1. Treelike System

then an invariant measure w satisfies

$$w(dx_k)\lambda(x_k, B_{k+1}) = \int_{B_{k-1}} w(dx_{k-1})\lambda(x_{k-1}, dx_k), \quad x_k \in B_k,\ 1 \le k \le n,$$

with $B_0 = 0 = B_{n+1}$ and $w(\{0\}) = 1$. Consequently,

$$w(dx_k) = \int_{B_1} \cdots \int_{B_{k-1}} \prod_{i=1}^{k} \frac{\lambda(x_{i-1}, dx_i)}{\lambda(x_i, B_{i+1})}, \quad x_k \in B_k,\ 1 \le k \le n.$$

In addition, we assume the service and routing rates are such that

$$\sup_{\mu} \phi_0(\mu) \int_{\mathbb{E}\setminus\{0\}} \frac{1}{\inf_{\mu \neq 0} \phi_x(\mu)} w(dx) < 1.$$

Then it follows by Theorems 10.4 and 10.5 that the spatial process X is ergodic and its stationary distribution is given by $\pi = c\sum_{n=0}^{\infty} \pi_n$, where π_n is as in Theorem 10.4 with w and Φ described above. □

10.4 Throughputs and Expected Sojourn Times

As in a network, important performance measures of a spatial queueing system are the speeds at which units move through it (throughput rates between sectors) and expected sojourn times of units in a sector. We now describe these quantities for an ergodic spatial queueing process X whose equilibrium distribution is given in Theorem 10.4.

To describe the movements of units between subsets of $\mathbb{E}$ at transitions of X, we will use the space–time point process N on $\mathbb{R}_+ \times \mathbb{E} \times \mathbb{E}$ defined, for $I \subset \mathbb{R}_+$, $A, B \in \mathcal{E}$, by

$$N(I \times A \times B) \equiv \sum_{t \in I} 1(X_t \neq X_{t-}, X_t = T_{AB}X_{t-}).$$

This is the number of times that units move from A to B in the time set I (the A and B may overlap). Then the average number of movements from A to B per unit time, called the *throughput from A to B*, is

$$\rho(A, B) \equiv \lim_{t\to\infty} t^{-1} N([0, t] \times A \times B).$$

Also, $\rho(A, B) = EN([0, 1] \times A \times B)$ when the process X is stationary. The *throughput of the set A* is defined by $\lambda(A) = \rho(A^c, A)$, which is the average number of units that enter A per unit time (A^c is the complement of A). It also equals the average number of units $\rho(A, A^c)$ that exit A per unit time since the process is ergodic.

By the ergodic theorem for Markov processes, we know that

$$\rho(A, B) = \int_{\mathbb{M}} \pi(d\mu) q(\mu, T_{AB}\mu), \tag{10.17}$$

where π is the stationary distribution for X. This throughput from A to B simplifies in some cases as follows. Here a is a positive constant.

Proposition 10.13. *Assume $\phi_0(\cdot) = a$ when X is open and, when X is closed, assume its service rates are of the form*

$$\phi_x(\mu) = a\Phi(\mu - \delta_x)/\Phi(\mu), \quad x \in \mathbb{E}, \mu \in \mathbb{M}.$$

If X is open with unlimited capacity, then

$$\rho(A, B) = a \int_A w(dx)\lambda(x, B), \quad A, B \in \mathcal{E}.$$

If X is closed with ν units, then

$$\rho(A, B) = ac_\nu c_{\nu-1}^{-1} \int_A w(dx)\lambda(x, B) \quad A, B \in \mathcal{E},$$

where c_ν is the normalizing constant for the equilibrium distribution of a closed system with ν units. This expression also applies when X is open with capacity ν, and the normalizing constant c_ν is for an open system with capacity ν.

Proof. As in the proof of Theorem 10.4, we write the stationary distribution of X as $\pi = c\sum_{n=0}^{\infty} \pi_n$, where c is the normalization constant. Using this π and the definition of the transition rates of X, it follows from (10.17) and the hypothesis that

$$\begin{aligned}\rho(A, B) &= c\sum_{n=0}^{\infty} \int_{\mathbb{E}^n} w(dx_1)\cdots w(dx_n)\Phi(\mu_{\mathbf{x}}) \prod_{z\in\mathbb{E}} \mu_{\mathbf{x}}(\{z\})! \\ &\quad\times \sum_{k=0}^{n} 1(x_k \in A)\lambda(x_k, B)\phi_{x_k}(\mu_{\mathbf{x}})/\mu_{\mathbf{x}}(\{x_k\}) \\ &= a\int_A w(dx)\lambda(x, B)c\sum_{n=1}^{\infty} \int_{\mathbb{E}^{n-1}} w(dy_1)\cdots w(dy_{n-1})\end{aligned}$$

$$\times \Phi(\mu_{\mathbf{y}}) \prod_{z \in \mathbb{E}} \mu_{\mathbf{y}}(\{z\})!$$

$$= a \int_A w(dx)\lambda(x, B) \sum_{n=1}^{\infty} c\pi_{n-1}(\mathbb{M}_{n-1}). \tag{10.18}$$

Now, the first assertion of the proposition follows since the last sum in (10.18) is 1 for the open infinite-capacity system. The second assertion follows since the last sum in (10.18) for the closed system with ν units is simply

$$c_\nu \pi_{\nu-1}(\mathbb{M}_{\nu-1}) = c_\nu / c_{\nu-1}.$$

Here $c = c_\nu$ and $c_{\nu-1}\pi_{\nu-1}(\mathbb{M}_{\nu-1}) = 1$. Similarly, for the open system with capacity ν, the last sum in (10.18) is

$$c_\nu \sum_{n=1}^{\nu} \pi_{n-1}(\mathbb{M}_{n-1}) = c_\nu / c_{\nu-1},$$

since $c_{\nu-1} \sum_{n=1}^{\nu} \pi_{n-1}(\mathbb{M}_{n-1}) = 1$. Applying this to (10.18) proves the third assertion. □

We now turn to expected queue lengths and sojourn times in a sector $A \in \mathcal{E}$. First note that the average number of units in A per unit time is

$$L(A) = \lim_{t \to \infty} t^{-1} \int_0^t X_s(A)\, ds = \int_{\mathbb{M}} \mu(A)\pi(d\mu), \quad \text{w.p.1}.$$

This follows by the ergodic theorem for Markov processes.

Hereafter, we assume that $w(A)$ and $w(A^c)$ are not 0. Consider the waiting times $W_1(A), W_2(A), \ldots$ of units in A, where $W_i(A)$ is the waiting (or sojourn) time in A of the ith unit to enter A. There is no restriction on the locations at which units enter or leave A; a unit may reside in several locations in A due to several jumps before it exits, and units need not exit A in the same order in which they entered. Then the average sojourn or waiting time of units in their visit to A is

$$W(A) = \lim_{n \to \infty} n^{-1} \sum_{i=1}^{n} W_i(A), \quad \text{w.p.1},$$

provided the limit exists.

The existence of these average waiting times is justified by the following Little laws that follow from Theorems 5.1 and 5.2, which are for ergodic Markovian systems that are recurrently empty. In this case, recurrently empty means that $P\{X_t(A) = 0\} > 0$ when X is stationary. But this condition is satisfied because of the assumption $w(A^c) > 0$ and

$$P\{X_t(A) = 0\} = c \sum_{n=0}^{\infty} \int_{(A^c)^n} w(dx_1) \cdots w(dx_n) \Phi(\mu_{\mathbf{x}}) \prod_{z \in \mathbb{E}} \mu_{\mathbf{x}}(\{z\})!.$$

Recall that $\lambda(A)$ is the rate at which units enter A. This is finite, and it is positive since $w(A)$ is. For simplicity, we assume $L(A)$ is finite. Note that $L(\cdot)$ and $\lambda(\cdot)$ are measures on $\mathbb{M}$, but $W(\cdot)$ is not. The first result below is for limiting averages and the second result is for expected values.

Theorem 10.14. *The average waiting time $W(A)$ exists and $L(A) = \lambda(A)W(A)$.*

Theorem 10.15. *Suppose the process X is stationary. Let $W(A)$ denote the expected sojourn time in A with respect to the Palm probability of the stationary process X conditioned that a unit enters A at time 0. Then $W(A)$ exists and $L(A) = \lambda(A)W(A)$. Furthermore, the $L(A)$ and $\lambda(A)$ defined above as limiting averages are also the expectations*

$$L(A) = EX_t(A), \quad \lambda(A) = EN([0,1] \times A^c \times A).$$

For an open system, one might be interested in the total time a unit spends in A in all of its visits there before it exits the system. This multiple-visit waiting time is different from the single-visit waiting time discussed above. To analyze such waiting times, we will consider the average waiting time of a unit in a set A while it is in a larger set $B \supset A$. This is defined by

$$W(A|B) = \lim_{n\to\infty} n^{-1} \sum_{i=1}^{n} W_i(A|B), \quad \text{w.p.1},$$

where $W_i(A|B)$ is the time the ith unit entering B spends in A before exiting B. A unit may have several visits to A while in B, and so $W_i(A|B)$ is the sum of all these waits in A. Note that $W(A|B)$ may be positive since each unit entering B has a positive probability of entering A (if the probability of moving from $B\backslash A$ to A is 0, then the ergodicity of the process ensures that there is a positive probability that a unit may enter A directly from B^c).

The general Little laws that justify the preceding results also apply to yield the following Little law for $W_B(A|B)$. This result is an analogue of Theorem 10.14; there is also an obvious analogue of Theorem 10.15.

Theorem 10.16. *The average waiting time $W(A|B)$ exists and*

$$L(A) = \lambda(B)W(A|B).$$

Furthermore, $W(A|B) = \lambda(A)\lambda(B)^{-1}W(A|B)$.

10.5 Poisson Flows in Open Systems

For an open, infinite-capacity stationary Jackson network process, the flows of departures and some other flows between nodes are Poisson processes. We now present analogous results for spatial queueing systems.

Throughout this section, we assume that $\{X_t : t \in \mathbb{R}\}$ is an open, infinite-capacity spatial queueing process defined on the entire time axis $\mathbb{R}$. Assume the process is ergodic and stationary. As in the previous section, let $N(I \times A \times B)$ denote the number of units that move from A into B at transitions of X in the time set $I \subset \mathbb{R}$.

We begin by characterizing the flow of arrivals into the system. This flow is described by the space–time point process N_0 defined on $\mathbb{R} \times \mathbb{E}$ by

$$N_0(I \times B) \equiv N(I \times \{0\} \times B), \quad I \in \mathcal{R}, B \in \mathcal{E}.$$

This is the number of arrivals into the set B from the outside location 0 during the time period I. Our interest is in determining when N_0 is a Poisson process. This means that the arrival flow into a fixed $B \in \mathcal{E}$ is a Poisson process $N_0(\cdot \times B)$ on the time axis $\mathbb{R}$ with rate $\lambda(0, B)$ and, these Poisson flows into disjoint sets in $\mathcal{E}$ are independent.

We say that *the future of N_0 is independent of the past of X*, denoted by $(N_0)_+ \perp X_-$, if for each $t \in \mathbb{R}$, the family of random variables

$$\{N_0(I \times B) : I \subset [t, \infty), \; B \in \mathcal{E}\}$$

is independent of $\{X_s : s \leq t\}$. This property is necessary for N_0 to have independent increments. In the following results, a is a positive constant.

Theorem 10.17. *The arrival process N_0 is a Poisson process such that $(N_0)_+ \perp X_-$ and*

$$EN_0(I \times B) = a|I|\lambda(0, B), \quad I \in \mathcal{R}, B \in \mathcal{E} \tag{10.19}$$

if and only if $\phi_0(\cdot) = a$.

The proof of this result is similar to the proof of Theorem 4.21. The equivalence follows basically since conditioned on X, the N_0 is a Poisson process with

$$E[N_0(I \times B)|X] = \int_I \phi_0(X_t)\lambda(0, B)\, dt.$$

Next, we consider the flow of units exiting the system that is depicted by the space–time point process $N_{\mathbb{E}}$ defined by

$$N_{\mathbb{E}}(I \times A) \equiv N(I \times A \times \{0\}), \quad I \in \mathcal{R}, A \in \mathcal{E}.$$

This is the number of units that exit the system from the set A during the time set I. The result above for arrivals has the following analogue for departures.

Theorem 10.18. *The following statements are equivalent.*
(i) *The exit process $N_{\mathbb{E}}$ is a Poisson process such that $(N_{\mathbb{E}})_- \perp X_+$ and*

$$EN_0(I \times A) = c|I| \int_A w(dx)\lambda(x, \{0\}).$$

(ii) $\phi_0(\cdot) = a$.
(iii) *For each $\mu \in \mathbb{M}$ and $A \in \mathcal{E}$,*

$$\bar{\alpha}(\mu, A) \equiv \int_{\mathbb{M}} \pi(d\eta) \frac{q(\eta, d\mu)}{\pi(d\mu)} 1(\mu \in T_{A0}\eta) = a \int_A w(dx)\lambda(x, \{0\}).$$

Proof. The equivalence of (i) and (iii) follows by a slight variation of Theorem 4.12 (here $N_{\mathbb{E}}$ in reverse time has a compensator with rate $\bar{\alpha}(\bar{X}_t, A)$, where $\bar{X}_t$ is the time reversal of X).

To prove (ii) is equivalent to (iii), consider the function $\bar{\alpha}$. Note that $1(\mu \in T_{A0}\eta) = 1$ if and only if $\eta = \mu + \delta_x$ for some $x \in A$. Using this observation and the definitions of q, π (recall (10.1), (10.5)), and then applying Φ-balance, we have

$$\begin{aligned}\bar{\alpha}(\mu, A) &= \int_A w(dx)\lambda(x, \{0\})\frac{\Phi(\mu + \delta_x)}{\Phi(\mu)}\phi_x(\mu + \delta_x)\\ &= \phi_0(\mu)\int_A w(dx)\lambda(x, \{0\}).\end{aligned}$$

In light of this observation, (ii) is equivalent to (iii). □

From the preceding results, it follows that the exit process $N_{\mathbb{E}}$ is Poisson provided the arrival process N_0 is Poisson. We now show that some flows inside the space $\mathbb{E}$ may also be Poisson processes. Suppose the subset $S \subset \mathbb{E}$ is such that each unit exiting S never returns to S (a set with "single" visits by units). This holds if the routing process ξ_t on $\mathbb{E}$ with rates $\lambda(x, B)$ has the property that whenever it exits S it must enter the outside node 0 before it can return to S again. Let N_S be the point process N restricted to $\mathbb{R} \times S \times S^c$. Then $N_S(I \times A \times B)$ is the number of units that move from $A \subset S$ to $B \subset S^c$ at transitions of X in the time set I. The next result gives sufficient conditions under which N_S is a Poisson process. This means that, for any fixed $A \subset S$ and $B \subset S^c$, the number of movements of units from any A to B over time is a Poisson process, and such Poisson processes are independent for disjoint pairs of sets.

We will assume that the service rates on S and S^c are independent in the following sense (here $\mu_B(\cdot) = \mu(B \cap \cdot)$, the restriction of μ to B):

- $\phi_0(\cdot) = 1$.
- $\phi_x(\mu) = \psi_x(\mu_S)$, for $x \in S$, for some function ψ_x.
- $\phi_y(\mu) = \psi'_y(\mu_{S^c})$, for $0 \neq y \in S^c$, for some function ψ'_y.
- $\{\psi_x : x \in S \cup \{0\}\}$ are Ψ-balanced, where $\psi_0(\cdot) = 1$.
- $\{\psi'_y : y \in S^c\}$ are Ψ'-balanced.

These assumptions imply that all the ϕ_x's are balanced by

$$\Phi(\mu) = \Psi(\mu_S)\Psi'(\mu_{S^c}).$$

Finally, let $\{Y_t : t \geq 0\}$ denote the process X restricted to the space S; that is, $Y_t = \mu_S$ if $X_t = \mu$.

Theorem 10.19. *Under the preceding assumptions, N_S is a Poisson process such that $(N_S)_- \perp Y_+$ and, for $I \subset \mathbb{R}$, $A \subset S$, and $B \subset S^c$,*

$$EN_S(I \times A \times B) = |I| \int_A w(dx)\lambda(x, B). \tag{10.20}$$

PROOF. Clearly Y is an open spatial process on S, and its routing rates are $\lambda_Y(x, \{0\}) \equiv \lambda(x, S^c)$ and

$$\lambda_Y(x, B) \equiv \lambda(x, B), \quad x \in S, B \subset S.$$

Recall that the process X has a measure w that satisfies the traffic equations (10.3). Under the assumption that a unit exiting S cannot return to S, the equations (10.3) are

$$\int_A w(dx)\lambda_Y(x, S \cup \{0\}) = \int_{S\cup\{0\}} w(dy)\lambda_Y(y, A), \quad A \in S \cup \{0\}.$$

But these are the traffic equations for Y, which are therefore satisfied by the truncation w_S of w. Thus, Y is an ergodic spatial queueing process, and its traffic equations have a solution w_S.

Now, by Theorem 10.19 and the assumption $\psi_0(\cdot) \equiv 1$, we know that the exit process M_S for Y is a Poisson process with

$$EM_S(I \times A) = \int_A w(dx)\lambda_Y(x, \{0\}).$$

Also, $(M_S)_- \perp Y_+$.

Next, recall that the routing of units is independent, and the probability that a unit moves from $x \in S$ to $B \subset S^c$, given that it moves into S^c, is $\lambda(x, B)/\lambda(x, S^c)$. Let (T_n, Y_n) $(n \in Z)$ denote the point locations of M_S, and let U_n $(n \in Z)$ be random elements of S^c that are conditionally independent given M_S and

$$P\{U_n \in B | M_S\} = \lambda(Y_n, B)/\lambda(Y_n, S^c).$$

Then it follows that

$$N_S(I \times A \times B) = \sum_n 1(T_n \in I, Y_n \in A, U_n \in B).$$

In other words, N_S is a marked point process of M_S with the location-dependent marks $\{U_n\}$. Since M_S is Poisson, it follows by a basic property of marked Poisson processes (see Theorem 9.12) that N_S is also Poisson with mean given by (10.20). Furthermore, $(N_S)_- \perp Y_+$ is a consequence of $(M_S)_- \perp Y_+$. □

10.6 Systems with Multiclass Units

Although the results above are presented for systems with homogeneous units, they also apply to the following types of systems with multiclass units.

Consider the system we have been studying in this chapter with the modification that each unit is labeled by a "mark" (sometimes called a customer type or class) from a set of marks. A mark represents auxiliary information about the unit that goes into determining its routing and service rates. Then each unit is identified by a pair $x = (z, \alpha)$, where z is its location in the space where it resides and α is its mark. Let $\mathbb{E}$ denote the measurable space of all such pairs. Using the notation in Section 10.1, we represent the state of the system by a counting measure $\mu(\cdot) = \sum_{k=1}^n \delta_{x_k}(\cdot)$, where the "point" $x_k = (z_k, \alpha_k)$ is now the location and mark of unit k.

As before, we assume that the units move in the space $\mathbb{E}$ according to a routing kernel $\lambda(x, A)$. A movement of a unit from $x = (z, \alpha)$ to $x' = (z', \alpha')$ means that the unit's location changes from z to z' and its mark changes from α to α'; the new location or mark may be the same as the previous one. Also, the service rate $\phi_x(\mu)$ is a function of the mark as well as the location, but we assume it is still Φ-balanced. Under these conventions, this measure valued process X is a spatial queueing process on the set $\mathbb{E}$ of pairs (z, α). Hence all the results above apply to it.

A large family of spatial systems are multiclass Jackson and Whittle networks we discussed in Section 3.1. When the class labels are discrete, then these systems can still be modeled as networks, but nondiscrete labels require a spatial model. Here is an example.

Example 10.20. *Multiclass Whittle Network.* Consider a closed Whittle network as in Example 10.3 with the additional feature that each unit carries a real-valued mark or class label. Then each unit is identified by a pair $x = (j, \alpha)$ in $\mathbb{E} \equiv \{1, \dots, m\} \times \mathbb{R}$. Suppose the routing kernel has the form

$$\lambda((j, \alpha), \{k\} \times A) = p_{jk} p_j(\alpha, A),$$

where p_{jk} is the probability that a unit moves from node j to node k and $p_j(\alpha, A)$ is the probability that a unit moving from j changes its mark from α to some value in A. Assume this kernel λ has an invariant measure $w(\{j\} \times A)$.

For instance, suppose p_{jk} is irreducible with stationary distribution p_j. Also, for each j, assume $p_j(\alpha, A)$ is an ergodic, reversible kernel with stationary distribution $\pi_j(A)$ and $\pi_j(d\alpha)p_j(\alpha, d\alpha)$ is independent of j. Then $w(\{j\} \times A) = p_j \pi_j(A)$ is an invariant measure of the kernel λ.

Next, assume the service rates at $x = (j, \alpha)$ are

$$\phi_x(\mu) = \psi_x(\mu)\mu(\{(j, \alpha)\})/\mu(\{j\} \times \mathbb{R}),$$

where $\psi_x(\mu)$ is the total service rate and $\mu(\{(j, \alpha)\})/\mu(\{j\} \times \mathbb{R})$ is the portion of units at node j with mark α. Assume the rates ψ_x are Ψ-balanced. Then the service rates are balanced by

$$\Phi(\mu) = \Psi(\mu) \prod_{j=1}^{m} \mu(\{j\} \times \mathbb{R})! \prod_{\alpha \in \mathbb{R}} \frac{1}{\mu(\{(j, \alpha)\})!}.$$

The results in this chapter apply to this spatial model. □

In some spatial systems, there may be periods of time during which customers do not receive services, possibly due to customer preferences or auxiliary travel time between service locations. These dead periods can often be incorporated by using marks as follows.

Example 10.21. *Customer Dead Periods and Auxiliary Travel Times.* We will consider the spatial queueing system with the variation that whenever a unit completes a service, it may incur a dead period before it proceeds to its next service location. To model this, let z_0 denote an auxiliary location at which a unit resides during a dead period. We assign a mark $\alpha = y$ to a unit if it is in location z_0 and its previous location was y, and we assign a mark $\alpha = 1$ to a unit if it is not at z_0 (it is therefore at a usual service location). Each unit is therefore identified by a pair $x = (z, \alpha)$, where z is the unit's location and α is its mark. Let $\mathbb{E}$ denote the set of all such pairs.

We assume the routing rates are

$$\begin{aligned}
\lambda((z, 1), \{z_0\} \times \{z\}) &= p(z, \{z_0\}),\\
\lambda((z, 1), A \times \{1\}) &= p(z, A),\\
\lambda((z_0, y), A \times \{1\}) &= p_0(y, A).
\end{aligned}$$

Here $p(z, \{z_0\})$ is the probability that a unit departing from a service location z goes to z_0 to incur a dead period; $p(z, A)$ is the usual routing probability kernel; and $p_0(y, A)$ is the routing probability kernel when exiting from node z_0 and y is the previous location. Here $p_0(y, \{z_0\}) = 0$. The $p_0(y, A)$ need not be related to $p(y, A)$, but a natural choice is to set

$$p_0(y, A) = p(y, A)/(1 - p(y, \{z_0\})).$$

Under these assumptions, whenever a unit finishes a service at z, it either goes to another service location in A with probability $p(z, A)$, or, with probability $p(z, \{z_0\})$, it goes to z_0; and after its sojourn there, it goes to another location according to the probability $p_0(z, A)$. We assume that a unit's sojourn time in location z_0 (its dead period) is exponentially distributed with rate γ_z, where z is its previous service location.

The service rates of the system are

$$\phi_x(\mu) = \begin{cases} \gamma_y & \text{if } x = (z_0, y)\\ \psi_z(\mu) & \text{if } x = (z, 1), \end{cases}$$

where $\psi_z(\mu)$ is the usual service rate at z. Assume that $\psi_z(\mu)$ are Ψ-balanced. Then $\phi_x(\mu)$ are balanced by $\Phi(\mu) = \Psi(\mu)\prod_y \gamma_y^{-\mu(\{y\})}$.

The traffic equations for the routing kernel are

$$w(\{z_0\} \times dy) = w(dy \times \{1\})p(y, \{z_0\})$$

$$w(dz \times \{1\}) = \int w(dy \times \{1\})p(y, dz) + \int w(\{z_0\} \times dy)p_0(y, dz).$$

The primary unknown in these equations is the measure $w_1(\cdot) = w(\cdot \times \{1\})$. Then substituting the first equation in the second one, we have

$$w_1(dz) = \int w_1(dy)[p(y, dz) + p(y, \{z_0\})p_0(y, dz)].$$

Solving this equation for w_1 determines w.

This completes the formulation of a spatial process that incorporates exponentially-distributed customer dead periods. Dead periods with phase-type distributions, or with other dependencies, can be formulated similarly using additional customer locations and more complex marks. □

10.7 Bibliographical Notes

This chapter is based on the Doctoral dissertation of Huang (1996). Spatial queueing models combine certain features of Jackson and Whittle models in Chapter 1 and the space–time models in Chapter 9. The term "spatial queueing" is also associated with other models in which the servers are mobile; examples are polling systems and the model described in Çinlar (1995). References on point processes are in Chapter 4, and the theory of Markov processes on general state spaces is reviewed in Meyn and Tweedie (1993).

References

Akyildiz, I. F. and H. von Brand (1989). Exact solutions for open, closed and mixed queueing networks with rejection blocking. *Theoret. Comput. Sci.*, **64**, 203–219.

Alexopoulos, C., El-Tannir, A. and R. F. Serfozo (1999). Partition-reversible Markov processes. *Operations Res.*, **47**, 125–130.

Asmussen, S. (1987). *Applied Probability and Queues.* John Wiley & Sons, New York.

Baccelli, F. and P. Brémaud (1994). *Elements of Queueing Theory.* Springer–Verlag, New York.

Baccelli, F. and S. Foss (1994). Ergodicity of Jackson-type queueing networks. *Queueing Systems Theory Appl.*, **17**, 5–72.

Baskett, F., Chandy, K. M., Muntz, R. R. and F. G. Palacios (1975). Open, closed and mixed networks of queues with different classes of customers. *J. Assoc. Comput. Mach.*, **22**, 248–260.

Bertozzi, A. and J. McKenna (1993). Multidimensional residues, generating functions, and their application to queueing networks. *SIAM Rev.*, **35**, 239–268.

Borovkov, A. A. and R. Schassberger (1994). Ergodicity of Jackson networks with batch arrivals. *J. Appl. Probab.*, **31**, 847–853.

Boucherie, R. and N. van Dijk (1990a). Spatial birth–death processes with multiple changes and applications to batch service networks and clustering processes. *Adv. in Appl. Probab.*, **22**, 433–455.

Boucherie, R. and N. van Dijk (1990b). Product forms for queueing networks with state dependent multiple job transitions. *Adv. in Appl. Probab.*, **23**, 152–187.

Boxma, O. J. (1979). On a tandem queueing model with identical service times at both counters. *Adv. in Appl. Probab.*, **11**, 616–659.

Bramson, M. (1998). State space collapse with application to heavy traffic limits for multiclass queueing networks. *Queueing Systems Theory Appl.*, **30**, 80–148.

Brandt, A., Franken, P. and B. Lisek (1990). *Stationary Stochastic Models.* John Wiley & Sons, New York.

Brandt, A. and G. Last (1995). *Marked Point Processes on the Real Line: The Dynamic Approach*. Springer–Verlag, New York.

Brémaud, P., Kannurpatti, R. and R. Mazumdar (1992). Event and time averages: A review. *Adv. in Appl. Probab.*, **24**, 377–411.

Brown, M. (1970). A property of Poisson processes and its application to macroscopic equilibrium of particle systems. *Ann. Math. Statist*, **41**, 1935–1941.

Brumelle, S. L. (1971). On the relation between customer and time averages in queues. *J. Appl. Probab.*, **8**, 508–520.

Burke, P. J. (1956). The output of a queuing system. *Operations Res.*, **4**, 699–704.

Burman, D. K., Lehoczky, J. P. and Y. Lim (1984). Insensitivity of blocking probabilities in a circuit–switched network. *J. Appl. Probab.*, **21**, 853–859.

Buzacott, J. A. and J. G. Shanthikumar (1993). *Stochastic Models of Manufacturing Systems*. Prentice–Hall, Englewood Cliffs, New Jersey.

Buzacott, J. A. and D. D. Yao (1986). On queueing network models of flexible manufacturing systems. *Queueing Systems Theory Appl.*, **1**, 5–27.

Buzen, J. P. (1973). Computational algorithms for closed queueing networks with exponential servers. *Comm. Assoc. Comput. Mach.*, **16**, 527–531.

Campbell, N. R. (1909). The study of discontinuous phenomena. *Proc. Cambridge Philos. Soc.*, **15**, 117–136.

Campbell, N. R. (1910). Discontinuities in light emmission. *Proc. Cambridge Philos. Soc.*, **15**, 310–328.

Chang, K., Serfozo, R. F. and W. Szczotka (1998). Treelike queueing networks: asymptotic stationarity and heavy traffic. *Ann. Appl. Probab.*, **8**, 541–568.

Chao, X. (1995). Networks of queues with customers, signals and arbitrary service time distributions. *Operations Res.*, **43**, 537–544.

Chao, X. and M. Miyazawa (1998). On quasi–reversibility and local balance: An alternative derivation of the product–form results. *Operations Res.*, **46**, 927–933.

Chao, X., Miyazawa, M., Serfozo, R. F. and H. Takada (1998). Markov network processes with product form stationary distributions. *Queueing Systems Theory Appl.*, **28**, 377–401.

Chao, X. and M. Pinedo (1993). On generalized networks of queues with positive and negative arrivals. *Probab. Engrg. Inform. Sci.*, **7**, 301–334.

Chao, X., Pinedo, M. and M. Miyazawa (1999). *Queueing Networks: Negative Customers, Signals and Product Form*. John Wiley & Sons, New York.

Chao, X., Pinedo, M. and D. Shaw (1996). Networks of queues with batch services and customer coalescence. *J. Appl. Probab.*, **33**, 858–869.

Chen, H., Kella, O. and G. Weiss (1997). Fluid approximations for a processor–sharing queue. *Queueing Systems Theory Appl.*, **27**, 99–125.

Choudhury, G. L., Leung, K. K. and W. Whitt (1995). Calculating normalization constants of closed queuing networks by numerically inverting their generating functions. *J. Assoc. Comput. Mach.*, **42**, 935–970.

Çinlar, E. (1975). *Introduction to Stochastic Processes*. Prentic–Hall, Englewood Cliffs, New Jersey.

Çinlar, E. (1995). An introduction to spatial queues. *Advances in Queueing*, 103–118. Probab. Stochastics Ser., CRC, Boca Raton, FL.

Cox, D. R. (1955). A use of complex probabilities in the theory of stochastic processes. *Proc. Cambridge Philos. Soc.*, **51**, 313–319.

Cramér, H, and M. R. Leadbetter (1967). *Stationary and Related Stochastic Processes. Sample Function Properties and Their Applications*. John Wiley & Sons, New York.

Daduna, H. (1988/89). Simultaneous busy periods for nodes in a stochastic network. *Performance Evaluation*, **9**, 103–109.

Dai, J. G. (1995). On positive Harris recurrence of multi–class queueing networks: a unified approach via fluid limit models. *Ann. Appl. Probab.*, **5**, 49–77.

Dai, J. G. and J. M. Harrison (1992). Reflected Brownian motion in an orthant: numerical methods for steady–state analysis. *Ann. Appl. Probab.*, **4**, 968–1012.

Daley, D. J. and D. Vere-Jones (1988). *An Introduction to the Theory of Point Processes.* Springer–Verlag, New York.

Derman, C. (1955). Some contributions to the theory of denumerable Markov chains. *Trans. Amer. Math. Soc.*, **79**, 541–555.

Disney, R. and P. C. Kiessler (1987). *Traffic Processes in Queueing Networks: A Markov Renewal Approach.* Johns Hopkins University Press, Baltimore.

Economou, A. and K. Fakinos (1998). Product form stationary distributions for queueing networks with blocking and rerouting. *Queueing Systems Theory Appl.*, **30**, 251–260.

El-Taha, M. and S. Stidham Jr. (1999). *Sample-Path Analysis of Queueing Systems.* Kluwer Academic Publishers, Boston.

Fishman, G. S. (1996). *Monte Carlo: Concepts, Algorithms, and Applications.* Springer–Verlag, New York.

Foley, R. D. (1982). The nonhomogeneous $M/G/\infty$ queue. *Opsearch*, **19**, 40–48.

Foley, R. D. and P. C. Kiessler (1989). Positive correlations in a three-node Jackson queueing network. *Adv. in Appl. Probab.*, **21**, 241–242.

Foss, S. (1991). Ergodicity of queueing networks. *Sib. Math. J.*, **32**, 184–203.

Franken, P., Köenig, D., Arndt, U. and V. Schmidt (1981). *Queues and Point Processes.* Akademie Verlag, Berlin.

Gelenbe, E. (1991). Product-form queueing networks with negative and positive customers. *J. Appl. Probab.*, **28**, 656–663.

Gelenbe, E. (1993). G-networks with triggered customer movements. *J. Appl. Probab.*, **30**, 931–942.

Gelenbe, E. and G. Pujolle (1987). *Introduction to Queueing Networks.* Translated from the French by J. C. C. Nelson. John Wiley & Sons, New York.

Gelenbe, E. and R. Schassberger (1992). Stability of G-networks. *Probab. Engrg. Inform. Sci.*, **6**, 3271–3276.

Gerasimov, A. I. (1995). On normalizing constants in multiclass queueing networks. *Operations Res.*, **43**, 704–711.

Gordon, W. J. and G. F. Newell (1967). Closed queueing systems with exponential servers. *Operations Res.*, **15**, 254–265.

Glasserman, P., Sigman, K. and D. D. Yao, eds. (1996). *Stochastic networks. Stability and rare events.* Papers from the workshop held at Columbia University, New York, November 3–4, 1995. Lecture Notes in Statistics, 117. Springer–Verlag, New York.

Glynn, P. W. and W. Whitt (1988). Ordinary CLT and WLLN versions of $L = \lambda W$. *Math Oper. Res*, **13**, 674–692.

Gross, D. and C. M. Harris (1985). *Fundamentals of Queueing Theory.* John Wiley & Sons, New York.

Harrison, J. M. and M. I. Reiman (1981). Reflected Brownian motion on an orthant. *Ann. Probab.*, **9**, 302–308.

Harrison, P. G. (1985). On normalizing constants in queueing networks. *Operations Res.*, **33**, 464–468.

Haverkort, B. R. (1998). *Performance of Computer Communication Systems.* John Wiley & Sons, New York.

Henderson, W. (1993). Queueing networks with negative customers and negative queueing lengths. *J. Appl. Probab.*, **30**, 931–942.

Henderson, W., Pearce, C. E. M., Taylor, P. G. and N. M. van Dijk (1990). Closed queueing networks with batch services. *Queueing Systems Theory Appl.*, **6**, 59–70.

Henderson, W., Pearce, C. E. M., Pollett, P. K. and P. G. Taylor (1992). Connecting internally balanced quasi-reversible Markov processes. *Adv. in Appl. Probab.*, **24**, 934–959.

Henderson, W. and P. G. Taylor (1990). Product form in networks of queues with batch arrivals and batch services. *Queueing Systems Theory Appl.*, **6**, 71–78.

Henderson, W., Northcote, B. S. and P. G. Taylor (1994a). State-dependent signaling in queueing networks. *Adv. in Appl. Probab.*, **24**, 934–959.

Henderson, W., Northcote, B. S. and P. G. Taylor (1994b). Geometric equilibrium distributions for queues with interactive batch departures. *Annals of Operations Research*, **48**, 463–492.

Henderson, W., Northcote, B. S. and P. G. Taylor (1995). Triggered batch movements in queueing networks. *Queueing Systems Theory Appl.*, **21**, 125–141.

Heyman, D. P. and S. Stidham Jr. (1980). The relation between customer and time averages in queues. *Operations Res.*, **28**, 983–994.

Hordijk, A. and N. van Dijk (1984). Networks of queues. I. Job-local-balance and the adjoint process. II. General routing and service characteristics. Modelling and performance evaluation methodology (Paris, 1983), 151–205, *Lecture Notes in Control and Inform. Sci.*, 60, Springer–Verlag, New York.

Horn, R. A., and C. R. Johnson (1994). *Topics in Matrix Analysis*. Cambridge University Press, Cambridge.

Hostinsky, B. and J. Potocek (1935). Chaines de Markoff inverses. *Bull. Intern. Acad. Technique Sci.*, **36**, 64–67.

Huang, X. (1996). *Spatial Queueing Systems and Reversible Markov Processes*. Doctoral dissertation, Industrial and Systems Engineering, Georgia Institute of Technology, Atlanta, GA.

Jackson, J. R. (1957). Networks of waiting lines. *Operations Res.*, **5**, 518–552.

Kallenberg, O. (1983). *Random Measures*. 3rd ed. Akademie Verlag, Berlin; Academic Press, New York.

Karr, A. (1991). *Point Processes and Their Statistical Inference*. 2nd ed. Marcel Dekker, New York.

Keilson, J. and L. D. Servi (1994). Networks of non-homogeneous $M/G/\infty$ systems. *Studies in Applied Probability; J. Appl. Probab.*, **31A**, 157–168.

Kelly, F. P. (1975). Networks of queues with customers of different types. *J. Appl. Probab.*, **12**, 542–554.

Kelly, F. P. (1976). Networks of queues. *Adv. in Appl. Probab.*, **8**,416–432.

Kelly, F. P. (1979). *Reversibility and Stochastic Networks*. John Wiley & Sons, New York.

Kelly, F. P. and P. K. Pollett (1983). Sojourn times in closed queueing networks. *Adv. in Appl. Probab.*, **15**, 638–656.

Kelly F. P. and R. J. Williams, eds. (1995). *Stochastic Networks*. The IMA Volumes in Mathematics and its Applications, **71**, Springer–Verlag, New York.

Kemeny, J. G. and J. L. Snell (1976). *Finite Markov Chains*. Reprinting of the 1960 original. Springer–Verlag, New York.

Kolmogorov, A. N. (1936). Zur Theorie der Markoffschen Ketten. *Math. Ann.*, **112**, 155–160.

Kingman, J. F. C. (1969). Markov population processes. *J. Appl. Probab.*, **6**, 1–18.

Knessl, C. and D. Tier (1995). Applications of singular perturbation methods in queueing. *Advances in queueing*, 3311-336, Probab. Stochastics Ser., CRC, Boca Raton, FL.

König, D. and V. Schmidt (1990). Extended and conditional versions of the PASTA property. *Adv. Appl. Probab.*, **22**, 510–512.

Kook, K. and R. F. Serfozo (1993). Travel and sojourn simes in stochastic setworks. *Ann. Appl. Probab.*, **3**, 228–252.

Krengel, U. (1985). *Ergodic Theorems*. de Gruyter Studies in Mathematics, 6. Walter de Gruyter, New York.

Kühn, P. (1979). Approximate analysis of general queueing networks by decomposition. *IEEE Transactions on Communications*, **27**, 113–126.

Kulkarni, V. G. (1995). *Modeling and Analysis of Stochastic Systems*. Chapman and Hall, London.

Kumar, S, and P. R. Kumar (1994). Performance bounds for queueing networks and scheduling policies. *IEEE Trans. Automat. Control*, **39**, 1600–1611.

Liggett, T. M. (1997). Stochastic models of interacting systems. *Ann. Probab.*, **25**, 1–29.

Letac, G. (1974). Transience and recurrence of an interesting Markov chain. *J. Appl. Probab.*, **11**, 818–824.

Leung, K. K., Massey, W. A. and W. Whitt (1994). Traffic models for wireless communication networks. *IEEE J. Selec. Areas Commun.*, **12**, 1353–1364.

Little, J. D. C. (1961). A proof for the queueing formula: $L = \lambda W$. *Operations Res.*, **9**, 383–387.

Malyshev, V. A. and A. V. Yakovlev (1996). Condensation in large closed Jackson networks. *Ann. Appl. Probab.*, **6**, 92–115.

Mandelbaum, A. and G. Pats (1998). State-dependent stochastic networks I. Approximations and applications with continuous diffusion limits. *Ann. Appl. Probab.*, **8**, 569–646.

Massey, W. A. and W. Whitt (1993). Networks of infinite-server queues with nonstationary Poisson input. *Queueing Systems Theory Appl.*, **13**, 183–250.

Massey, W. A. and W. Whitt (1994). A stochastic model to capture space and time dynamics in wireless communications systems. *Prob. Engin. Inform. Sci.*, **8**, 541–569.

McKenna, J. and D. Mitra (1982). Asymptotic expansions and integral representations of moments of queue lengths in closed Markovian networks. *J. Assoc. Comput. Mach.*, **31**, 346–360.

Melamed, B. (1982a). Sojourn times in queueing networks. *Math. Oper. Res.*, **7**, 223–244.

Melamed, B. (1982b). On Markov jump processes imbedded at jump epochs and their queueing-theoretic applications. *Math. Oper. Res.*, **7**, 111–128.

Melamed, B. and W. Whitt (1990). On arrivals that see time averages. *Operations Res.*, **38**, 156–172.

Meyn, S. and R. L. Tweedie (1993). *Stationary Markov Chains and Stochastic Stability*. Springer–Verlag, New York.

Miyazawa, M. (1994). Palm calculus for a process with a stationary random measure and its applications to fluid queues. *Queueing Systems Theory Appl.*, **17**, 183–211.

Miyazawa, M. (1994). Rate conservation laws: a survey. *Queueing Systems Theory Appl.*, **15**, 1–58.

Miyazawa, M. (1995). Note on generalizations of Mecke's formula and extensions of $H = \lambda G$. *J. Appl. Probab.*, **32**,105–122.

Morse, P. M. (1958). *Queues, Inventories and Maintenance*. John Wiley & Sons, New York.

Neuts, M. F. (1994). *Matrix-Geometric Solutions in Stochastic Models*. An algorithmic approach. Corrected reprint of the 1981 original. Dover Publications, Inc., New York.

Perros, H. G. (1994). *Queueing Networks With Blocking. Exact and Approximate Solutions*. The Clarendon Press, Oxford University Press, New York.

Pollett, P. K. (1986). Connecting reversible Markov processes. *Adv. in Appl. Probab.*, **18**, 880–900.

Reich, E. (1957). Waiting times when queues are in tandem. *Ann. Math. Statist.*, **28**, 768–773.

Reiser, M. and S. S. Lavenberg (1980). Mean-value analysis of closed multichain queueing networks. *J. Assoc. Comput. Mach.*, **27**, 313–322.

Resnick, S. I. (1992). *Adventures in Stochastic Processes*. Birkhäuser, Boston.

Révész, P. (1968). *The Laws of Large Numbers*. Academic Press, New York.

Rieman, M. I. (1984). Open queueing networks in heavy traffic. *Math. Oper. Res.*, **39**, 441–458.

Rolski, T. (1981). *Stationary Random Processes Associated with Point Processes*. Lecture Notes in Statistics, **5**. Springer–Verlag, New York.

Rolski, T. and S. Stidham Jr. (1983). Continuous versions of the queueing formulas $L = \lambda W$ and $H = \lambda G$. *Operations Res. Lett.*, **2**, 211–215.

Rozanov, Yu. A. (1967). *Stationary Random Processes*. Translated from the Russian by A. Feinstein. Holden-Day, San Francisco.

Ross, K. W., Tsang, D. H. K. and J. Wang (1994). Monte Carlo summation and integration applied to multiclass queueing networks. *J. Assoc. Comput. Mach.*, **41**, 1110–1135.

Ross, S. (1983). *Stochastic Processes*. John Wiley & Sons, New York.

Schassberger, R. (1978). The insensitivity of stationary probabilities in networks of queues. *Adv. in Appl. Probab.*, **10**, 906–912.

Schassberger, R. and H. Daduna (1983). The time for a round trip in a cycle of exponential queues. *J. Assoc. Comput. Mach.*, **30**, 146–150.

Schassberger, R. and H. Daduna (1987). Sojourn times in queuing networks with multiserver modes. *J. Appl. Probab.*, **24**, 511–521.

Schmidt, V. and R. F. Serfozo(1995). Campbell's formula and queueing applications. *Advances in Queueing*, 225–242, Probab. Stochastics Ser., CRC, Boca Raton, FL.

Serfozo, R. F. (1985). Partitions of point processes: multivariate Poisson approximations. *Stoch. Proc. Appl.*, **20**, 281–294.

Serfozo, R. F. (1989). Poisson functionals of Markov processes and queueing networks. *Adv. in Appl. Probab.*, **21**, 595–611.

Serfozo, R. F. (1990). Point processes. In *Stochastic Models*, eds. Heyman, D. P. and M. J. Sobel, Elsevier Science/North Holland, Amsterdam, 1–93.

Serfozo, R. F. (1993). Queueing networks with dependent nodes and concurrent movements. *Queueing Systems Theory Appl.*, **13**, 143–182.

Serfozo, R. F. (1994). Little laws for utility processes and waiting times in queues. *Queueing Systems Theory Appl.*, **17**, 137–181.

Serfozo, R. F. and B. Yang (1998). Markov network processes with string transitions. *Ann. Appl. Probab.*, **8**, 793–821.

Sevcik, K. C. and I. Mitrani (1981). The distribution of queuing network states at input and output instants. *J. Assoc. Comput. Mach.*, **28**, 358–371.

Sigman, K. (1995). *Stationary Marked Point Processes: An Intuitive Approach*. Chapman and Hall, New York.

Sigman, K. (1991). A note on a sample-path rate conservation law and its relationship with $H = \lambda G$. *Adv. in Appl. Probab.*, **23**, 662–665.

Simon, B. and R. D. Foley (1979). Some results on sojourn times in acyclic Jackson networks. *Mgmt. Sci.*, **25**, 1027–1034.

Stecke, K. E. and J. J. Solberg (1985). The optimality of unbalancing both workloads and machine group size in closed queueing networks of multi-server queues. *Operations Res.*, **33**, 882–910.

Stidham Jr., S. (1972). $L = \lambda W$: a discounted analogue and a new proof. *Operations Res.*, **20**, 1115–1126.

Stidham Jr., S. (1974). A last word on $L = \lambda W$. *Operations Res.*, **22**, 417–421.

Stidham Jr., S. and M. El-Taha (1989). Sample-path analysis of processes with imbedded point processes. *Queueing Systems Theory Appl.*, **5**, 131–166.

Suomela, P. (1979). Invariant measures of time-reversible Markov chains. *J. Appl. Probab.*, **16**, 226–229.

Towsley, D. (1980). Queuing network models with state-dependent routing. *J. Assoc. Comput. Mach.*, **27**, 323–337.

van Dijk, N. M. (1993). *Queueing Networks and Product Forms: A Systems Approach*. John Wiley & Sons, New York.

Walrand, J. (1988). *Introduction to Queueing Networks*. Prentice–Hall, Englewood Cliffs, New Jersey.

Walrand, J., and P. Varaiya (1980). Sojourn times and the overtaking condition in Jacksonian networks. *Adv. in Appl. Probab.*, **12**, 1000–1018.

Watanabe, S. (1964). On discontinuous additive functionals and Lévy measures of a Markov process. *Japanese J. Math.*, **34**, 53–70.

Whitt, W. (1983). The queueing network analyzer. *AT&T Bell Labs. Tech. J.*, **62**, 2779–2815.

Whitt, W. (1991). A review of $L = \lambda W$ and extensions. *Queueing Systems Theory Appl.*, **9**, 235–268.

Whittle, P. (1968). Equilibrium distributions for an open migration process. *J. Appl. Probab.*, **5**, 567–571.

Whittle, P. (1985). Partial balance and insensitivity. *J. Appl. Probab.* **22**, 168–176.

Whittle, P. (1986a). Partial balance, insensitivity and weak coupling. *Adv. in Appl. Probab.*, **18**, 706–723.

Whittle, P. (1986b). *Systems in Stochastic Equilibrium*. John Wiley & Sons, New York.

Williams, R. J. (1998). Diffusion approximations for open multiclass queueing networks: sufficient conditions involving state space collapse. *Queueing Systems Theory Appl.*, **30**, 27–88.

Wolff, R. A. (1982). Poisson arrivals see time averages. *Operations Res.*, **30**, 223–231.

Zirbel, C. L. and E. Çinlar (1996). Dispersion of particle systems in Brownian flows. *Adv. in Appl. Probab.*, **28**, 53–74.

Index

Applications of Mathematics

(continued from page ii)

33 Embrechts/Klüppelberg/Mikosch, **Modelling Extremal Events** (1997)
34 Duflo, **Random Iterative Models** (1997)
35 Kushner/Yin, **Stochastic Approximation Algorithms and Applications** (1997)
36 Musiela/Rutkowski, **Martingale Methods in Financial Modeling: Theory and Application** (1997)
37 Yin/Zhang, **Continuous-Time Markov Chains and Applications** (1998)
38 Dembo/Zeitouni, **Large Deviations Techniques and Applications**, Second Ed. (1998)
39 Karatzas/Shreve, **Methods of Mathematical Finance** (1998)
40 Fayolle/Iasnogorodski/Malyshev, **Random Walks in the Quarter Plane** (1999)
41 Aven/Jensen, **Stochastic Models in Reliability** (1999)
42 Hernández-Lerma/Lasserre, **Further Topics on Discrete-Time Markov Control Processes** (1999)
43 Yong/Zhou, **Stochastic Controls: Hamiltonian Systems and HJB Equations** (1999)
44 Serfozo, **Introduction to Stochastic Networks** (1999)